The Metrology Handbook

The Metrology Handbook

Edited by Sarah Hudson

STATES
ACADEMIC PRESS
www.statesacademicpress.com

States Academic Press,
109 South 5th Street,
Brooklyn, NY 11249, USA

ISBN: 978-1-63989-520-5

Cataloging-in-Publication Data

The metrology handbook / edited by Sarah Hudson.
p. cm.
Includes bibliographical references and index.
ISBN 978-1-63989-520-5
1. Metrology. 2. Weights and measures. 3. Measurement. I. Hudson, Sarah.
QC88 .M48 2022
389.1--dc23

For information on all States Academic Press publications visit our website at www.statesacademicpress.com

STATES
ACADEMIC PRESS

Contents

Preface

Every book is a source of knowledge and this one is no exception. The idea that led to the conceptualization of this book was the fact that the world is advancing rapidly; which makes it crucial to document the progress in every field. I am aware that a lot of data is already available, yet, there is a lot more to learn. Hence, I accepted the responsibility of editing this book and contributing my knowledge to the community.

Metrology is the scientific study of measurement that establishes a common understanding of units, which is crucial in linking human activities. The three overlapping activities of metrology are the definition, realization and traceability of the units of measurement. There are three basic sub-fields of metrology in which these activities are used. They are scientific or fundamental metrology, technical or industrial metrology, and legal metrology. Study of scientific metrology is focused on the units of measurement, the development of new measurement methods, the realisation of measurement standards, and the transfer of traceability from these standards to users in a society. Technical metrology studies are based on the application of measurement to manufacturing and other processes and their use in society. Legal metrology is based on the activities that result from statuatory requirements, measurements, units of measurement, measuring instruments and their methods of measurement. This book provides significant information of this discipline to help develop a good understanding of metrology and related fields. Its objective is to give a general view of the different areas of this discipline, and its applications. Researchers and students in this field will be assisted by this book.

While editing this book, I had multiple visions for it. Then I finally narrowed down to make every chapter a sole standing text explaining a particular topic, so that they can be used independently. However, the umbrella subject sinews them into a common theme. This makes the book a unique platform of knowledge.

I would like to give the major credit of this book to the experts from every corner of the world, who took the time to share their expertise with us. Also, I owe the completion of this book to the never-ending support of my family, who supported me throughout the project.

Editor

1

Silent Speech Recognition by Surface Electromyography

Andrzej B. Dobrucki, Piotr Pruchnicki,
Przemysław Plaskota, Piotr Staroniewicz,
Stefan Brachmański and Maciej Walczyński

Abstract

For some time, new methods based on a different than acoustic signal analysis are used for speech recognition. The purpose of nonacoustic signals is to allow silent communication. One of these methods based on the electromyography signal is generated by the human speech articulation system. This article presents a device for electromyographic (EMG) signal acquisition and the first measurements from its use.

Keywords: Electromyography, Speech recognition

1. Introduction

Speech recognition is a very important part of computer–human communication system. The most frequently used methods for speech recognition are based on the analysis of an acoustic signal. This signal can be represented as a real-time signal or previously recorded data with a human speech signal.

The use of the acoustic signal analysis is not always possible. In a noisy environment, it is not possible to separate the speech signal from background noise. In this case, the limitations of technical capabilities are present. On the other hand, in a very quiet environment, it is not possible to use voice communication because of the need to maintain silence. In both cases, other signals generated by a speech articulation system can be useful.

It is possible to use electromyographic (EMG) signals produced by muscles of the human speech system [1, 2]. These signals are characterized by low voltage levels and therefore are difficult to record. Another problem is their low resistance to external interference due to the need for large gains [3].

For a practical use of EMG signals, it is necessary to record and gain EMG signals. In the next part of this article, the assumptions and implementation of recording devices are presented. Also, the results of the first attempts to use the device are described.

2. System concept

2.1. EMG signal sensor

The designed hardware part of the SVR (subvocal recognition) system should comply with some specific assumptions because of the specific characteristics of signals:

- high quality of the signal path due to work with low- level signals – characterized by low noise, high dynamics, and immunity to external interference,
- flexibility in the configuration of the signal path – variable usage of electrodes, different types of electrodes, and the possibility of simultaneous use of several signal paths,
- mobility – portable character of the system, small size, and weight.

Based on the assumptions outlined above, a block diagram of the system has been developed (**Figure 1**). Also the necessary requirements for its parts have been defined.

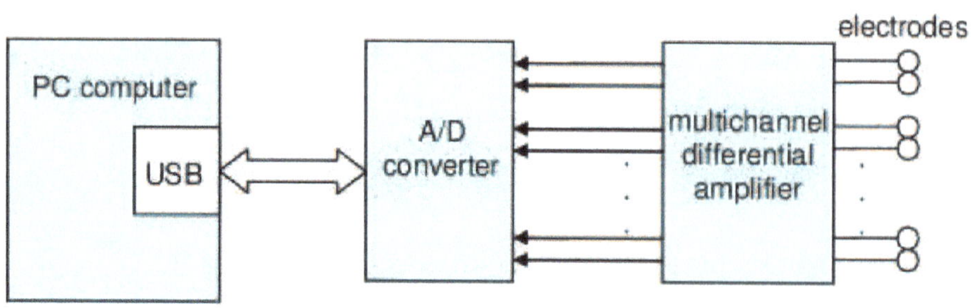

Figure 1. Block diagram of the EMG sensor.

Properties of electrodes that have been used in the system were selected primarily on the basis of SENIAM (Surface ElectroMyoGraphy for the Non-Invasive Assessment of Muscles) recommendations [4]. According to them, there are bipolar electrodes. Due to the small size of the analyzed muscle, the electrodes must be placed close to those muscles. Also, the electrodes must be small.

Electrical impulses collected by the electrodes need to be amplified around 500–1000 times. For this purpose, it is necessary to use specially designed low-noise amplifiers. Due to the use of bipolar electrodes, the amplifier operates in a differential regime that helps significantly to reduce the impact of external noise in the recorded signal.

For further analysis of EMG signals, it is necessary to replace electrical EMG signals with a digital form. This process allows to carry on digital processing of EMG signals on computer devices. This transformation process of analog signals into their digital form requires the use of a signal acquisition card (DAQ card). Due to the requirement of system mobility, the optimal solution is to use an external adapter with a USB interface. It enables the use of easy-to-carry laptop, and besides, there is no need to use additional power supplies since enough energy is supplied by the USB port. An important element of the system is a PC. On this device, the programs for SVR based on electromyography signals are installed.

2.2. Electrodes

The elements used to collect the electromyography signals are electrodes. The device uses various types of electrodes in order to allow finding the best solution. The self-adhesive electrodes (**Figure 2**) provide the most flexibility for an individual subject. The use of this kind of electrodes allows placing them almost anywhere. Initial experiments have shown that the adhesion of such electrodes is only possible on smooth and dry skin. The presence of hairs on the skin, sweat, and wrinkles often prevents the attachment of electrodes.

Figure 2. Self-adhesive electrodes.

An alternative solution is to use cup-shaped metal electrodes (**Figure 3**). They also contain isolated electrodes, which make them any deployment. They require using a special adhesives conductive gel that is used for fixing the electrode and increases the conductivity of the connection. The electrodes may be made of different materials, such as silver, silver coated with silver chloride (Ag/AgCl), gold, iron, or tin. Due to the very good properties – low and stable in time voltage – the most commonly used are silver electrodes coated with silver chloride.

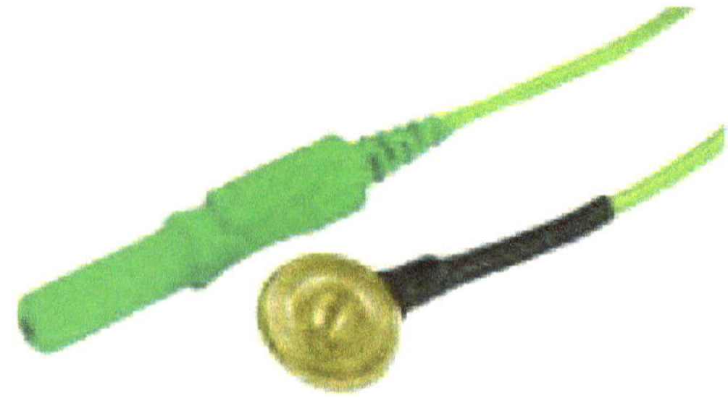

Figure 3. Cup-shaped electrodes.

The main advantage of individual electrodes is the possibility of putting them on any place, although in certain circumstances, it may be a significant drawback. That happens when the repeatable electrode placement is needed in order to couple each other. A good solution in such a situation is the use of bar-type bipolar electrodes (**Figure 4**). These electrodes are compact and are lead shielded. The distance between the electrodes is strongly determined by their structure.

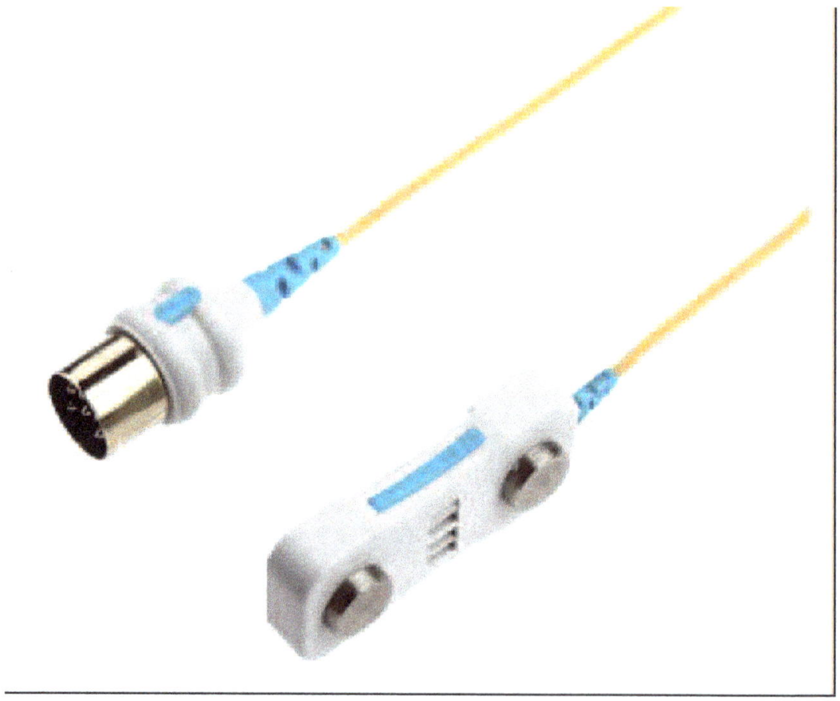

Figure 4. Bar-type electrodes.

During the measurement process, in addition to EMG signal electrodes, it is necessary to use a reference electrode. The purpose of the additional electrode is to determine the reference potential of the speaker. Without this additional electrode, the amount of noise (especially from the power network) in the acquisition increases greatly, which prevents the measurement completely. During the measurements, the reference electrode is connected to the patient in a place where there is no muscular activity.

2.3. Amplifier

To amplify the electrical EMG signals from the electrodes, specially designed amplifiers are used [5]. The development of the designed systems allows for a perfect fit of the amplifier for the SVR purpose.

Amplifier design guidelines are defined on the basis of the solutions used in the world as follows:

- The frequency band should be in the range of 10–500 Hz; hence, the filters should be used in order to limit this frequency range.
- The total amplification factor has to be very high (up to 1000 V/V).
- The self-noise level in reference to the input signal is assumed less than 1 μV.
- In order to suppress interference from the environment, it is very important to get the largest possible value for CMRR (common-mode rejection ratio). It is assumed that CMRR is >95 dB value.
- Input impedance of EMG amplifiers is usually high and is at least of 10 MΩ. Because of high resistance of the human skin and nonideal coupling of the electrodes with the skin as well as adipose tissue, the impedance of EMG electrodes is significant and it can be as much as several dozens of kiloohms. Due to a high input impedance value of the amplifier, it is possible to obtain maximum values of EMG voltages.
- The need of decrease of external distortions induces the minimum length of a cable connecting electrodes and amplifiers; thus, the amplifiers should be placed as near as possible and their size must be also as small as possible.

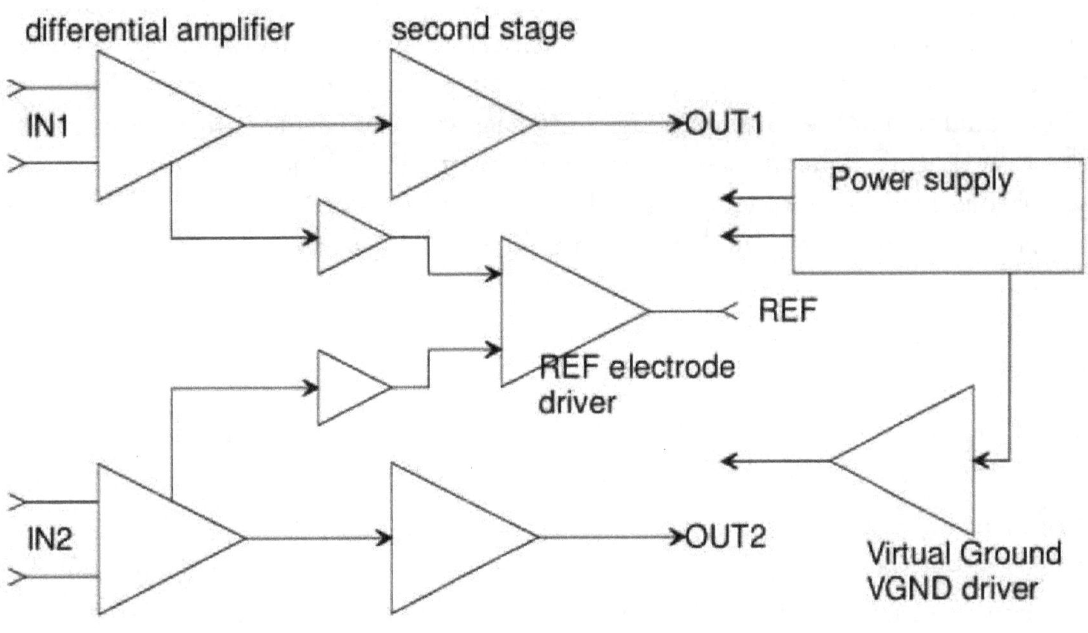

Figure 5. EMG amplifier.

- Power supply for the amplifier is on the board of the DAQ card; so the additional power unit is not necessary to use.

The scheme of the EMG amplifier is presented in **Figure 5**.

2.4. Driver of the reference electrode

The task of the reference electrode is fixing an adequate potential of the patient's body and ensuring returns to the bias currents of the inputs of a differential amplifier. Using the driver improves the damping of the common component for the amplifier (CMRR) of a value from 10 to 50 dB [6, 7]. The operation of this system consists of a continuous comparison of average voltage from measuring electrodes with the virtual ground's voltage. If a difference occurs, it is amplified with an opposite sign and put to the reference electrode. In this way, the system still maintains the proper value of the DC component of the measuring electrodes. The reference electrode driver (RLD) is used to raise the CMRR of the instrumentation amplifier. With this higher signal-to-noise ratio (SNR), the differential signal obtained is ensured to possess only relevant information and a minimum of interference currents or irrelevant data. The idea behind the RLD is to maintain a known voltage potential in the human subject that is directly related to the system board ground. This method then reduces the common-mode DC offset previously found in the system and thereby attempts to cancel any different DC offsets that individual channels or probes may experience. The actual method of the RLD is quite unique. A feedback network is created that depends on the averaged inputs from the combined instrumentation amplifier floating grounds and a GROUND signal originating from the human. This signal is then sent through an inverting gain stage that completes the feedback loop, which effectively counteracts any potential changes in the subject.

2.5. Virtual ground source

Because all electronic circuits of the power amplifier are supplied with a single voltage, it was necessary to produce an artificial "ground" potential. DC voltage of such a virtual ground should be equal to half the supply voltage. The easiest way to achieve this is to use a simple resistive divider (this version was used in the construction of the first amplifier). Unfortunately, the change of current from the divider causes changes of the voltage. Therefore, in the second version the buffer amplifier has been added, which holds the potential of virtual ground unchanged.

2.6. ADC card (DAQ)

Because of the narrow bandwidth of collected signals and a large level of signal after amplification, requirements for the DAQ card parameters are not critical.

Mobility of the whole system involves the use of an external card with a USB interface. A precise sampling runs with high dynamics and requires a resolution of not less than 16 bits. Because of the research reason, the originally formed multicard system must provide a sampling for at least eight independent channels. The sampling rate is not critical. At the top of the analyzed signal frequency of 500 Hz and the provided sample for the antialiasing filter slope, there is quite an adequate sampling frequency equal to 4 kHz. Even while working with eight channels, the resultant frequency for the whole card will only be 32 kHz.

A desirable feature of the card used is switched ranges. They allow for a flexible adjustment range of the amplitude of the sampled signal and better use of dynamics DAC.

A software part of this project is written in C++ and runs on the Linux operating system. Due to the availability of drivers, the application was tested on Linux Debian, and due to the licensing requirements, it was decided that the program is a console application. The target of the software is data acquisition, signal parameterization, and classification.

3. EMG measurement by using cup-shaped and double (glued) electrodes

The measurements were made by using adhesive (**Figure 2**) cup-shaped (**Figure 3**) and double electrodes (Figure 4). The reference electrode was used as one of the glued electrodes (**Figure 2**). An example of the reference disc electrode placement is shown in **Figure 6**.

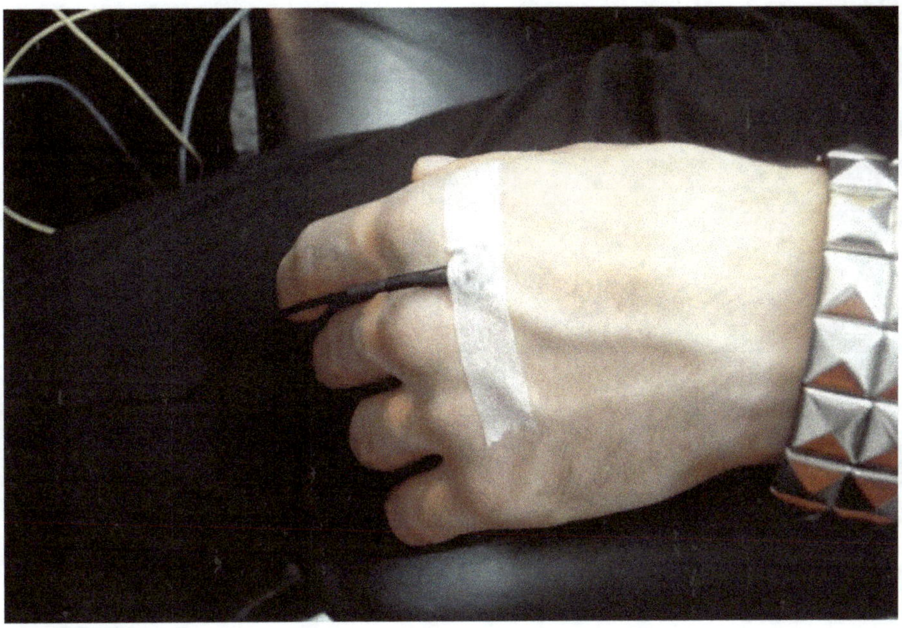

Figure 6. Example of the reference disc electrode placement.

3.1. Measurement system

The measurement system consists of electrodes, a preamplifier of an EMG signal, an AD converter, and a multichannel recorder. As an AD converter, a multichannel recorder has been used, or a sound card with a software for digital signal processing.

A few extra features have been checked before finalizing the configuration of the measurement system. To record the results of the measurements, five different measurement configurations were used. Three of them have been reviewed negatively. The main problems concern the interference from the power supply, noise from FireWire (not always), and demodulation of the EMG signal cables from a nearby radio station transmitter. The use of battery power, the reduction of EMG wires length, and the use of the USB bus made it possible to eliminate the problems listed above.

The EMG signals were stored as a wave format file with a 48-kHz sampling frequency, with 16-bit resolution. The EMG signal dynamics reach about 20 dB, so that the signal parameters are sufficient. An easy analysis of the recorded signals is guaranteed by using a wave format file. During the measurements, the recorded signal can be observed, and these signals can be displayed in real time. It is also possible to insert some markers for easy highlighting in the next repetition of words. **Figure 7** shows some examples of the results of the measurements.

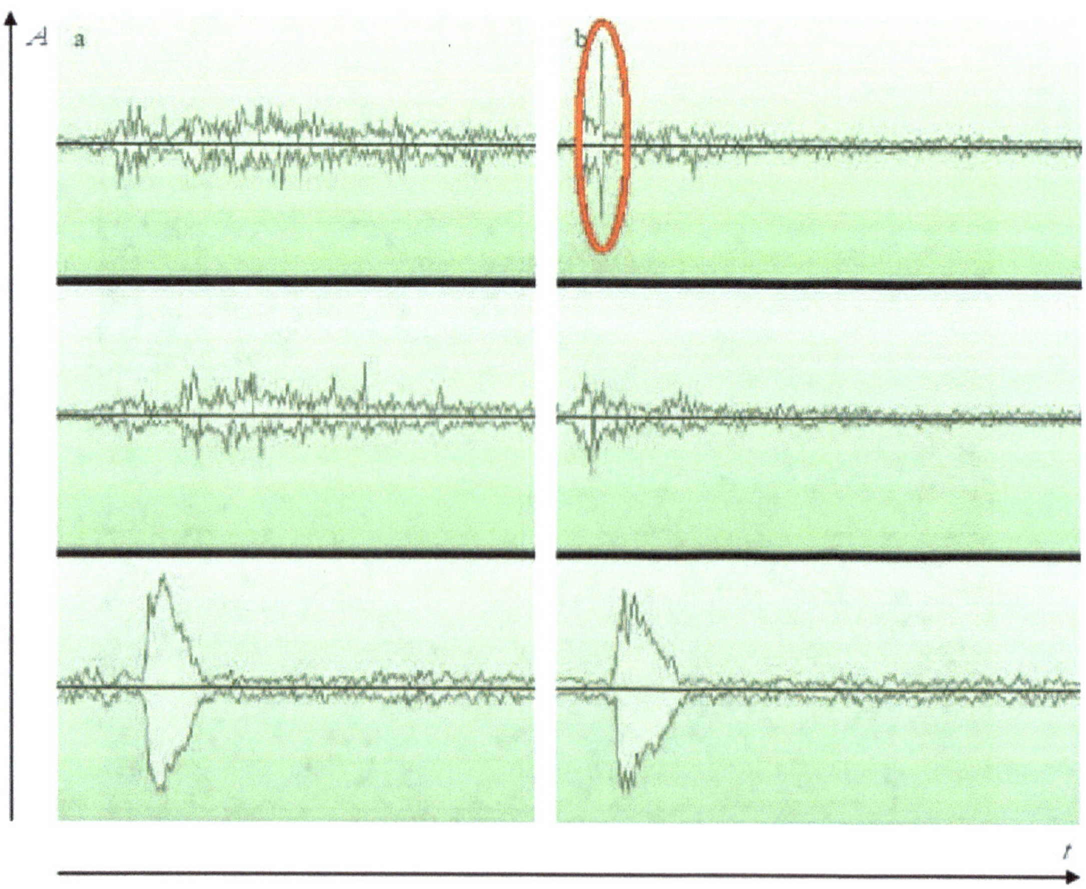

Figure 7. Sample measurement results for a person with a normal EMG signal (vowel "a," 1 s duration): a) a valid signal and b) a signal with interference (marked). From the top: an EMG signal from the bar-type electrode, an EMG signal from the cup-shaped electrode, an acoustic signal recorded by using a microphone. A – amplitude (arbitrary units), t – time.

3.2. Measurements

The measurements were provided for a group of 14 people (11 men and 3 women), aged 20–25 years. A typical location for electrodes is shown in **Figure 8**. The electrodes' location differed slightly between individuals; the differences resulted primarily from the differences in the anatomy of the human body. In order to improve the conductivity between the skin and the electrodes, an electroconductive paste was used. The EMG signals were recorded with and without the conductive paste. The use of the paste in some cases slightly improved the conductivity (the recorded EMG signal was greater); the other did not affect the value of the EMG signal. There were no cases of deterioration of the EMG signals after applying the paste.

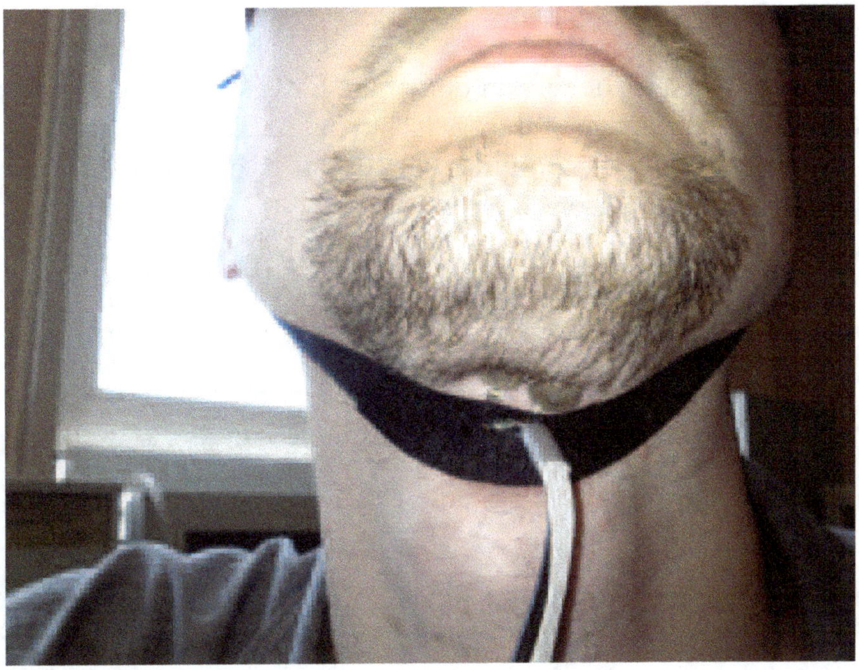

Figure 8. Example of cup-shaped electrodes' location.

To achieve the best repeatability of the measured signals, the research process was performed with the following procedure: 1) application of the sensor to the neck; 2) signal check: if bad – a change of sensor position is needed, if good – recording the signal. The electrodes were attached by using a hook and loop tape, ensuring the pressure stability over the entire process of measurements.

During the measurements, it was found that the EMG signals fundamentally differ within the group. For eight subjects, the correct values of the EMG signals were obtained immediately, while for six people, the correct values could not be obtained immediately. Three of these people learned to speak words in a short time, so that the EMG signal had the correct value. The other three were not able to obtain the EMG signal even after some time.

During the measurements, 10 repetitions of the following elements of the dictionary were recorded: the vowels a, e, i, o, u, y; the words: "stop", "start", "dalej" (further), "v levo" (left), "v pravo" (right), "pauza" (pause), "enter", "tak" (yes), "nie" (no).

To get the correct number of repetitions and the corresponding interval between different words, the graphical presentation on the screen was used. With this solution, the signals did not overlap, and it was easy to get the required number of repetitions.

3.3. Measurements summary

For higher values of EMG signals during speaking, it is necessary to use more sensitive sensors. The use of the conductive paste did not increase the value of the signal. The EMG signal has the highest value depending on the subjects and positions of the electrode. Finding the best location of electrodes requires several attempts. For some people, the correct EMG signal could not be obtained, but some of these people could learn to speak such words, and thus the EMG signals were correct. The use of rigid location of the sensor helps to improve the correct EMG signal quality.

The first experiments related to the registration of the EMG signals were conducted with the use of the first introductory version of the electronic circuit. The characteristics of this device were wires of 1.5 m, connecting the amplifier with the surface electrodes, which were placed on the skin of the investigated person.

In the process of measuring, some problems were observed, as follows:

- high level hum of 50 Hz – the level of this noise was strongly dependent on the wires position,
- change of a noise level during the measurement due to changing of the wires position, and
- signal from the FM transmitter (analog radio program), depending on the position in the building.

All the presented problems are connected with a large length of wires connecting the electrodes with the amplifier. The EMG signals amplifier has a very large amplification and input impedance. It is very sensitive to violations from the electric–magnetic fields around the device.

The best method to minimize these type of violations is the reduction of the wire length from the electrode to the amplifier. To realize this aim, the miniaturization of the device is necessary. The first step is to fasten the electrodes directly to the printed circuit board, and the second step is to place the whole device on the neck of the measured person.

4. State of the art of EMG speech recognition

The early research in the field of electromyography in speech technology was basically carried out only for proper speech prosthetic (parameters comparison between healthy people with prosthesis).

First significant trials to use EMG for automatic speech recognition were done in 1985–1986 in the USA and Japan. Morse et al. [8] applied four-channel surface electrodes with a sampling frequency of 5120 samples/channel/s. After the limitation of amplitude and frequency band to the range of 100–1000 Hz, 20 repetitions of isolated words (digits in English) for two voices were examined. After the use of ML (maximum likelihood), the scores at the level of 60% were obtained. In a later research, the same team obtained similar results with the use of neural networks [9].

In a parallel study, Japanese scientists Sugie et al. [10] with three-channel surface electrodes with a sampling frequency of 1250 samples/channel/s succeeded in real-time phone recognition experiments. For three voices and 50 Japanese syllables, they reached a proper classification at the level of 64%.

Among research centers where most advanced works on subvocal speech (based on EMG) are done, the NASA Ames Research Center should be mentioned [11, 12]. Jorgensen et al. use wavelet transform or linear predictive coding as the parameterization method, and they use artificial neural networks, SVM (support vector machines), and HMM (hidden Markov models) as classifiers. For the limited set of five voice commands, they obtained the scores of the order of 90%. For the set of English phones and female voices, the scores were lower – at the level of 50%. Among applications designed by NASA, beside voice commands controls (e.g., for remote steering of application or robot), also speech recognition in difficult conditions should be mentioned (e.g., acoustic – with high noise level or with speech pronounced when the gas mask is on).

Research carried out by Schultz et al. [13–15] at Karlsruhe University can be considered as the most crucial in the last decade because they tried to overcome the speaker dependence on one hand and the limitation of the dataset on the other. The collected signal database (EMG-PIT) was recorded for 78 speakers with the use of six electrodes placed on the speaker's face and neck. For the signal parameterization, the parameters of time domain were applied (mean value in frame, energy, crossing zero density). For the sample of 14 speakers (10 sentences each), they obtained 47.15% of WER (word error rate).

Manabe et al. [16, 17] carrying out research at the Japanese NTT DoCoMo used a novel technique of setting of the EMG electrodes. In the proposed technique, three electrodes are set on the speaker's fingers which he keeps on his face. In their preliminary tests with the neural network and signals energy, they reached around a 90% recognition score for five vowels. In their later works where more sophisticated signal parameterization techniques (MFCC, LPC, etc.) and HMM as classifier were used, the score at the level of 60% for several comments in Japanese was obtained.

The research on EMG speech signal recognition was also carried out by Instituto de Investigacion en Ingenieria de Aragon and Dpt. De Informatica e Ingenieria de Sistemas [18]. The prepared system uses eight electrodes detecting signals of face muscles. For three voices and 30 syllables of Spanish, they reached a proper classification at the level of 69%.

In the research of Bu et al. [19] instead of standard bipolar electrodes put on a singular muscle, they applied a differential signal between two unipolar electrodes put on different muscles, which let them use less number of electrodes to obtain similar recognition results. In the classification tests of several Japanese phones, they reached the score of 90.6%.

The parameterization and classification techniques used for the EMG speech signal recognition are presented in **Table 1**. Despite the fact that there are crucial differences between acoustic and EMG signals, most of them are used also in automatic acoustic speech and speaker recognition systems.

Parameterization	Positions	Classification	Positions
WT (Wavelet Transform)	[11, 12]	HMM (Hidden Markov Models)	[11, 13, 14, 16]
STFT (short time Fourier transform)	[12, 13, 15, 18]	ANN (Artificial Neural Network)	[12, 19]
LPC (Linear Predictive Coding)	[12, 13]	SVM (Support Vector Machines)	[12]
Time-domain and zero-crossing parameters	[14, 15, 18,]	DTC (Decision Trees Classifier)	[14, 18]
Cepstral parameters	[13, 16–18]	LDA (Linear Discriminant Analysis)	[15]
Phonetic parameters	[14]		

Table 1. Parameterization and classification techniques used for the EMG speech signal recognition described in the literature.

5. Automatic recognition tests

The block diagram of general speech recognition procedures is presented in **Figure 9**. The digital EMG signal is given to the front end of automatic speech recognition system. The system also uses the signals stored before in the EMG signals database. The task of the front-end procedure is to obtain the parametric picture of the registered digital EMG signal that can be later recognized in the classification process. The classification consists of two stages: the training and recognition.

The preliminary tests were carried out for the EMG recordings of six Polish vowels [a, e, i, o, u, y] and eight isolated words [stop, start, dalej, v levo, v pravo, pauza, enter, tak] [vowels and words are given in the SAMPA (Speech Assessment Methods Phonetic Alphabet) notation]. The tested material was recorded with two types of electrodes (cup-shaped and self-adhesive) during one recording session of one speaker. Since the problem of speech recognition from the EMG signal is still very new, in the literature there is no commonly accepted set of best features for that kind of task. Most features applied and described in the literature are copied directly from nowadays well-developed techniques of speech recognition based on the acoustic signal. The EMG signal which has completely different physical origins than the acoustic one has also a completely different nature both in frequency (limited to substantially lower frequency range) and time domain (longer time intervals caused by preparation and relaxation of muscles for each utterance than for acoustic signal). Since the selection of most efficient features set is a big task, it demands longer studies and tests. In the preliminary tests, the feature vector was the combination of spectral MFCC (mel frequency cepstral coefficients) and LPC (linear predictive coefficients) features. In the following test results, the 76 parameters were applied (average and deviation for each): 13 MFCC, 10 LPC, 5 spectral moments, and 10 area method of moments of MFCC. MFCC and LPC are the most common features applied in speech technology; additionally the spectral variability was accounted for by using the FFT (fast Fourier transform) moment and its modification for MFCC.

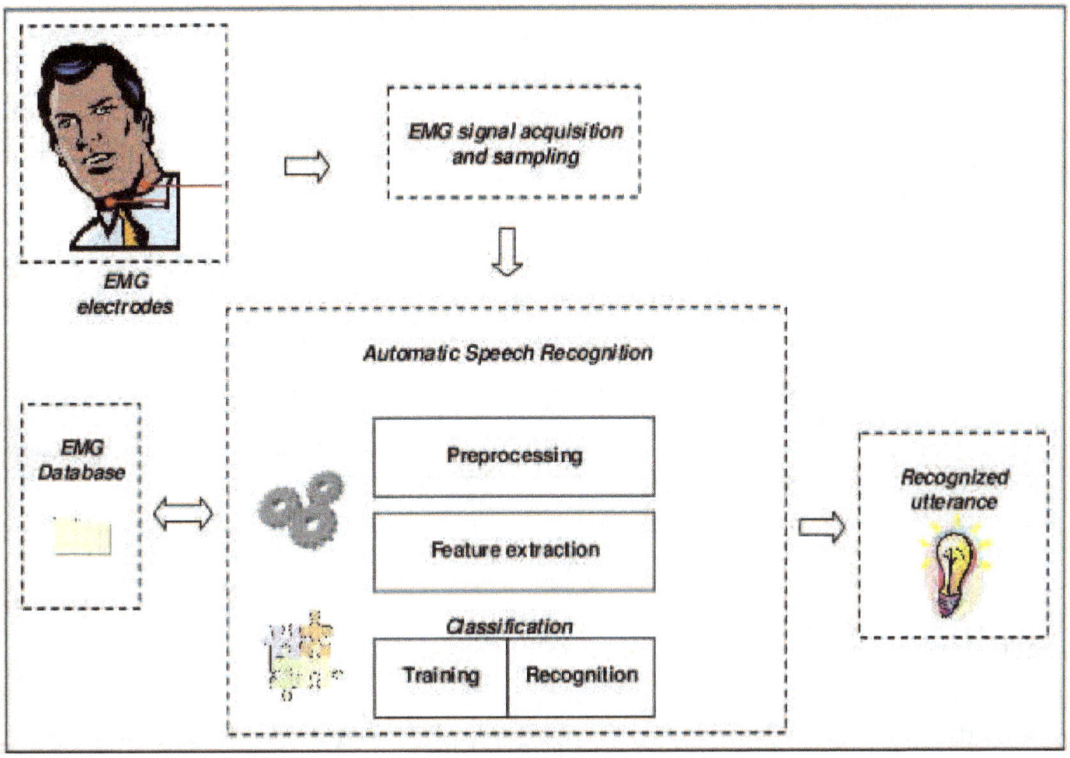

Figure 9. Block diagram of speech recognition procedures.

The window size for feature vector was set up experimentally. The tests were carried out for recognition of six Polish vowels with the Bayes network classification. Due to a small number of instances, instead of dividing them into training and testing sets, a cross-validation was applied. **Table 2** presents the results of correctly classified instances in the function of window size (represented in samples and corresponding time intervals). Kappa is a chance-corrected measure of agreement between the classifications and the true classes. It is calculated by taking the agreement expected by chance away from the observed agreement and dividing by the maximum possible agreement. The windows overlap was set to 0.5.

Window size [samples]	Window size [ms]	Correctly classified instances	Kappa
128	8	90.70%	0.89
256	16	93.02%	0.92
512	32	97.67%	0.97
1024	64	93.02%	0.92

Table 2. Results of correctly classified instances in the function of window size.

Since the best results were obtained for the window size of 32 ms, the following tests were carried out for this value.

Table 3 presents the overall results of correctly classified utterances for vowels, words, and three tested classifiers (Bayes network, Naive Bayes, and multilayer perceptron). Some conclusions from the obtained results can be drawn.

Classifier	Correctly classified instances			
	Vowels		Words	
	Cup-shaped electrodes	Self-adhesive electrodes	Cup-shaped electrodes	Self-adhesive electrodes
Bayes network	97.67%	90.70%	97.40%	96.10%
Naive Bayes	95.35%	83.72%	93.51%	92.21%
Multilayer perceptron	93.02%	79.07%	80.52%	84.42%

Table 3. Overall results for vowel and word recognition.

6. Conclusions

The level of the EMG signals obtained from the applied electrodes is sufficient for further analysis for most cases.

Subjects, who during the first trial did not have the correct EMG signal, may learn to speak such words to be able to record the signal. After a few minutes of training, it was possible to speak the words correctly, so the EMG signals could be recorded. It should be noted that the way of speaking words was associated with an unnatural speaking out, with very expressive facial expressions.

Subjects involved in the research can be divided into three groups:

- People who were possible to speak in such the way that the registered EMG signals could be done at once. The expressive facial expressions allow to record signals with a higher amplitude.
- People who were able to produce the sufficient for recording EMG signals after training. The expressive facial expressions allow to capture data with normal values.
- People for whom EMG signals could not be registered even after training.

The great difficulty in the measurement process was to obtain the undisturbed EMG signals. Each movement of the speaker was reflected in the EMG signal parameters. Particularly, a strong interference was generated by the reflex of swallowing saliva – the signal is two to three times greater than the speech signal. It is important to ensure the adequate measurement conditions – separate from the power supply, and other signals that can be picked up by the electrodes.

The best scores during automatic recognition tests were obtained for the Bayes network, but the differences between chosen classifiers are considerably insignificant. Very similar results were obtained for words and vowels (in vocal speech signal recognition, the vowel recognition would gain better results than word recognition).

The difference between electrodes indicates that the influence of electrodes type or their position can have a significant impact on final scores (recording for both types of electrodes were carried out during the same recording session). The difference was quite big in the case of vowels recognition (difference around 10%) and considerably insignificant in case of words recognition. Considerably high results of correctly classified instances are partly caused by a session-dependent and a speaker-dependent case.

Author details

Andrzej B. Dobrucki*, Piotr Pruchnicki, Przemysław Plaskota, Piotr Staroniewicz, Stefan Brachmański and Maciej Walczyński

*Address all correspondence to: andrzej.dobrucki@pwr.edu.pl

Department of Acoustics and Multimedia, Faculty of Electronics, Wrocław University of Technology, Wrocław, Poland

References

[1] Basmajian JV, De Luca CJ. Muscles Alive, Their Function Revealed by Electromyography. Baltimore: Williams Wilkins; 1985.

[2] De Luca CJ, Knaflitz M. Surface Electromyography: What's New?. Torino: C.L.U.T.; 1992.

[3] Konrad P. The ABC of EMG. A Practical Introduction to Kinesiological Electromyography. USA: Noraxon U.S.A. Inc. ; 2005.

[4] Hermens HJ, Freriks B. SENIAM 5. The State of the Art on Sensors and Sensor Placement Procedures for Surface ElectroMyoGraphy: A Proposal for Sensor Placement Procedures. The Netherlands; 1997.

[5] Winter DA. Biomechanics and Motor Control of Human Movement. New York: John Wiley & Sons; 1990.

[6] MettingVanRijn AC, Peper A, Grimbergen CA. Amplifiers for bioelectric events: a design with a minimal numbers of parts. Medical & Biological Engineering & Computing. 1994, 32: pp. 305–310.

[7] Benning M, Boyd S, Cochrane A, Uddenberg D. The Experimental Portable EEG/EMG Amplifier, University of Victoria Faculty of Engineering, ELEC 499A Report; 2003.

[8] Morse MS, O'Brien EM. Research summary of a scheme to ascertain the availability of speech information in the myoelectric signals of neck and head muscles using surface electrodes. Computers in Biology and Medicine. 1986, 16, no. 6: pp. 399–410.

[9] Jorgensen C, Dusan S. Speech interfaces based upon surface electromyography. Speech Communication. 2010, 52: pp. 354–366.

[10] Sugie N, Tsunoda K. A speech prosthesis employing a speech synthesizer – vowel discrimination from perioral muscle activities and vowel production. IEEE Transactions on Biomedical Engineering. 1985, 32, no. 7: pp. 485–490.

[11] Betts BJ, Binsted K, Jorgensen C. Small-vocabulary speech recognition using surface electromyography. Interacting with Computers. 2006, 18: pp. 1242–1259.

[12] Jorgensen C, Binsted K. Web browser control using EMG based sub vocal speech recognition. In: Proc. of the 38th Hawaii International Conference on System Sciences; 2005.

[13] Maier-Hein L, Metze F, Schultz T, Waibel A. Session independent non-audible speech recognition using surface electromyography. In: IEEE Workshop on Automatic Speech Recognition and Understanding, San Juan, 27–27 November, pp. 331–336; 2005.

[14] Schultz T, Wand M. Modeling coarticulation in EMG-based continuous speech recognition. Speech Communication. 2010, 52: pp. 341–353.

[15] Szu-Chen J, Schultz T, Walliczek M, Kraft F, WaibelA. Towards continuous speech recognition using surface electromyography. In: Proc. Interspeech 2006, pp. 573–576, Pittsburgh, Pennsylvania, September 17–21; 2006.

[16] Manabe H, Zhang Z. Multi-stream HMM for EMG-based speech recognition. In: Proc. of the 26th Annual International Conference of the IEEE EMBS, San Francisco, CA, USA, September 1–5; 2004.

[17] Zhang Z, Manabe H, Horikoshi T, Ohya T. Robust methods for EMG signal processing for audio-EMG-based multi-modal speech recognition. In: COST278 and ISCA Tutorial and Research Workshop on Robustness Issues in Conversational Interaction, University of East Anglia, Norwich, UK, August 30–31; 2004.

[18] Lopez-Larraz E, MozosO M, Antelis JM, Minguez J. Syllable-based speech recognition using EMG. Conference Proceedings of the IEEE Engineering in Medicine and Biology Society. 2010, 1: pp. 4699–4702.

[19] Bu N, Tsuji T, Arita J, Ohga M. Phoneme Classification for Speech Synthesizer using Differential EMG Signals between Muscles. In: Proc. of the 2005 IEEE Engineering in Medicine and Biology 27th Annual Conference, Shanghai, China, September 1–4; 2005.

[20] Staroniewicz P, Brachmański S, Dobrucki AB. Subvocal speech recognition based on electromyographic signal [in Polish: Rozpoznawanie mowy subwokalnej w oparciu o

sygnał elektromiograficzny], Przegląd Telekomunikacyjny, Wiadomości Telekomunikacyjne No. 6; 2013, pp. 571–574.

[21] Mendes J, Robson RR, Labidi S, Barros AK. Subvocal speech recognition based on EMG signal using independent component analysis and neural network. In: Congress on Image and Signal Processing; 2008.

2

Self-Calibration of Two-Dimensional Precision Metrology Systems

Chuxiong Hu and Yu Zhu

Abstract

In modern industry, there usually exist high-precision stages, such as reticle stages and wafer stages in VLSI lithography tools, metrology stages in coordinate measurement machines, motion stages in CNC machine tools, etc. These stages need high-precision measurement/metrology systems for monitoring its XY movement. As the metrology systems are quite accurate, we often cannot find a standard tool with better accuracy to implement a traditional calibration process for systematic measurement error (i.e., stage error) determination and measurement accuracy compensation. Subsequently, self-calibration technology is developed to meet this challenge and to solve the calibration problem. In this chapter, we study the self-calibration of two-dimensional precision metrology systems and present a holistic self-calibration strategy. This strategy utilizes three measurement views of an artifact plate with mark positions not precisely known on the un-calibrated two-dimensional metrology stage and constructs relevant symmetry, transitivity, and redundancy of the stage error of the metrology stage. The misalignment errors of all measurement views, especially including those of the translation view, are totally determined by detailed mathematical manipulations. Then, a redundant equation group is synthesized, and a least-square–based robust estimation law is employed to calculate out the stage error even under the existence of random measurement noise. Furthermore, as the determination of the misalignment error components of the measurement views is rather complicated but important in previous and the proposed methods, this chapter also significantly analyzes the necessity of this costly computation. The proposed approach is investigated by simulation computation, and the simulation results prove that the proposed determination scheme can calculate out the stage error rather exactly without random measurement noise. Furthermore, when there exist various random measurement noises, the calibration accuracy of the proposed strategy is also investigated, and the results illustrate that the standard deviations of the calibration error are consistently with the same level of those of the random measurement noises. All these results verify that the proposed scheme can realize the stage error rather accurately even under the existence of random measurement noise. The practical procedure for performing a standard self-calibration is also introduced for engineers to facilitate actual implementation.

Keywords: Self-calibration, Metrology system, Stage error, Measurement noise, VLSI lithography

---

1. Introduction

Modern precision applications such as VLSI lithography tools, CNC machine tools, and coordinate measurement machines usually need multi-dimensional stages those are capable of manufacture accuracy at micro/nanometer levels [1, 2]. As automatic servo systems, these stages have metrology/measurement systems for position information and motion feedback [3–5]. Because of the unavoidable surface non-flatness, non-orthogonality, scale difference, etc., in the metrology systems, there exists systematic measurement error (i.e., stage error), which is the difference between the actual metrology system and the ideal metrology system [6]. The common way to determine the stage error map in Cartesian space needs traditional calibration technology [7, 8] with a rather standard measurement plate or scale for direct "high-precision calibrates low-precision". If the stage error can be figured out, the measurement accuracy of the metrology system in the stage could be determined and compensated, which will improve the positioning accuracy and repeatability of practical motion systems [9, 10]. However, in practical ultra-precision applications such as nano-lithography, engineers usually cannot find a standard measurement tool with better accuracy than the stage's metrology system to perform a traditional calibration process. Therefore, the idea of self-calibration is developed. The basic principle is to utilize an artifact plate with mark positions not precisely known and place the plate on the metrology stage with different locations or orientations to construct different measurement views [11]. In the following **Figure 1**, one example of the measurement view is shown where the pedestal is the un-calibrated metrology system and the grid plate is the artifact plate.

The key point of the self-calibration way is that the positions of the marks on the plate are fixed whatever the measurement view is, and researchers could utilize the measurement information of the marks in different views to construct equation about the stage error for final determination. Following this way, self-calibration has attracted attention of researchers and engineers and has been applied to certain special but important applications [12, 13], such as nano-positioners [14], profiling stages [15], scanning probe microscope [16], and coordinate measuring machines [17]. For example, Takac [18] concerned the self-calibration problem of one-dimensional stages with the utilization of an un-calibrated artifact plate to obtain congruency via transitivity and provided a calibration that makes a set of tool graduation markers with identical spacing. Raugh proposed a rigorous mathematical method for two-dimensional self-calibration under the assumption that the stage error map in Cartesian space can be expressed as a finite polynomial [19]. However, the scheme is complicated with extensive computation and even may be unstable under the existence of random measurement noise. To reduce the computation of Raugh's algorithm, for the two-dimensional self-calibration, Takac et al. [9] developed a transitive algorithm based on direct point-to-point comparison, which is intuitive and simple, but quite sensitive to random measurement noise due to that the handling of the rotation and translation of each measurement view is oversimplified.

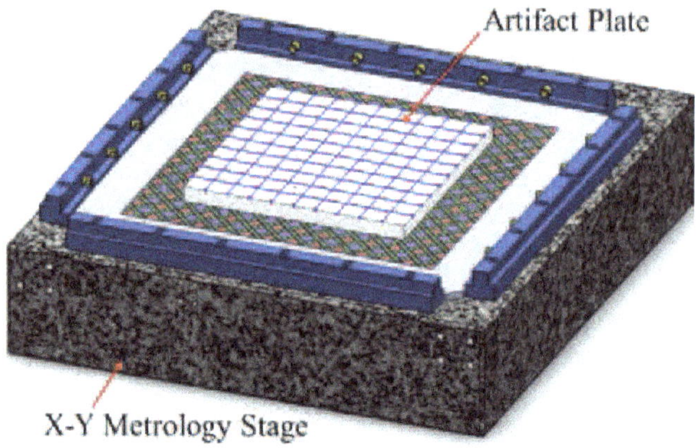

Figure 1. An artifact plate on a two-dimensional metrology stage.

In [20], Ye proposed a discrete Fourier transform-based algorithm for two-dimensional self-calibration, which is numerically robust. He also validated that the algorithm provides an exact self-calibration on the discrete sample sites when there was no random measurement noise, and only introduced calibration error of about the same size as the random measurement noise when there exists random measurement noise. The development of this algorithm is inspired by the achievement of Takac and Raugh and is considered as a standardized process by related researchers and engineers [16]. However, the computation for the misalignment error components of translation view is complex due to that the information of the rotation view is not sufficiently utilized. Moreover, discrete Fourier transform is widely used in the algorithm, which further increases the difficulty on understanding and implementation for researchers and engineers [21].

In this chapter, considering the complexity of existing algorithms, the self-calibration of X–Y precision metrology stages is studied with simplicity and effectiveness orientation. The method sufficiently utilizes the three properties of stage error detailed in [20] but abandons the Fourier transform way to provide an easily understood self-calibration algorithm. Specifically, with the three measurement views of the plate on the un-calibrated stage, the measurement information of the plate's mark positions in all views is utilized to construct symmetry, transitivity, and redundancy of the stage error. Consequently, a least-square-based self-calibration algorithm is proposed to effectively realize the stage error accurately even under the existence of random measurement noise. The computation process for the components of misalignment error of each measurement view is also provided in detail, which could be utilized as basis for synthesis of other self-calibration strategies. Furthermore, as the determination of the misalignment error components of the measurement views is rather complicated but important in previous and the proposed methods, this chapter also significantly analyzes the necessity of this costly computation. The proposed self-calibration algorithm is investigated by computer simulation, and the results validate that the stage error could be determined quite exactly when there is no random measurement noise. Furthermore, when there exists random measurement noise, the calibration accuracy also could be guaranteed, and the calibration error is at the same level as the random measurement noises, which illustrates that the proposed scheme possesses certain performance robustness. The proposed scheme actually develops a well-understood solution to two-dimensional self-calibration problem, which could facilitate practical implementation.

2. Problem formulation of two-dimensional self-calibration

2.1. Stage error

In this section, the expression of stage error would be presented. In a two-dimensional metrology/measurement system of the stage, define (x, y) as the true location of the sample site in the Cartesian grid, and **G**(x, y) as the stage error at (x, y), which is the difference between the actual metrology/measurement system and the ideal metrology/measurement system. The field to be calibrated is assumed to be L × L, and the origin of the X–Y axis is at the center of this area. Define

$$\mathbf{G}(x,y) = G_x(x,y)\mathbf{e}_x + G_y(x,y)\mathbf{e}_y \tag{1}$$

where $\mathbf{e}_x$ and $\mathbf{e}_y$ are the unit vectors of the two-dimensional stage axis; $G_x(x, y)$ and $G_y(x, y)$ are functions of the continuous X–Y field L × L. The sample sites are set to be as an N × N square array covering the field L × L, and N is odd. Then, the sample site locations in the Cartesian grid can be expressed as follows:

$$x_m = m\Delta\,,\ y_n = n\Delta \tag{2}$$

where m= $-\frac{N-1}{2}, -\frac{N-3}{2}, \cdots, \frac{N-1}{2}$, $n = -\frac{N-1}{2}, -\frac{N-3}{2}, \cdots, \frac{N-1}{2}$, and $\Delta = L/(N-1)$, which is called sample site interval. For notation simplicity, we denote as follows:

$$\begin{aligned} \mathbf{G}_{m,n} &\equiv G_{x,m,n}\mathbf{e}_x + G_{y,m,n}\mathbf{e}_y \\ G_{x,m,n} &\equiv G_x(x_m, y_n) \\ G_{y,m,n} &\equiv G_y(x_m, y_n) \end{aligned} \tag{3}$$

Figure 2(1) shows an example of the stage error, $G_{m,n}$ which leads the distortion of the actual measurement system. The objective of self-calibration in this chapter was to calculate $\mathbf{G}_{m,n}$ out which can be directly utilized to compensate the measurement accuracy. According to the presentation of [20], there are three properties for $\mathbf{G}_{m,n}$ those essentially define the ideal calibration coordinates.

- $\mathbf{G}_{m,n}$ has no translation, i.e.,

$$\sum_{m,n} G_{x,m,n} = \sum_{m,n} G_{y,m,n} = 0 \tag{4}$$

- $\mathbf{G}_{m,n}$ has no rotation, i.e.,

$$\sum_{m,n} (G_{y,m,n} x_m - G_{x,m,n} y_n) = 0 \tag{5}$$

- $\mathbf{G}_{m,n}$ has no magnification, i.e.,

$$\sum_{m,n} (G_{x,m,n} x_m + G_{y,m,n} y_n) = 0 \tag{6}$$

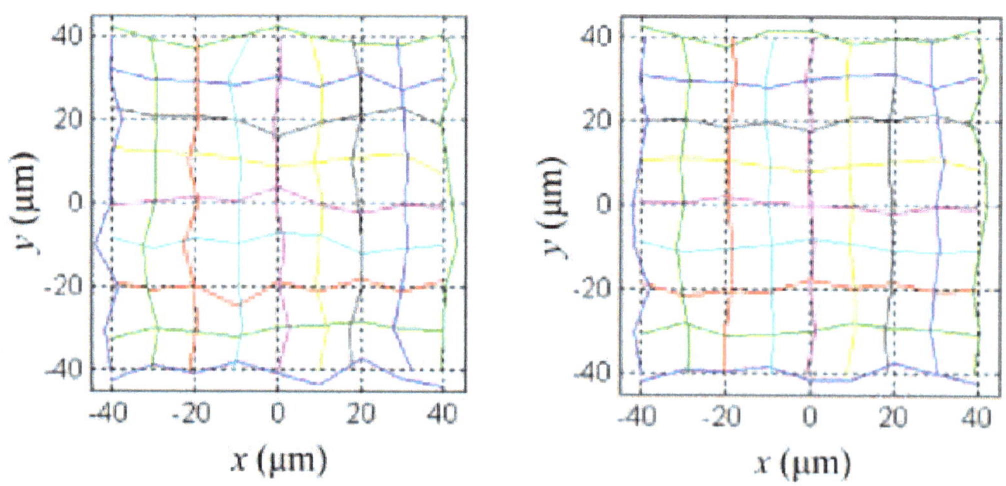

Figure 2. (1) An example of $\mathbf{G}_{m,n}$; (2) an example of $\mathbf{A}_{m,n}$.

Meanwhile, two dimensionless parameters O and R are defined as the X–Y non-orthogonality and the X–Y scale difference of $\mathbf{G}_{m,n}$, respectively. Consequently, $\mathbf{G}_{m,n}$ be described as follows [20]:

$$\begin{aligned} G_{x,m,n} &= Oy_n + Rx_m + F_{x,m,n} \\ G_{y,m,n} &= Ox_m - Ry_n + F_{y,m,n} \end{aligned} \tag{7}$$

Therefore, the determination of $\mathbf{G}_{m,n}$ can be completed by calculating out the first-order components O and R, and the residual error $\mathbf{F}_{m,n}$. It must be noted that the origin of X–Y axis is the center of the sample array and the following properties of $\mathbf{F}_{m,n}$ could be obtained [20]:

$$\begin{aligned}
&\sum_{m,n} F_{x,m,n} = \sum_{m,n} F_{x,m,n} x_m = \sum_{m,n} F_{x,m,n} y_n = 0 \\
&\sum_{m,n} F_{y,m,n} = \sum_{m,n} F_{y,m,n} x_m = \sum_{m,n} F_{y,m,n} y_n = 0
\end{aligned} \tag{8}$$

2.2. Artifact error

As we state previously, an artifact plate is needed as a critical device for the self-calibration. The used artifact plate has a square N × N mark array with the same size of the stage sample site array, and the origin of the plate X–Y axis is located at the center of the mark array. The plate axis is fixed on the plate and will move with the plate during its motion. The nominal mark locations in the plate coordinate system are the same as the sample site locations in the stage coordinate system. It is well known that the artifact plate cannot be perfect, and each actual mark at (m,n) deviates from its nominal location by $A_{m,n}$. Herein, $A_{m,n}$ is named as artifact error with one example illustrated in **Figure 2(2)**, and

$$\begin{aligned}
\mathbf{A}_{m,n} &\equiv A_{x,m,n}\mathbf{e}_{px} + A_{y,m,n}\mathbf{e}_{py} \\
A_{x,m,n} &\equiv A_x(x_m, y_n) \\
A_{y,m,n} &\equiv A_y(x_m, y_n)
\end{aligned} \tag{9}$$

where m= $-\frac{N-1}{2}, -\frac{N-3}{2}, \ldots, \frac{N-1}{2}$, $n = -\frac{N-1}{2}, -\frac{N-3}{2}, \ldots, \frac{N-1}{2}$, $\mathbf{e}_{px}$ and $\mathbf{e}_{py}$ are the unit vectors of the plate axes. It should be noted that each mark on the artifact plate is with an identification number (m,n). The mark's identification number is fixed with the mark and does not change when the plate moves on the stage. The mark number is utilized to identify each physical mark of the plate, when different measurement views are conducted and compared. Similar to the stage error $\mathbf{G}_{m,n,}$ the artifact error $\mathbf{A}_{m,n}$ also has two properties as follows [20]:

- $\mathbf{A}_{m,n}$ has no translation, i.e.,

$$\sum_{m,n} A_{x,m,n} = \sum_{m,n} A_{y,m,n} = 0 \tag{10}$$

- $\mathbf{A}_{m,n}$ has no rotation, i.e.,

$$\sum_{m,n} (A_{y,m,n} x_m - A_{x,m,n} y_n) = 0 \tag{11}$$

3. Fundamental of X–Y self-calibration

The self-calibration requires performing independent measurement views with different permutations of the artifact plate on the un-calibrated metrology stage as illustrated in

Figure 3, where the grid is the artifact plate and the colorful plane out of the grid is the metrology stage. The first view is the original orientation of the artifact plate referred to as View 0, in which the X-axis and Y-axis of the plate are aligned with those of the stage as closely as possible. In View 1, the artifact plate is rotated 90° around the origin from View 0 on the metrology stage. In View 2, the artifact plate is translated from the original View 0 by one sampling site + Δ along X-axis. In the following, we measure the rigid artifact plate marks when the plate is placed on the metrology stage in three measurement views. Through these measurement data, the stage error map can be determined by comparing different measurement views, and the misalignment error can be directly figured out with mathematical manipulations in detail.

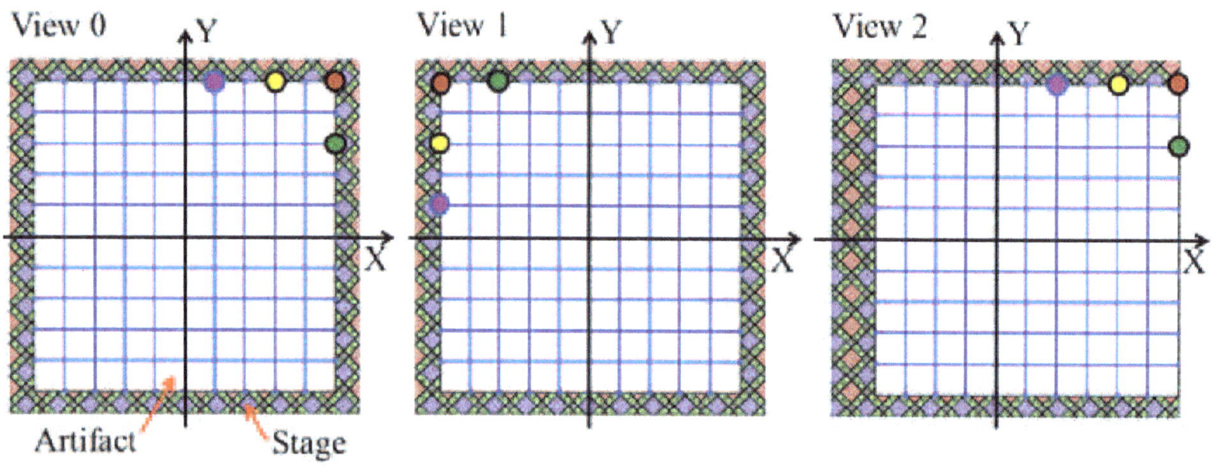

Figure 3. Three measurement views for X–Y self-calibration.

3.1. View 0

In the first view, the artifact plate is placed on the un-calibrated metrology stage with the corresponding axes aligned roughly as shown in View 0 of **Figure 3**. The deviation of mark (m,n) from the measurement value to its nominal position in the stage coordinate system is denoted as $\mathbf{v}_{0,m,n}$, where subscript 0 represents View 0. Define

$$\mathbf{v}_{0,m,n} = v_{0,x,m,n}\mathbf{e}_x + v_{0,y,m,n}\mathbf{e}_y \tag{12}$$

It must be noted that the alignment between the stage axes and plate axes cannot be perfect, i.e., there inevitably exists a small offset between their origins and a small rotation between their orientations. Therefore,

$$\begin{aligned} v_{0,x,m,n} &= G_{x,m,n} + A_{x,m,n} + E_{0,x,m,n} + r_{0,x,m,n} \\ v_{0,y,m,n} &= G_{y,m,n} + A_{y,m,n} + E_{0,y,m,n} + r_{0,y,m,n} \end{aligned} \tag{13}$$

where the misalignment error $\mathbf{E}_{0,m,n}$ is denoted as follows:

$$\begin{aligned} E_{0,x,m,n} &= -\theta_0 y_n + t_{0x} \\ E_{0,y,m,n} &= \theta_0 x_m + t_{0y} \end{aligned} \tag{14}$$

and θ_0 and $\mathbf{t}_0$ are the rotation and offset of the misalignment. $r_{0,y,m,n}$ and $r_{0,y,m,n}$are the random measurement noises. To attenuate the effects of $r_{0,x,m,n}$and $r_{0,y,m,n}$, repeated measurements are constructed with exactly the same mark position of the artifact plate. Resultantly, $V_{0,x,m,n}$ and $V_{0,y,m,n}$ are the mean values of a number of repeated measurements of $V_{0,x,m,n}$ and $V_{0,y,m,n}$, and the random measurement noise is consequently assumed to be completely attenuated, i.e.,

$$\begin{aligned} V_{0,x,m,n} &= G_{x,m,n} + A_{x,m,n} + E_{0,x,m,n} \\ V_{0,y,m,n} &= G_{y,m,n} + A_{y,m,n} + E_{0,y,m,n} \end{aligned} \tag{15}$$

Substituting Eqs. (7) and (14) into Eq. (15) leads to the following:

$$\begin{aligned} V_{0,x,m,n} &= F_{x,m,n} + Oy_n + Rx_m + A_{x,m,n} - \theta_0 y_n + t_{0x} \\ V_{0,y,m,n} &= F_{y,m,n} + Ox_m - Ry_n + A_{y,m,n} + \theta_0 x_m + t_{0y} \end{aligned} \tag{16}$$

Noting Eqs. (8), (10), (11), and (16), t_{0x}, t_{0y}, and θ_0 of the misalignment error $\mathbf{E}_{0,m,n}$ are determined [20], i.e.,

$$t_{0x} = \frac{\sum_{m,n} V_{0,x,m,n}}{N^2}, t_{0y} = \frac{\sum_{m,n} V_{0,y,m,n}}{N^2}, \theta_0 = \frac{\sum_{m,n}(V_{0,y,m,n}x_m - V_{0,x,m,n}y_n)}{\sum_{m,n}(x_m^2 + y_n^2)} \tag{17}$$

After computing $\mathbf{t}_0$ and θ_0, the misalignment error $\mathbf{E}_{0,m,n}$ is determined. Then, define

$$\begin{aligned} U_{0,x,m,n} &= V_{0,x,m,n} - t_{0x} + \theta_0 y_n \\ U_{0,y,m,n} &= V_{0,y,m,n} - t_{0y} - \theta_0 x_m \end{aligned} \tag{18}$$

It must be pointed out that, the transformation from $\mathbf{V}_{0,m,n}$ to $\mathbf{U}_{0,m,n}$ is commonly referred to as multiple-point alignment. Combining Eqs. (16) and (18) leads to the following

$$\begin{aligned} F_{x,m,n} + A_{x,m,n} &= U_{0,x,m,n} - Oy_n - Rx_m \\ F_{y,m,n} + A_{y,m,n} &= U_{0,y,m,n} - Ox_m + Ry_n \end{aligned} \tag{19}$$

3.2. View 1

In the second measurement view, the artifact plate is rotated 90° around the origin from View 0 as shown in View 1 of **Figure 3**. In this view, the plate X-axis is aligned along the stage Y-direction with same direction, and plate Y-axis is aligned along the stage X-direction but with converse direction. The deviation of mark (m,n) from the measurement value to its nominal position in the stage coordinate system is denoted as $\mathbf{V}_{1,m,n}$, where subscript 1 represents View 1. It must be noted that although located at a different position relative to the stage, the mark (m,n) in View 1 is the same physical mark (m,n) in View 0. Then,

$$\mathbf{v}_{1,m,n} = v_{1,x,m,n}\mathbf{e}_x + v_{1,y,m,n}\mathbf{e}_y \tag{20}$$

and

$$\begin{aligned} V_{1,x,m,n} &= G_{x,-n,m} - A_{y,m,n} + E_{1,x,m,n} \\ V_{1,y,m,n} &= G_{y,-n,m} + A_{x,m,n} + E_{1,y,m,n} \end{aligned} \tag{21}$$

where $\mathbf{V}_{1,m,n}$ is the mean value of a number of measurements of $\mathbf{V}_{1,m,n}$, and $\mathbf{E}_{1,m,n}$ is the misalignment error. With similar procedure of above subsection, we define

$$\begin{aligned} U_{1,x,m,n} &= V_{1,x,m,n} - \frac{\sum_{m,n} V_{1,x,m,n}}{N^2} + \frac{\sum_{m,n}(-V_{1,y,m,n}y_n - V_{1,x,m,n}x_m)}{\sum_{m,n}(x_m^2 + y_n^2)} x_m \\ U_{1,y,m,n} &= V_{1,y,m,n} - \frac{\sum_{m,n} V_{1,y,m,n}}{N^2} + \frac{\sum_{m,n}(-V_{1,y,m,n}y_n - V_{1,x,m,n}x_m)}{\sum_{m,n}(x_m^2 + y_n^2)} y_n \end{aligned} \tag{22}$$

Consequently, Eqs. (21) and (22) yield

$$\begin{aligned} F_{x,-n,m} - A_{y,m,n} &= U_{1,x,m,n} - Ox_m + Ry_n \\ F_{y,-n,m} + A_{x,m,n} &= U_{1,y,m,n} + Oy_n + Rx_m \end{aligned} \tag{23}$$

Comparing Eq. (23) of View 1 with Eq. (19) of View 0, with the same procedure in [20], the stage error components O and R can be calculated out as follows:

$$O=\frac{1}{2}\left[\frac{\sum_{m,n}(U_{0,x,m,n}y_n+U_{0,y,m,n}x_m)}{\sum_{m,n}(x_m^2+y_n^2)}+\frac{\sum_{m,n}(U_{1,x,m,n}x_m-U_{1,y,m,n}y_n)}{\sum_{m,n}(x_m^2+y_n^2)}\right]$$
$$R=\frac{1}{2}\left[\frac{\sum_{m,n}(U_{0,x,m,n}x_m-U_{0,y,m,n}y_n)}{\sum_{m,n}(x_m^2+y_n^2)}+\frac{\sum_{m,n}(-U_{1,x,m,n}y_n-U_{1,y,m,n}x_m)}{\sum_{m,n}(x_m^2+y_n^2)}\right] \tag{24}$$

After calculating out O and R, Eqs. (19) and (23) can lead to the following

$$\begin{aligned} F_{x,m,n}-F_{y,-n,m}&=P_{m,n}\\ F_{y,m,n}+F_{x,-n,m}&=Q_{m,n} \end{aligned} \tag{25}$$

where

$$\begin{aligned} P_{m,n}&=U_{0,x,m,n}-U_{1,y,m,n}-2Oy_n-2Rx_m\\ Q_{m,n}&=U_{0,y,m,n}+U_{1,x,m,n}-2Ox_m+2Ry_n \end{aligned} \tag{26}$$

3.3. View 2

As the first-order components O and R have been determined by View 0 and View 1, we try to determine $\mathbf{F}_{m,n}$ in the third view. In this view, the artifact plate is translated by roughly one sample site interval $+\Delta$ from View 0 along X-axis relative to the stage, while the stage axes and plate axes are roughly aligned, which is shown in View 2 of **Figure 3**. The deviation of mark (m,n) from the measurement value to its nominal position in the stage coordinate system is denoted as $\mathbf{v}_{2,m,n}$, where subscript 2 represents View 2. Different from View 0 and View 1, the subscript m here is from $-\frac{N-1}{2}$ to $-\frac{N-3}{2}$, due to that the far right column, i.e., $m=-\frac{N-1}{2}$, is outside the initial N × N stage sample sites. It also should be noted that although located at a different position relative to the stage (by about one sample site interval translation), the mark (m,n) here is the same physical mark (m,n) in View 0. As only the (N − 1) × N sites in View 2 within the initial N × N sites can be used, the properties of no translation and no rotation cannot be used.

Similar to Views 0 and 1, we can obtain

$$\mathbf{v}_{2,m,n}=v_{2,x,m,n}\mathbf{e}_x+v_{2,y,m,n}\mathbf{e}_y \tag{27}$$

and

$$
\begin{aligned}
V_{2,x,m,n} &= G_{x,m+1,n} + A_{x,m,n} + E_{2,x,m,n} \\
V_{2,y,m,n} &= G_{y,m+1,n} + A_{y,m,n} + E_{2,y,m,n} \\
E_{2,x,m,n} &= -\theta_2 y_n + t_{2x} \\
E_{2,y,m,n} &= \theta_2 x_m + t_{2y}
\end{aligned}
\tag{28}
$$

where m= $-\frac{N-1}{2}$, $-\frac{N-3}{2}$, ..., $\frac{N-1}{2}$, $\mathbf{V}_{2,m,n}$, is the mean value of a number of measurements of, $\mathbf{V}_{2,m,n}$, θ_2 and $\mathbf{t}_2$ are the rotation and offset of the misalignment. To be consistent with the previous two views, $\mathbf{U}_{2,m,n}$ is used instead of $\mathbf{V}_{2,m,n}$, i.e.,

$$
U_{2,x,m,n} = V_{2,x,m,n},\ U_{2,y,m,n} = V_{2,y,m,n} \tag{29}
$$

Combining Eqs. (28) and (29) and noting $x_{m+1} = x_m + \Delta$ lead to the following

$$
\begin{aligned}
F_{x,m+1,n} + A_{x,m,n} &= U_{2,x,m,n} - Oy_n - Rx_m + \xi_x - \xi_\theta y_n \\
F_{y,m+1,n} + A_{y,m,n} &= U_{2,y,m,n} - Ox_m + Ry_n + \xi_y + \xi_\theta x_m
\end{aligned}
\tag{30}
$$

where $\xi_x = -(t_{2x} + R\Delta)$, $\xi_y = -(t_{2y} + O\Delta)$, $\xi_\theta = -\theta_2$. Subtracting Eq. (19) from (30) leads to the following

$$
\begin{aligned}
F_{x,m+1,n} - F_{x,m,n} &= U_{2,x,m,n} - U_{0,x,m,n} + \xi_x - \xi_\theta y_n \\
F_{y,m+1,n} - F_{y,m,n} &= U_{2,y,m,n} - U_{0,y,m,n} + \xi_y + \xi_\theta x_m
\end{aligned}
\tag{31}
$$

Unlike the cases in View 0 and View 1, the misalignment error of View 2 is not identified simply by applying the properties of $\mathbf{G}_{m,n}$, $\mathbf{F}_{m,n}$ and $\mathbf{A}_{m,n}$, as View 2 provides no measurement data for the first column of the sampling sites. Certain algebraic manipulation is needed and is explained in Appendix A, B, and C, where the full procedures to determine the misalignment error components of View 2, i.e., ξ_θ, ξ_x, and ξ_y, are separately presented in detail. It should be pointed out that the determination process of ξ_θ, ξ_x, and ξ_y also can be used as foundation for research of other self-calibration schemes.

4. Self-calibration algorithm for X–Y metrology systems

In this section, we try to synthesize a self-calibration algorithm for X–Y precision metrology stages with accuracy and simplicity orientation. The proposed method should be well understood and robust to meet the challenge of random measurement noise, which are both important for engineers to implement in practical applications. In the following, two schemes are provided.

4.1. Scheme I

Combining Eqs. (25), (31), and (8), the following equation group for $F_{x,m,n}$ and $F_{y,m,n}$ can be obtained:

$$
\begin{aligned}
&F_{x,m,n} - F_{y,-n,m} = U_{0,x,m,n} - U_{1,y,m,n} - 2Oy_n - 2Rx_m \\
&F_{y,m,n} + F_{x,-n,m} = U_{0,y,m,n} + U_{1,x,m,n} - 2Ox_m + 2Ry_n \\
&F_{x,m+1,n} - F_{x,m,n} = U_{2,x,m,n} - U_{0,x,m,n} + \xi_x - \xi_\theta y_n \\
&F_{y,m+1,n} - F_{y,m,n} = U_{2,y,m,n} - U_{0,y,m,n} + \xi_y + \xi_\theta x_m \\
&\sum_{m,n} F_{x,m,n} = \sum_{m,n} F_{x,m,n} x_m = \sum_{m,n} F_{x,m,n} y_n = 0 \\
&\sum_{m,n} F_{y,m,n} = \sum_{m,n} F_{y,m,n} x_m = \sum_{m,n} F_{y,m,n} y_n = 0
\end{aligned}
\tag{32}
$$

Eq. (32) actually supplies linear equations for calculation of $F_{x,m,n}$ and $F_{y,m,n}$ with certain redundancy. Therefore, a least-square solution for $\mathbf{F}_{m,n}$ can be synthesized with robustness to meet the challenge of random measurement noise. To provide a more explicit illustration, we take $N = 11$ as an example, and Eq. (32) can then be assembled into matrix form, i.e., $ME_F = S$, where M is a relational matrix of dimension 468 × 242, E_F is a vector of dimension 242 × 1, and S is a vector of dimension 468 × 1. This assembly of equations under matrix form is as follows:

$$ME_F = S \tag{33}$$

where

$$
\begin{aligned}
E_F = [&F_{x,-5,-5}, F_{x,-4,-5}, F_{x,-3,-5}, \ldots, F_{x,5,-5}, \\
&F_{x,-5,-4}, F_{x,-4,-4}, F_{x,-3,-4}, \ldots, F_{x,5,-4}, \\
&\vdots \\
&F_{x,-5,5}, F_{x,-4,5}, F_{x,-3,5}, \ldots, F_{x,5,5}, \\
&F_{y,-5,-5}, F_{y,-4,-5}, F_{y,-3,-5}, \ldots, F_{y,5,-5}, \\
&F_{y,-5,-4}, F_{y,-4,-4}, F_{y,-3,-4}, \ldots, F_{y,5,-4}, \\
&\vdots \\
&F_{y,-5,5}, F_{y,-4,5}, F_{y,-3,5}, \ldots, F_{y,5,5}]^T_{242\times 1}
\end{aligned}
\tag{34}
$$

We have stated previously that three measurement views are required to generate equations with data redundancy. Consequently, a least-square method can be synthesized to provide a solution as follows

$$E_F = (M^T M)^{-1} M^T S \tag{35}$$

which possesses certain robustness to random measurement noise. Therefore, $\mathbf{F}_{m,n}$ is obtained which leads to the final determination of $G_{m,n}$.

4.2. Scheme II

As presented above, with the determination of the misalignment errors detailed in the Appendix, the stage error can be calculated out. However, it also can be observed from the Appendix that the computation process is complicated to some extent. Herein, this chapter also analyses the necessity of this costly computation and provides another alternative as follows [22].

As ξ_θ can be calculated out in the Appendix, define:

$$\begin{aligned} L_{x,m,n} &= U_{2,x,m,n} - U_{0,x,m,n} - \xi_\theta y_n \\ L_{y,m,n} &= U_{2,y,m,n} - U_{0,y,m,n} + \xi_\theta x_m \end{aligned} \tag{36}$$

Combining (31) and (36) leads to the following

$$\begin{aligned} &F_{x,m+2,n} - 2F_{x,m+1,n} + F_{x,m,n} = L_{x,m+1,n} - L_{x,m,n} \\ &F_{x,m+1,n+1} - F_{x,m+1,n} - F_{x,m,n+1} + F_{x,m,n} = L_{x,m,n+1} - L_{x,m,n} \\ &F_{y,m+2,n} - 2F_{y,m+1,n} + F_{y,m,n} = L_{y,m+1,n} - L_{y,m,n} \\ &F_{y,m+1,n+1} - F_{y,m+1,n} - F_{y,m,n+1} + F_{y,m,n} = L_{y,m,n+1} - L_{y,m,n} \end{aligned} \tag{37}$$

Consequently, the equations for $\mathbf{F}_{m,n}$ from (8), (25), and (37) can be grouped as follows

$$\begin{aligned} &\sum_{m,n} F_{x,m,n} = \sum_{m,n} F_{x,m,n} x_m = \sum_{m,n} F_{x,m,n} y_n = 0 \\ &\sum_{m,n} F_{y,m,n} = \sum_{m,n} F_{y,m,n} x_m = \sum_{m,n} F_{y,m,n} y_n = 0 \\ &F_{x,m,n} - F_{y,-n,m} = P_{m,n} \\ &F_{y,m,n} + F_{x,-n,m} = Q_{m,n} \\ &F_{x,m+2,n} - 2F_{x,m+1,n} + F_{x,m,n} = L_{x,m+1,n} - L_{x,m,n} \\ &F_{x,m+1,n+1} - F_{x,m+1,n} - F_{x,m,n+1} + F_{x,m,n} = L_{x,m,n+1} - L_{x,m,n} \\ &F_{y,m+2,n} - 2F_{y,m+1,n} + F_{y,m,n} = L_{y,m+1,n} - L_{y,m,n} \\ &F_{y,m+1,n+1} - F_{y,m+1,n} - F_{y,m,n+1} + F_{y,m,n} = L_{y,m,n+1} - L_{y,m,n} \end{aligned} \tag{38}$$

Eq. (38) also provides linear equations for determination of $F_{x,m,n}$ and $F_{y,m,n}$ with certain redundancy. Then, a least-square solution for $\mathbf{F}_{m,n}$ can be synthesized. In detail, Eq. (38) can be assembled into matrix form as follows

$$\Gamma E_F = S \tag{39}$$

where $\mathbf{\Gamma}$ is a relational matrix of dimension $D_1 \times D_2$, E_F is a vector of dimension $D_2 \times 1$, and S is a vector of dimension $D_1 \times 1$. Noting the subscripts of (38), we can calculate out D_1 and D_2 as: $D_1 = 6 + N^2 + N^2 + (N-2) \times N + (N-1)^2 + (N-2) \times N + (N-1)^2 = 6N^2 - 8N + 8$, $D_2 = N^2 + N^2 = 2N^2$. For example, assuming $N = 11$, Γ is of dimension 646×242 which can be determined by (38), while

$$\begin{aligned}
E_F = [&F_{x,-5,-5}, F_{x,-4,-5}, F_{x,-3,-5}, \ldots, F_{x,5,-5},\\
&F_{x,-5,-4}, F_{x,-4,-4}, F_{x,-3,-4}, \ldots, F_{x,5,-4},\\
&\vdots\\
&F_{x,-5,5}, F_{x,-4,5}, F_{x,-3,5}, \ldots, F_{x,5,5},\\
&F_{y,-5,-5}, F_{y,-4,-5}, F_{y,-3,-5}, \ldots, F_{y,5,-5},\\
&F_{y,-5,-4}, F_{y,-4,-4}, F_{y,-3,-4}, \ldots, F_{y,5,-4},\\
&\vdots\\
&F_{y,-5,-3}, F_{y,-4,-3}, F_{y,-3,-3}, \ldots, F_{y,5,-3},\\
S = [&0,0,0,0,0,0, P_{-5,-5}, P_{-4,-5}, \ldots, P_{5,5},\\
&Q_{-5,-5}, Q_{-4,-5}, Q_{-3,-5}, Q_{-2,-5}, \ldots, Q_{5,5},\\
&L_{x,-4,-5} - L_{x,-5,-5}, \ldots, L_{x,5,5} - L_{x,4,5},\\
&\vdots\\
&L_{y,-5,-4} - L_{x,-5,-5}, \ldots, L_{x,5,5} - L_{x,5,4}]^T_{646\times 1}
\end{aligned} \tag{40}$$

As the system has requisite data redundancy, it can be solved using a least-square method to calculate $F_{x,m,n}$ and $F_{y,m,n}$, producing a solution like

$$E_F = (\Gamma^T \Gamma)^{-1} \Gamma^T S \tag{41}$$

Therefore, $\mathbf{F}_{m,n}$ is obtained with certain robustness which finally leads to the determination of $\mathbf{G}_{m,n}$ by (7) as O and R are all known.

It should be pointed out that the proposed scheme does not calculate out the misalignment error components ξ_x and ξ_y, while these calculations are important in previous published strategies [15, 23, 24]. The proposed scheme does not need this costly computation, which would be meaningful for simplification of the calculation process to some extent. In the reminder of this chapter, as Scheme I is a holistic method and Scheme II is a simplified alternative, the simulation test and self-calibration procedure are all presented just for Scheme I. Readers could follow the following presentation to conduct similar issues for Scheme II.

5. Computer simulation

In the following, the proposed self-calibration scheme would be tested by simulation. The stage error is assumed to be on a 11 × 11 sample site array with the sample site internal Δ = 10 mm. And the stage error is generated by the command "normrnd" of MATLAB software with mean of 0 and standard deviation of 0.3, plus certain sine/cosine functions which may be partial characteristics of the stage error data, i.e.,

$$\begin{aligned} G_{x,m,n} &= normrnd(0,0.3,11,11) + 0.5sin(3m \times \frac{2\pi}{11}) \\ G_{y,m,n} &= normrnd(0,0.3,11,11) + 0.5cos(3n \times \frac{2\pi}{11}) \end{aligned} \tag{42}$$

The above data are utilized as the nominal $G_{m,n}$ shown in **Figure 4**, where the red vector lines are the stage error which have been zoomed in for 5000 times. It must be noted that $G_{x,m,n}$ and $G_{y,m,n}$ are both generated by *normrnd* (0, 0.3, 11, 11), respectively. To satisfy relevant constrains such as Eq. (4), little data modification has been made. In addition, the artifact error including $\mathbf{A}_{x,m,n}$and $\mathbf{A}_{y,m,n}$ is generated with mean of 0 and standard deviation of 0.5 μm.

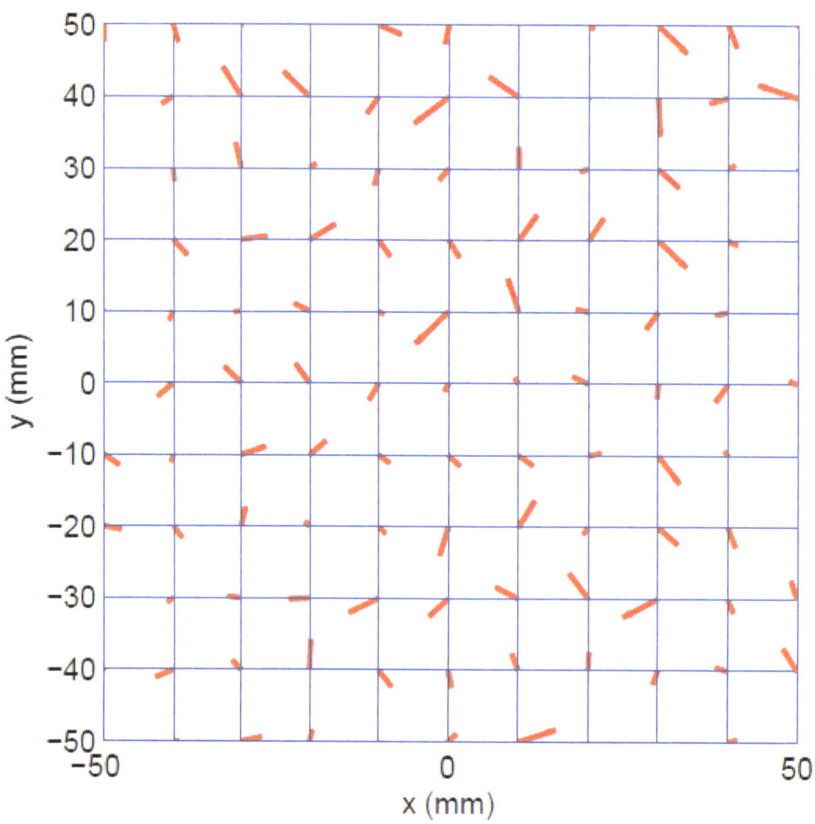

Figure 4. Nominal $\mathbf{G}_{m,n}$ × 5000 with max($\mathbf{G}_{m,n}$) = = 1.0832 μm, min($\mathbf{G}_{m,n}$) = − 1.1448 μm, and std($\mathbf{G}_{m,n}$) = 0.4611 μm.

5.1. Case I: simulation without random measurement noise

In this subsection, the random measurement noise is assumed to be perfectly compensated by the mean value of a number of repeated measurements. The recalculated stage error $\widehat{\mathbf{G}}_{m,n}$, i.e., $\widehat{G}_{x,m,n}$ and $\widehat{G}_{y,m,n}$, can be calculated out through the proposed self-calibration strategy. **Figure 5** illustrates the calibration error E_G including E_{Gx} and E_{Gy}, i.e., $\boldsymbol{E}_G = \boldsymbol{G}_{m,n} - \widehat{\mathbf{G}}_{m,n}$, $E_{Gx} = G_{x,m,n} - \widehat{G}_{x,m,n}$and $E_{Gy} = G_{y,m,n} - \widehat{G}_{y,m,n}$. The calibration error is shown as the red vector lines those have been zoomed in for 5×10^{17} times. The maximum value $max(\cdot)$, the minimum value $min(\cdot)$ and the standard deviation $std(\cdot)$of $\boldsymbol{G}_{m,n}$ $\widehat{\mathbf{G}}_{m,n}$ and E_G are all listed in **Table 1**. It can be observed that the reconstructed stage error $\widehat{\mathbf{G}}_{m,n}$ is always quite close to the actual stage error $\widehat{\mathbf{G}}_{m,n}$, and the calibration errors are all below 1.6×10^{-14} μm , which validate that the proposed self-calibration algorithm can accurately determine the stage error without the existence of random measurement noise.

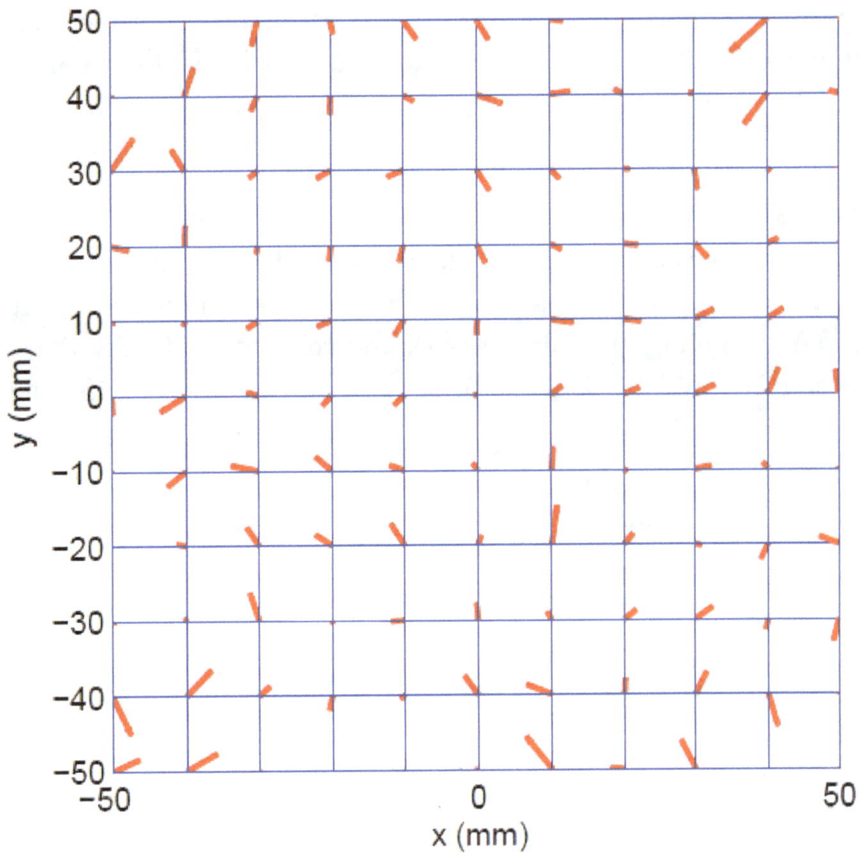

Figure 5. Calibration error $E_{Gy}(\mu m)s \times 5 \times 10^{17}$ with $max(E_G) = 1.5127 \times 10^{-14}$μm, $min(E_G) = -1.1546 \times 10^{-14}$μm, and $std(E_G) = 4.6547 \times 10^{-15}$μm (without random measurement noise).

	max (·)	min (·)	std (·)
$G_{mn}(\mu m)$	1.0832	−1.1448	0.4611
$G_{x,m,n}(\mu m)$	1.0832	−1.1448	0.4659
$G_{y,m,n}(\mu m)$	0.9753	−1.0870	0.4583
$\widehat{G}_{m,n}(\mu m)$	1.0832	−1.1448	0.4611
$\widehat{G}_{x,m,n}(\mu m)$	1.0831	−1.1448	0.4659
$\widehat{G}_{y,m,n}(\mu m)$	0.9753	−1.0870	0.4583
$E_G(\mu m)$	1.5127×10^{-14}	-1.1546×10^{-14}	4.6547×10^{-15}
$E_{Gx}(\mu m)$	1.2046×10^{-14}	-1.0214×10^{-14}	4.3670×10^{-15}
$E_{Gy}(\mu m)$	1.5127×10^{-14}	-1.1546×10^{-14}	4.9431×10^{-15}

Table 1. Calculation performance indexes (without random measurement noise).

5.2. Case II: simulation with random measurement noise of standard deviation 0.05 μm

The nominal stage error and artifact error in this subsection keep same with those of Section 5.1, except that in all site's measurements, we add independent random Gaussian measurement noises. The random measurement noise is generated with mean of 0 and standard deviation of $0.05\mu m$. Following the proposed self-calibration scheme, the reconstructed stage error $\widehat{\mathbf{G}}_{m,n}$, i.e., $\widehat{G}_{x,m,n}$ and $\widehat{G}_{y,m,n}$ also can be computed out. The calibration error E_G, which is the difference between the recalculated and the actual stage error, is shown in **Figure 6**, where the red vector lines as E_G have been zoomed in for 5×10^4 times. The maximum value $max(\cdot)$, the minimum value $max(\cdot)$, and the standard deviation of $G_{x,m,n}$, $G_{y,m,n}$, E_{Gx}, and E_{Gy} are all listed in **Table 2**. It can be seen that the proposed self-calibration method

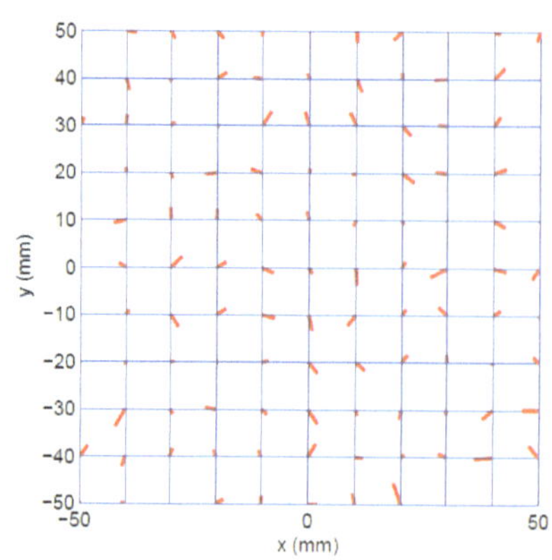

Figure 6. Calibration error $E_G \times 5 \times 10^4$ with $max(E_G) = 0.0899$ μm, $min(E_G) = -0.0793$ μm and $std(E_G) = 0.0334$ μm (with random measurement noise std = 0.05 μm).

can determine the stage error rather exactly even there exists certain random measurement noise —when the measurement noise is with standard deviation of 0.05 μm , the corresponding calibration error is with a standard deviation of 0.0334 μm which is at the same level as the random measurement noise.

We also further test the calculation robustness of the proposed scheme to various random measurement noises those are continuously generated by arbitrary 15 times. The standard deviations of the resultant calibration errors are shown in **Figure 7**, where the 15 standard deviations are all below 0.05 μm . These results consistently validate that the proposed self-calibration algorithm can meet the challenge of random measurement noise effectively.

	max (·)	min (·)	std (·)
$\hat{G}_{m,n}(\mu m)$	1.0919	−1.1598	0.4645
$\hat{G}_{x,m,n}(\mu m)$	1.0919	−1.1598	0.4693
$\hat{G}_{y,m,n}(\mu m)$	0.9744	−1.0835	0.4617
$E_G(\mu m)$	0.0899	−0.0793	0.0334
$E_{Gx}(\mu m)$	0.0552	−0.0793	0.0334
$E_{Gy}(\mu m)$	0.0899	−0.0759	0.0336

Table 2. Calculation performance indexes (with random measurement noise std = 0.05 μm).

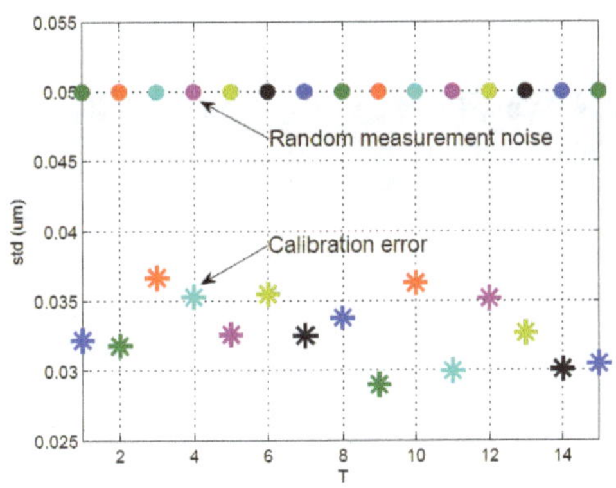

Figure 7. Standard deviation of calibration error for arbitrary 15 times (with random measurement noise std = 0.05 μm).

5.3. Case III: simulation with random measurement noise of standard deviation 5×10^{-4} μm

In this subsection, we implement other simulation to test the consistency of the proposed scheme to random measurement noise. The random measurement noise is generated with mean of 0 and standard deviation of 0.0005 μm which is rather small. **Figure 8** shows the resultant calibration error E_G under the existence of random measurement noise, where the red vector lines is E_G which have been zoomed in for 5×10^6 times. The maximum value $max(\cdot)$, the minimum value $min(\cdot)$, and the standard deviation $std(\cdot)$ of $G_{x,m,n}$, $G_{y,m,n}$, E_{Gx} and E_{Gy} are all listed in **Table 3**. Seen from this table, the proposed method can determine the stage error when there exists random measurement noise with standard deviation of 0.0005 μm --- —the standard deviation of calibration error is about 0.000327 μm , again the same level as the random measurement noise. The algorithm's robustness is further tested for arbitrary 15 times with standard deviation of 0.0005 μm , and **Figure 9** shows the standard deviations of the calibration errors. It can be seen from the figure that the standard deviations are all <0.0005 μm . All these results consistently validate that the proposed self-calibration algorithm can compute the stage error rather accurately no matter what the random measurement noise is.

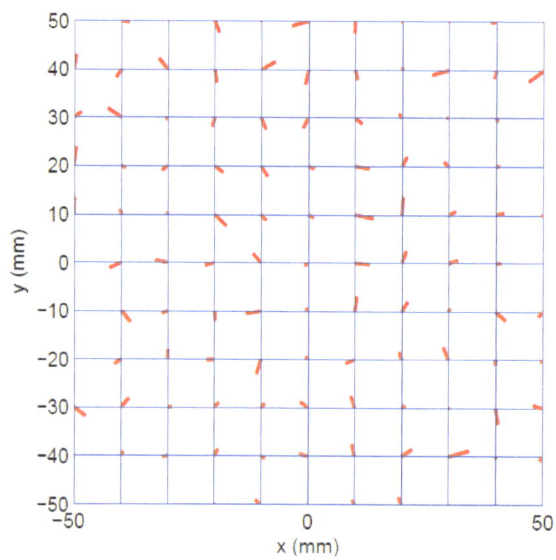

Figure 8. Calibration error $E_G \times 5 \times 10^6$ with max(E_G) = 8.8410 × 10^{-4} μm, min(E_G) = −7.4194 × 10^{-4} μm and std(E_G) = 3.2684 × 10^{-4} μm (with random measurement noise std = 5 × 10^{-4} μm).

	max (·)	min (·)	std (·)
$\hat{G}_{m,n}(\mu m)$	1.0833	−1.1441	0.4612
$\hat{G}_{x,m,n}(\mu m)$	1.0833	−1.1441	0.4659
$\hat{G}_{y,m,n}(\mu m)$	0.9756	−1.0868	0.4583
$E_G(\mu m)$	8.8410×10^{-4}	-7.4194×10^{-4}	3.2684×10^{-4}
$E_{Gx}(\mu m)$	8.8410×10^{-4}	-7.4194×10^{-4}	3.2804×10^{-4}
$E_{Gy}(\mu m)$	7.8703×10^{-4}	-6.6555×10^{-4}	3.2700×10^{-4}

Table 3. Calculation performance indexes (with random measurement noise std = 5 × 10^{-4} μm).

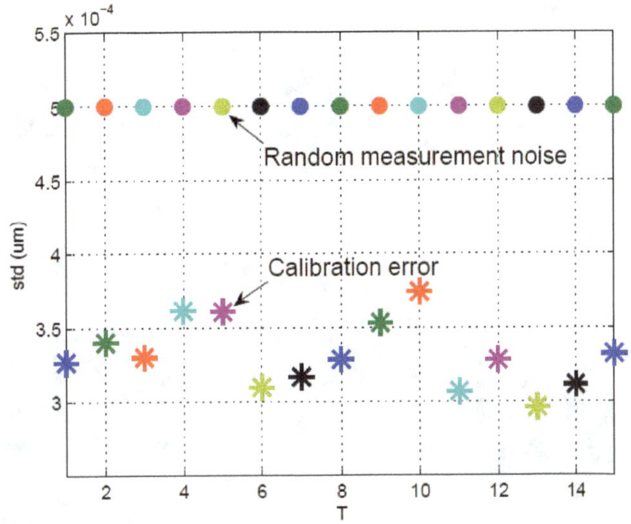

Figure 9. Standard deviation of calibration error for arbitrary 15 times (with random measurement noise std = 5×10^{-4} μm).

6. Standard self-calibration procedure

In this section, the procedure of performing a standard self-calibration following the proposed scheme is provided to facilitate practical implementations. To illustrate the procedure more clearly, a self-calibration system including an artifact plate, a two-dimensional stage, and a mark alignment system is developed and shown in **Figure 10**. Herein, the mark alignment system is developed to precisely obtain the measurement information of each mark position. The actual steps are listed in the following.

- Step 0: Place the artifact plate on the 2-D metrology stage as shown in View 0 of **Figure 1**. Utilize the metrology stage to measure the marks of the artifact plate through the mark alignment system. Obtain $v_{0,x,m,n}$ and $v_{0,y,m,n}$ for K (e.g. $K = 10$) times, and get $U_{0,x,m,n}$ and $U_{0,y,m,n}$ by (18), respectively.
- Step 1: Place the artifact plate on the 2-D metrology stage as shown in View 1 of **Figure 1**. Use the metrology stage to measure the marks of the artifact plate, and get $v_{1,x,m,n}$ and $v_{1,y,m,n}$ for K (e.g. $K = 10$) times. Then, determine $U_{1,x,m,n}$ and $U_{1,y,m,n}$ by (22), and obtain O and R by (24).
- Step 2: Place the artifact plate on the 2-D metrology stage as shown in View 2 of **Figure 1**. Use the metrology stage to measure the marks of the artifact plate, and get $v_{2,x,m,n}$ and $v_{2,y,m,n}$ for K (e.g. $K = 10$) times. Compute the misalignment error components ξ_θ, ξ_x, ξ_y, and obtain (31).
- Step 3: Through above steps, Eq. (32) can be consequently obtained. Then, by (35), $F_{x,m,n}$ and $F_{y,m,n}$ can be figured out. Finally, $G_{x,m,n}$ and $G_{y,m,n}$ can be computed out by (7) as O and R are previously determined by Step 1.

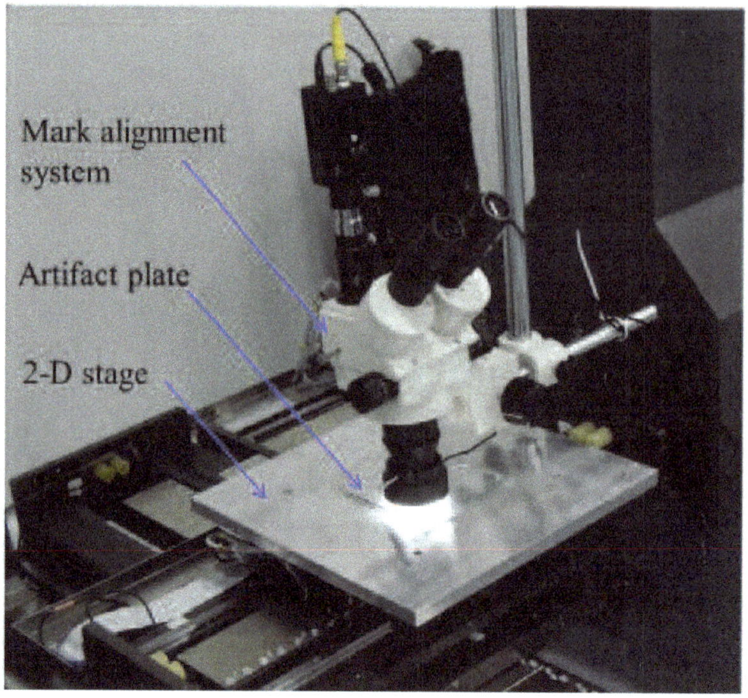

Figure 10. An example setup for 2-D self-calibration.

7. Conclusions

In this chapter, a least-square-based self-calibration strategy with simplicity and accuracy orientation has been proposed for two-dimensional precision metrology stages to address the measurement accuracy calibration problem. Three measurement views of an artifact plate on the un-calibrated metrology stage are utilized to construct symmetry, transitivity, and redundancy of the stage error, which is the basis for the synthesis of a least-square-based self-calibration algorithm. The misalignment error of each measurement view has been calculated out with explicit mathematical manipulations, while the necessity of these costly computations has been discussed. The proposed algorithm has been investigated by computer simulation, and the results show that the self-calibration strategy can reconstruct the stage error map rather precisely even there exist various random measurement noises. Finally, the procedure for performing the proposed self-calibration has been presented for engineers to facilitate practical implementation.

Acknowledgements

This work is supported by National Nature Science Foundation of China (Grant 51475262) and Autonomous Scientific Research Project of Tsinghua University (Grant 20151080363).

Appendix

Appendix A

Calculation of misalignment error component ξ_θ : Through Eq. (**31**), one can get

$$\begin{aligned}
&(F_{x,m+1,n}+F_{x,-(m+1),-n})-(F_{x,m,n}+F_{x,-m,-n}) \\
&=(F_{x,m+1,n}-F_{x,m,n})-(F_{x,-m,-n}-F_{x,-(m+1),-n}) \\
&=U_{2,x,m,n}-U_{0,x,m,n}+\xi_x-\xi_\theta y_n-(U_{2,x,-(m+1),-n}-U_{0,x,-(m+1),-n}+\xi_x-\xi_\theta y_{-n}) \\
&=(U_{2,x,m,n}-U_{0,x,m,n}-U_{2,x,-(m+1),-n}+U_{0,x,-(m+1),-n})-\xi_\theta(y_n-y_{-n})
\end{aligned} \tag{43}$$

Changing the index of Eq. (25), we can obtain

$$\begin{aligned}
F_{x,-m,-n}+F_{y,-n,m}&=Q_{-n,m} \\
F_{x,-n,m}-F_{y,-m,-n}&=P_{-n,m}
\end{aligned} \tag{44}$$

Combining Eqs. (25) and (44) leads to the following

$$\begin{aligned}
F_{x,m,n}+F_{x,-m,-n}&=C_{m,n}\equiv P_{m,n}+Q_{-n,m} \\
F_{y,m,n}+F_{y,-m,-n}&=D_{m,n}\equiv Q_{m,n}-P_{-n,m}
\end{aligned} \tag{45}$$

Then, noting Eqs. (43) and (45), we can obtain

$$(U_{2,x,m,n}-U_{0,x,m,n}-U_{2,x,-(m+1),-n}+U_{0,x,-(m+1),-n})-\xi_\theta(y_n-y_{-n})=C_{m+1,n}-C_{m,n} \tag{46}$$

Eq. (46) can be directly utilized to determine the value of ξ_θ.

Appendix B

Calculation of misalignment error component ξ_x : Summing Eq. (31) over n leads to the following

$$\begin{aligned}
H_{x,m+1}-H_{x,m}&=Z_{x,m}+N\xi_x \\
H_{y,m+1}-H_{y,m}&=Z_{y,m}+N\xi_y+N\xi_\theta x_m
\end{aligned} \tag{47}$$

where $m=-\frac{N-1}{2}, \cdots, \frac{N-3}{2}$, and

$$
\begin{aligned}
H_{x,m} &= \sum_n F_{x,m,n},\ Z_{x,m} = \sum_n (U_{2,x,m,n} - U_{0,x,m,n}) \\
H_{y,m} &= \sum_n F_{y,m,n},\ Z_{y,m} = \sum_n (U_{2,y,m,n} - U_{0,y,m,n})
\end{aligned} \tag{48}
$$

Considering $H_{x,m}$, with $j = -\frac{N-1}{2}$, we can obtain

$$
\begin{aligned}
&\sum_{m=-j}^{j} H_{x,m} x_m \\
&= H_{x,-j} x_{-j} + \{[H_{x,-j+1} - H_{x,-j}] + H_{x,-j}\} x_{-j+1} + \ldots \\
&\quad + \{[H_{x,j} - H_{x,j-1}] + [H_{x,j-1} - H_{x,j-2}] \ + [H_{x,j-2} - H_{x,j-3}] + \ldots + H_{x,-j}\} x_j \\
&= H_{x,-j} \sum_{m=-j}^{j} x_m + [H_{x,-j+1} - H_{x,-j}] \sum_{m=-j+1}^{j} x_m + \ldots + [H_{x,j} - H_{x,j-1}] \sum_{m=j}^{j} x_m
\end{aligned} \tag{49}
$$

Substituting (47) into (49), and noting $\sum_{m=-j}^{j} x_m = 0$, one can get

$$
\begin{aligned}
&\sum_{m=-j}^{j} H_{x,m} x_m \\
&= N\xi_x [\sum_{m=-j+1}^{j} x_m + \sum_{m=-j+2}^{j} x_m + \ldots + \sum_{m=j}^{j} x_m] \\
&\quad + [Z_{x,-j} \sum_{m=-j+1}^{j} x_m + Z_{x,-j+1} \sum_{m=-j+2}^{j} x_m + \ldots \ + Z_{x,j-1} \sum_{m=j}^{j} x_m]
\end{aligned} \tag{50}
$$

From Eq. (8), it can be seen that

$$
\sum_{m=-j}^{j} H_{x,m} x_m = \sum_{m,n} F_{x,m,n} x_m = 0 \tag{51}
$$

Appendix C

Calculation of misalignment error component ξ_y : Similar to $H_{x,m}$, the following equation for

$H_{y,m}$ with $j = -\frac{N-1}{2}$ can be constructed, i.e.,

$$\begin{aligned}
&\sum_{m=-j}^{j} H_{y,m} x_m \\
&= H_{y,-j} x_{-j} + \{[H_{y,-j+1} - H_{y,-j}] + H_{y,-j}\} x_{-j+1} + \ldots \\
&\quad + \{[H_{y,j} - H_{y,j-1}] + [H_{y,j-1} - H_{y,j-2}] + \ldots + H_{y,-j}\} x_j \\
&= H_{y,-j} \sum_{m=-j}^{j} x_m + [H_{y,-j+1} - H_{y,-j}] \sum_{m=-j+1}^{j} x_m \\
&\quad + \ldots + [H_{y,j} - H_{y,j-1}] \sum_{m=j}^{j} x_m
\end{aligned} \tag{52}$$

Substituting Eq. (47) into (52), and noting $\sum_{m=-j}^{j} x_m = 0$, one can obtain

$$\begin{aligned}
&\sum_{m=-j}^{j} H_{y,m} x_m \\
&= N\xi_y [\sum_{m=-j+1}^{j} x_m + \sum_{m=-j+2}^{j} x_m + \ldots + \sum_{m=j}^{j} x_m] \\
&\quad + N\xi_\theta [x_{-j} \sum_{m=-j+1}^{j} x_m + x_{-j+1} \sum_{m=-j+2}^{j} x_m + \ldots \ + x_{j-1} \sum_{m=j}^{j} x_m] \\
&\quad + [Z_{y,-j} \sum_{m=-j+1}^{j} x_m \ + Z_{y,-j+1} \sum_{m=-j+2}^{j} x_m + \ldots + Z_{y,j-1} \sum_{m=j}^{j} x_m]
\end{aligned} \tag{53}$$

From Eq. (8), it also can be seen that

$$\sum_{m=-j}^{j} H_{y,m} x_m = \sum_{m,n} F_{y,m,n} x_m = 0 \tag{54}$$

As ξ_θ has been calculated out in Appendix A, ξ_y can be figured out by Eqs. (53) and (54).

Author details

Chuxiong Hu[1,2*] and Yu Zhu[1,2]

*Address all correspondence to: cxhu@tsinghua.edu.cn

1 State Key Lab of Tribology, Department of Mechanical Engineering, Tsinghua University, Beijing, China

2 Beijing Key Lab of Precision/Ultra-precision Manufacturing Equipment and Control, Tsinghua University, Beijing, China

References

[1] M. R. Raugh. Auto calibration method suitable for use in electron beam lithography. United States, US Patent. March 7, 1984.

[2] X.-J. Wan, C.-H. Xiong, X.-F. Wang, X.-M. Zhang, and Y.-L. Xiong. A machining-feature-driven approach to locating scheme in multi-axis milling. International Journal of Machine Tools &; Manufacture. 2010, 50 (1): 42–50.

[3] C. Hu, B. Yao, and Q. Wang. Coordinated adaptive robust contouring control of an industrial biaxial precision gantry with cogging force compensations. IEEE Transactions on Industrial Electronics. 2010, 57 (5): 1746–1754.

[4] C. Hu, B. Yao, and Q. Wang. Integrated direct/indirect adaptive robust contouring control of a biaxial gantry with accurate parameter estimations. Automatica. 2010, 46 (4): 701–707.

[5] C. Hu, B. Yao, and Q. Wang. Global task coordinate frame-based contouring control of linear-motor-driven biaxial systems with accurate parameter estimations. IEEE Transactions on Industrial Electronics. 2011, 58 (11): 5195–5205.

[6] C. J. Evans, R. J. Hocken, and W. T. Estler. Self-calibration: reversal, redundancy, error separation, and 'absolute testing. Annals of the CIRP. 1996, 35 (2): 617–632.

[7] J. Kawagoe and T. Kawasaki. A new precision digital phase meter and its simple calibration method. IEEE Transactions on Instrumentation and Measurement. 2010, 59 (2): 396–403.

[8] S.-H. P. Won and F. Golnaraghi. A triaxial accelerometer calibration method using a mathematical model. IEEE Transactions on Instrumentation and Measurement. 2010, 59 (8): 2144–2153.

[9] M. T. Takac, J. Ye, M. R. Raugh, et al. Self-calibration in two dimensions: the experiment. In Proceedings of SPIE Metrology, Inspection, and Process Control for Microlithography, Bellingham, 1996, pp. 130–146.

[10] M. R. Raugh. Two-dimensional stage self-calibration: role of symmetry and invariant sets of points. Journal of Vacuum Science &; Technology B. 1997, 15 (6): 2139–2145.

[11] J. Ye. Errors in high-precision mask making and metrology. Ph.D. Thesis, Stanford University, 1996.

[12] U. Stumper. Uncertainties of VNA S-parameter measurements applying the TAN self-calibration method. IEEE Transactions on Instrumentation and Measurement. 2007, 56 (2): 2047–2052.

[13] T. Barry, G. Fuller, K. Hayatleh, and J. Lidgey. Self-calibrating infrared thermometer for low-temperature measurement. IEEE Transactions on Instrumentation and Measurement. 2011, 60 (6): 2047–2052.

[14] Y. H. Jeong, J. Dong, and P. M. Ferreira. Self-calibration of dual-actuated single-axis nanopositioners. Measurement Science and Technology. 2008, 19 (4): 1–13.

[15] S. Yoo and S. W. Kim. Self-calibration algorithm for testing out-of-plane errors of two-dimensional profiling stages. International Journal of Machine Tools &; Manufacture. 2004, 44: 767–774.

[16] M. Xu, T. Dziomba, G. Dai, and L. Koenders. Self-calibration of scanning probe microscope: mapping the errors of the instrument. Measurement Science and Technology. 2008, 19 (2): 1–6.

[17] Q. C. Dang, S. Yoo, and S.-W. Kim. Complete 3-D self-calibration of coordinate measuring machines. Annals of the CIRP. 2006, 55 (1): 1–4.

[18] M. T. Takac. Self-calibration in one dimension. In Proceedings of SPIE 13th Annual BACUS Symposium on Photomask Technology and Management, 1993, pp. 80–86.

[19] M. R. Raugh. Absolute two-dimensional sub-micron metrology for electron beam lithography: a theory of calibration with applications. Precision Engineering. 1985, 7: 3–13.

[20] J. Ye, M. T. Takac, C. N. Berglund, et al. An exact algorithm for self-calibration of precision metrology stages. Precision Engineering. 1997, 20 (1): 16–32.

[21] X. M. Lu. Real-time self-calibration and geometry error measurement in nm level multi-axis precision machines based on multi X–Y encoder integration. Ph.D. Thesis. University of New Mexico, Albuquerque, 2004.

[22] Y. Zhu, C. Hu, J. Hu, and K. Yang. Accuracy and simplicity oriented self-calibration approach for two-dimensional precision stages. IEEE Transactions on Industrial Electronics. 2013, 60 (6): 2264–2272.

[23] M. Xu, T. Dziomba, G. Dai, and L. Koenders. Self-calibration of scanning probe microscope: mapping the errors of the instrument. Measurement Science and Technology. 2008, 19 (2): 1–6.

[24] C. Hu, Y. Zhu, J. Hu, M. Zhang, and D. Xu. A holistic self-calibration algorithm for X–Y precision metrology systems. IEEE Transactions on Instrumentation and Measurement. 2012, 61 (9): 2492–2500.

3

THz Measurement Systems

Leopoldo Angrisani, Giovanni Cavallo,
Annalisa Liccardo, Gian Paolo Papari and
Antonello Andreone

Abstract

The terahertz (THz) frequency region is often defined as the last unexplored area of the electromagnetic spectrum. Over the past few years, the full access has been the objective of intense research efforts. Progress in this area has played an important role in opening up the possibility of using THz electromagnetic radiation (T-waves) in science and in real-world applications. T-waves are not perceptible by the human eye, are not ionizing, and have the ability to cross many non-conducting materials such as paper, fabrics, wood, plastic, and organic tissues. Moreover, the use of THz radiation allows non-destructive analysis of the materials under investigation both by study of their "fingerprint" via spectroscopic measurements and by high-resolution spatial imaging operations, exploiting the see-through capability of T-waves. Such technology can be applied in diverse areas, spanning from biology to chemical, pharmaceutical, environmental sciences, etc. In this chapter, we will present the typical architecture of measurement systems based on the THz technology, detailing what are the parameters that define their performance, the measurement methods, and the related errors and uncertainty, and focusing at the end on the use of time-domain spectroscopy for the evaluation of different material properties in this specific frequency region.

Keywords: THz measurements, THz technology, time domain spectroscopy, imaging, metrological characterization

1. Introduction to THz technology

Terahertz (THz) spectrum refers to the frequency domain ranging approximately from 100 GHz to 10 THz, corresponding to wavelengths from 3 mm to 30 μm (T-waves). The lower limit is the microwave region, where mobile and satellite systems operate, and the upper limit is the far infrared, widely used for optical communications. **Figure 1** shows the characteristics of the T-waves in the electromagnetic (EM) spectrum.

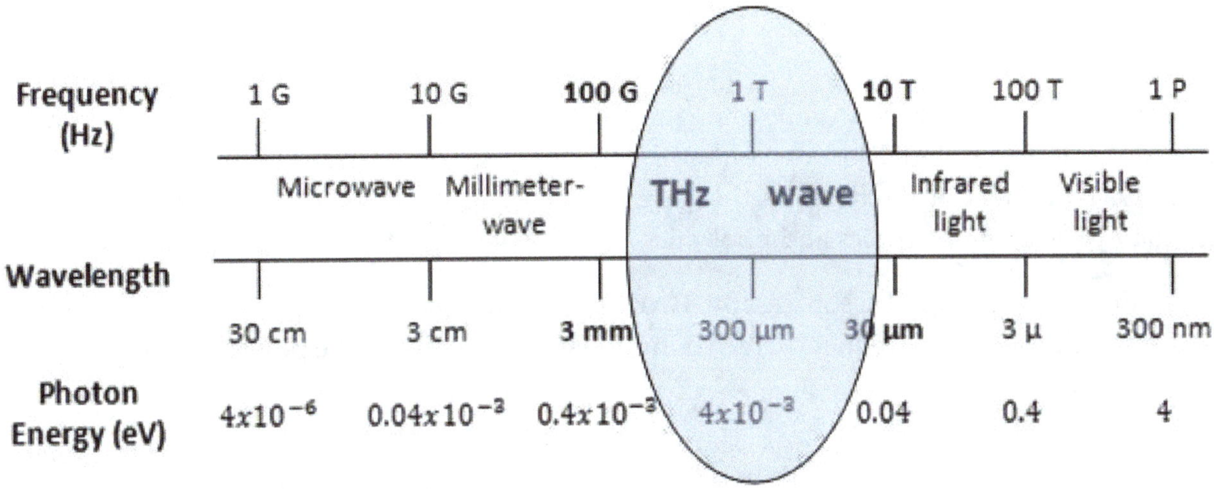

Figure 1. Position of the THz waves in the electromagnetic spectrum.

Terahertz frequency region is often defined as the last unexplored area of the EM spectrum, since it represents an area of convergence between electronics and photonics presently lacking a mature technology. Since T-waves are located between microwaves and far-infrared waves, there are two enabling technologies that can be considered for their full exploitation: electronics and photonics. Over the past 20 years, the full access and exploitation of this frequency window have been the objective of intense research efforts both in academia and in industry, in order to close the so-called THz gap. T-waves present many important properties potentially with a shattering impact both in science and in many real-world applications. First, they are not perceptible by the human eye, are not ionizing, and have the ability to cross many non-conducting common materials such as paper, fabrics, wood, plastic, and organic tissues [1, 2]. Then, in terms of energy, they give access to the rotational and vibrational modes of many molecules and macromolecules. These modes can be observed as absorption peaks in the THz spectra, providing the "fingerprint" of unknown compounds via spectroscopic measurements. Finally, the use of THz radiation allows contactless and non-destructive analysis of the materials under investigation, by spatial imaging operations [3–5] with resolution higher than micro- and millimeter waves. THz science can be applied in so many and different areas of interest, from biology to physical, chemical, pharmaceutical, and environmental research, within a broad range of industries including the medical, security, cultural heritage, manufacturing, and aerospace sectors.

The objective of this chapter is to provide readers who are not familiar with the basics of this breakthrough technology, a brief review of the typical architectures of measurement systems and the different sources and detectors that are commonly used, and looking at possible applications. Successively, they will provide details on the parameters that define the performance of a THz system, the measurement methods, and the related errors and uncertainty, focusing at the end on the use of time-domain spectroscopy for the evaluation of different material properties in this specific frequency region.

2. Overview of THz sources

Since its early beginning, one of the main hurdles for the development of THz technology was a lack of solid-state signal sources, rather than detectors. Since T-waves are located between microwaves and far-infrared waves, there are two enabling technologies that can be considered for their emission: electronics and photonics.

One can roughly group THz sources into two major operational families based on the emission mode and the operating frequency: the continuous wave (CW) and the pulsed (or time domain – TD) mode [1, 6].

Most of CW systems have been developed from the electronics side, in particular the microwave field. The typical way to generate THz emission, in fact, is by scaling the frequency by using frequency multipliers. These CW systems usually operate in the lower frequency range of the THz band with maximum frequencies around 0.8 THz. Still, there are some systems able to emit at frequencies as high as 5 THz, for example backward-wave oscillators and quantum cascade lasers. Nevertheless, CW systems could be realized from the photonics side too, since down-conversion is possible by mixing two lasers that work at different frequencies [6].

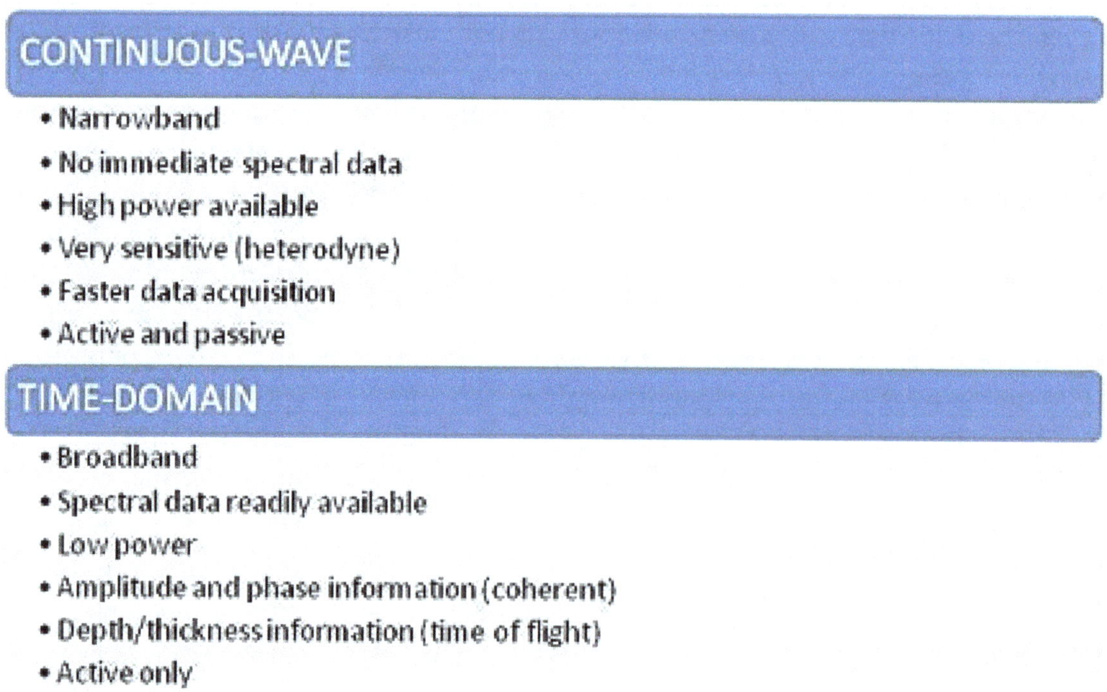

Figure 2. Characteristics of the two major operational families for the THz sources: continuous wave and time domain.

These systems are capable to operate at a single frequency, and their emission is continuous or modulated up to GHz frequencies. Therefore, CW systems are intrinsically narrowband and have often a limited tunability with a high spectral resolution (~100 MHz) very useful for gas-phase spectroscopy. Moreover, CW systems typically provide output power higher than pulsed sources. They can be passive or active; in the first case, the system detects the radiation emitted by the sample or body, whereas in the second case, the system illuminates the sample and detects the reflected or transmitted radiation. CW systems can be used in telecommunications applications and non-destructive evaluation (NDE) applications.

In pulsed or time-domain (TD) systems, the distinctive element is an optical-to-THz signal conversion technology, based on the generation and detection of an electromagnetic transient having duration of few picoseconds by means of ultrafast pulsed lasers [6]. The short pulse is composed of many frequencies, which can be accessed with a Fourier transform of the pulse. Contrary to what happens in CW systems, pulsed systems can be only active, are broadband in nature, and do not benefit of a continuous emission, so that they are ideal to study ultrafast phenomena and for general purpose spectroscopic applications.

In **Figure 2**, the major differences between CW and pulsed systems are summarized.

In the following, the most commonly used THz sources will be described, with some emphasis on those based on electro-optical conversion, since they form the base for the development of coherent systems operating in the time domain.

2.1. Backward-wave oscillator (BWO)

The backward-wave oscillator is one of the most important and successful sources based on up-conversion, allowing to extend the frequency of microwave sources to the THz range using harmonic generation. The mechanism of a BWO is very similar to a travelling wave amplifier, with the difference that it is a slow-wave structure, deliberately designed to provide feedback. In particular, a BWO seems a sophisticated high vacuum diode, where the cathode is heated by a low voltage heater and emits electrons accelerated by a high-voltage field travelling toward the anode. The electrons are collimated by a uniform external magnetic field and pass over the slow-wave structure like a comb. This mechanism produces the required quantity for the transfer of the kinetic energy of the electrons to an electromagnetic wave that builds up from noise fluctuations. The most important advantage of BWOs is their tunability; in fact, the tuning rate is ~10 MHz/V for low frequency devices, rising to ~100 MHz/V for devices operating at 1 THz or above. In addition, the output power of each BWO varies quite rapidly with frequency and the useful tuning range is approximately ±10% from the centre frequency [7].

2.2. Quantum cascade laser (QCL)

The quantum cascade laser belongs to the semiconductor-based THz sources. Over the years, the enormous progress in the field of nanotechnology is making QCL the most employed source in the CW family. In typical semiconductor lasers, light is generated by the recombination of electrons in the conduction band with holes in the valence band, separated by the gap of the active material. In QCLs, the presence of coupled quantum wells (QWs) splits the conduction band by quantum confinement into a number of distinct sub-bands. In fact, the structure of a QLC is composed of semiconducting layered QWs from hundreds to thousands (like InGaAs/AlGaAs) [8]. Applied electric field, lifetimes, and tunneling probabilities of each level are fundamental to obtain population inversion between two sub-bands in a series of identical repeat units. The output radiation frequency is defined by the energy spacing (or QW thickness) of the lasing sub-bands. The active regions are connected with injector/collector structures allowing electrical transport through injection of carriers into the upper laser level and extraction of carriers from the lower laser level [9, 10].

2.3. Stimulated terahertz amplitude radiation (STAR) emitters

Stimulated terahertz amplified radiation (STAR) is compact and technologically simple all solid-state emitters based on superconducting devices. Coherent electromagnetic waves are generated at THz frequencies because of the Josephson effect between one or more superconductor–insulator–superconductor (SIS) junctions. Biased by a DC voltage V, a Josephson junction is essentially a two-level system with energy difference of 2 eV. As Cooper pairs are tunneling, EM waves are emitted from the junction. The radiation from a single junction, however, is only about 1 pW, and the frequency is below THz, limited in conventional low Tc superconductors by the small superconducting energy gap. The radiation power can be enhanced by fabricating an artificial array of Josephson junctions. Nevertheless, the crucial aspect relies on the coherent emission of EM waves, which requires synchronized oscillations of individual elements. The synchronization looks easier to achieve for the so-called intrinsic Josephson junctions (IJJs) that are densely packed inside high-quality single crystals of a high transition temperature Tc superconductor, usually $Bi_2Sr_2CaCu_2O_{8+\delta}$ (BSCCO). IJJs are formed naturally in BSCCO, where Bi-Sr-O layers between the superconducting CuO_2 layers act as a non-conducting barrier of nanoscale thickness [11]. The main feature rendering the STAR emitters so attractive is the nature of the emitted radiation, which is fairly robust, reasonably intense (~ μW), and characterized by high spectral purity. Furthermore, the line width is so sharp that its observation is limited by the resolution of the conventional spectrometers (0.25 cm^{-1}). Then, the frequency of the THz radiation can be tuned considerably, approximately up to 10–15% of the central frequency, by varying the bias voltage applied to the N IJJ stack, even if its variable range of frequency strongly differs from sample to sample, depending on the preparation conditions [12].

2.4. Photoconductive antenna (PCA)

A photoconductive antenna is the most commonly used source in THz TDS systems. It consists of a semiconductor substrate where two metallic electrodes are deposited and separated by a gap of few microns. The substrate is typically a direct III-V semiconductor such as GaAs or low-temperature grown GaAs, sometimes doped silicon is adopted. In PCA, photocarriers are produced by the laser pulse and then accelerated by a bias field applied across the gap. In order to generate photo-induced free carriers, it is necessary that the photons of the laser pulse have an energy higher than the bandgap of the semiconductor. If the laser is focused at the gap between the electrodes, the photo-induced carriers are accelerated by the bias field across the gap, which generates a current. The amplitude of the current is a function of time, and thus, the derivative of the current respect time generates the THz pulse, in fact, for the pulsed nature of the laser beam [13, 14]:

$$E_{THz} = \frac{Ae}{4\pi\varepsilon_0 c^2 z}\frac{\partial N(t)}{\partial t}\mu E_b \quad (1)$$

where A is the area under beam illumination, e the electron charge, ε_0 is the vacuum permittivity, c is the light velocity, z is the pulse penetration inside the semiconductor, N is the photocarrier density, μ is the carrier mobility, and E_b is the bias field. The main contribution to the photocurrent comes from the electrons since their mobility is often higher than that of holes. In **Figure 3**, the photoconducting antenna is excited by a fs laser pulse. The dipole structure is biased at a voltage V_{bias} to increase the THz signal emitted from the device.

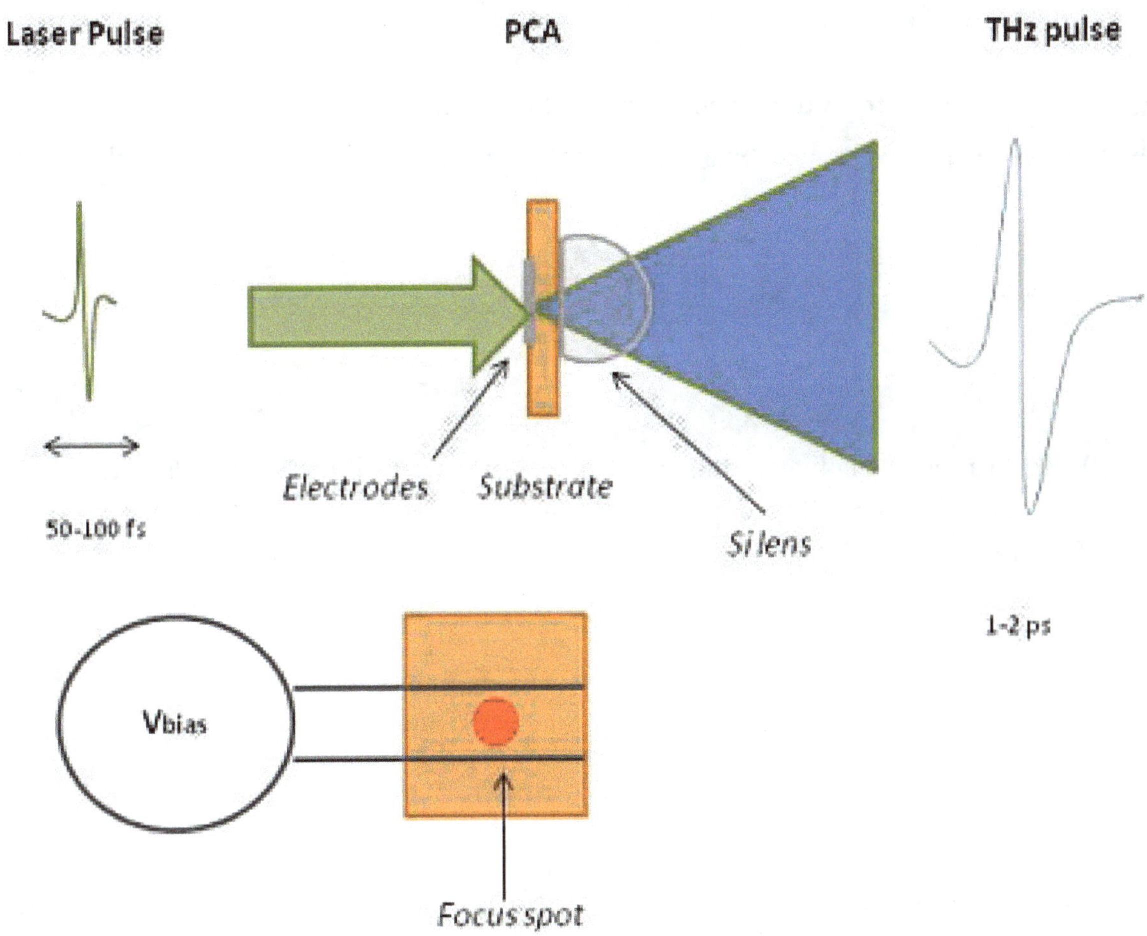

Figure 3. Sketch of the THz emission mechanism induced in a photoconductive antenna by a fs laser pulse.

Performances of PCAs can be appreciated considering some important and fundamental aspects as: semiconductor bandgap, carrier life and mobility, antenna gap and bias field. Finally, PCAs structures can be resonant or non-resonant. The first ones generate THz radiation around a central frequency, which depends on the gap distance, and the second ones have variable gap distances and provide broader frequency THz emission [13, 15].

2.5. Electro-optical conversion (EOC)

Another type of THz source is based on the electro-optical rectification. It is a nonlinear optical effect, and THz waves are generated as a result of a difference-frequency process between the frequency components contained in the femtosecond laser pulse, occurring in materials having a higher order susceptibility that is different from zero.

Mathematically, the polarization induced by the electric field associated with the optical pulse can be expressed in power series [16]:

$$\overline{P}(\overline{r},t) = \chi^{(1)}\overline{E}(\overline{r},t) + \chi^{(2)} : \overline{E}(\overline{r},t)\overline{E}(\overline{r},t) + \chi^{(3)} : \overline{E}(\overline{r},t)\overline{E}(\overline{r},t)\overline{E}(\overline{r},t) + \ldots \tag{2}$$

EO rectification comes from the second term in the previous equation. If the incident light is a plane wave, the polarization induced by the second-order susceptibility can be expressed as:

$$\overline{P}_{OR}(t) = 2\chi^{(2)} : \int_{\infty}^{0}\int_{\infty}^{0} \overline{E}(\omega+\Omega)\overline{E}'(\omega)e^{-j\Omega r}d\Omega d\omega \tag{3}$$

Here, Ω is the difference between two frequency components of the laser pulse, whereas $\chi^{(2)}$ is the second-order susceptibility, depending on the material crystalline structure. The radiated electric field due to the EO induced polarization can be expressed as follows [17, 18]:

$$\overline{E}_{THz} \propto \frac{\partial J(t)}{\partial t} = \frac{\partial^2 P(t)}{\partial t^2} = \chi^{(2)}\frac{\partial^2 E_{laser}(t)}{\partial t^2} \tag{4}$$

It is worth to emphasize that in EO rectification, no bias is necessary to realize THz generation (**Figure 4**). For a given material, the radiation efficiency and bandwidth are affected by factors such as thickness, laser pulse duration, absorption and dispersion, crystal orientation, and phase-matching conditions [19].

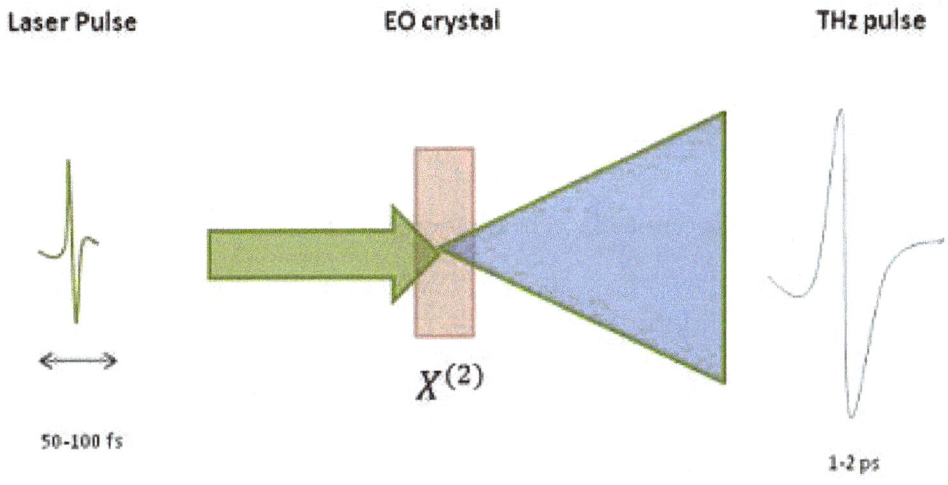

Figure 4. Sketch of the electro-optical mechanism.

2.6. Photomixing for the generation of THZ radiation

The term "photomixing" describes the generation of T-waves as a difference frequency in a nonlinear element. In the case of the THz region, it is necessary to use two IR or two visible laser photons, so that the laser frequency difference lies in this frequency range. Basically, a THz photomixer consists of two independently tunable sources (usually, solid-state diode lasers) lighting a photoconductive antenna PCA and yielding the desired difference frequency by heterodyning [20], as shown in **Figure 5**.

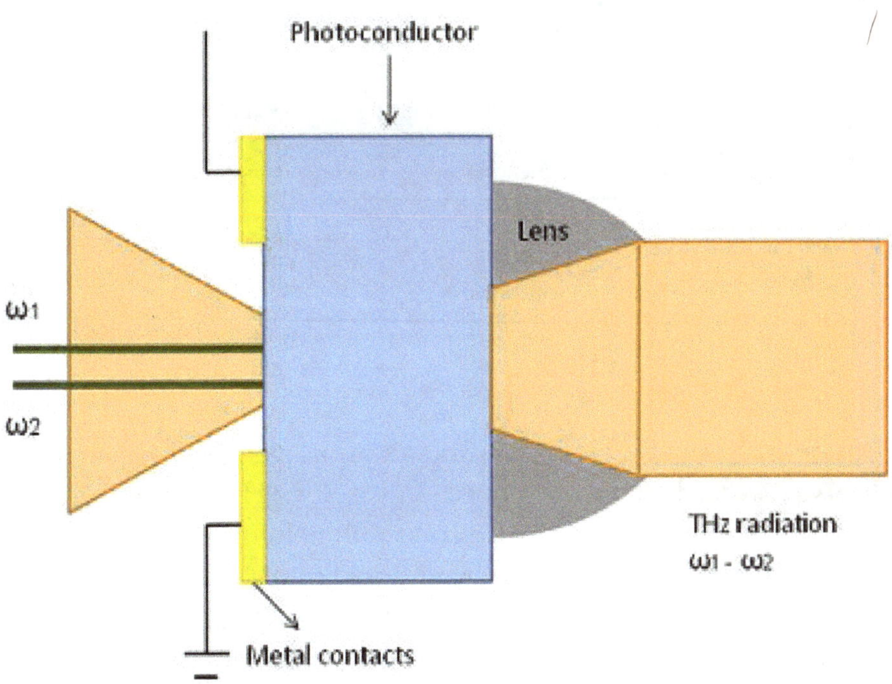

Figure 5. Operation of a THz photomixer.

The device serves as a terahertz coherent wave emitter. Changing the temperature or the operation current of the laser diodes, the value of the beat frequency can be straightforwardly regulated.

The band of the photomixer depends on the spectral width of the employed diodes; in particular, in order to enlarge the band or to move the band of the photomixer into higher frequencies, diodes with different central frequencies are necessary [21].

Table 1 summarizes the major strengths and weaknesses for the operation of the sources reported above in the THz region.

Source	Advantages	Drawbacks
BW oscillator	✓ High tunability	✗ Bulky
	✓ High output power	✗ Low frequency
QC laser	✓ High integration	✗ Small tunability
	✓ Moderate output power	✗ Low operation temperature
STAR emitter	✓ High integration	✗ Small output power
	✓ High spectral purity	✗ Low operation temperature
	✓ Moderate tunability	
Photoconductive antenna	✓ Broadband	✗ Small output power
		✗ Bulky
Electro-optical conversion	✓ Broadband	✗ Small output power
		✗ Bulky
Photomixing	✓ Moderate output power	✗ Bulky
	✓ High frequency	

Table 1. List of pros and cons for different THz sources.

3. Overview of THz detectors

In general, detectors are transducers converting an input signal into some convenient form that can be observed, recorded, and analyzed. In case of THz technology, the signal is an electromagnetic wave whose amplitude and phase both hold important information. According to the nature of EM wave, detectors can be grouped into two classes: incoherent or direct detectors that detect the amplitude only and coherent detectors that detect both amplitude and phase.

The performance of a THz detector depends on a number of parameters, some of them correlated among them. The most important are as follows: bandwidth (the spectral range over which the detector responds), responsivity (input–output gain of the detector system), noise characteristics (characterized by the noise equivalent power, NEP, that is the signal power required to yield a signal-to-noise ratio of unity at the output of the detector in a 1 Hz bandwidth), dynamic range, response time, and sensitive area.

3.1. Thermoelectric detectors

The photon energy of a THz wave is relatively weak and comparable to the phonon energy of a crystal; for this reason, it could be detected as thermal energy. In principle, the response of such thermoelectric detectors is very slow—in the order of milliseconds—because of the thermoelectric reaction of the crystal, but it is usually ultra-broadband.

The most commonly used thermoelectric detectors are as follows: Golay cell, pyroelectric, and bolometer.

The Golay cell is a gas cell detector in which an IR-absorptive gas is encapsulated, and its thermal expansion produced by THz waves is optically detected. The pyroelectric detector is a photovoltaic detector made with a dielectric material, exhibiting temperature-sensitive surface polarization and high sensitivity to THz imaging. Typical value of NEP is 100 pW/√Hz. Finally, the bolometer is a temperature-sensitive semiconductor, such as Si or Ge, and detects the THz radiation as the change in resistivity by the heating due to absorption of THz radiation. The bolometer could operate at cryogenic temperatures, ultimately minimizing the background noise level. The typical value of NEP is $10^{-6}\mathrm{W}/\sqrt{\mathrm{Hz}}$ at 0.3 K and $1\mathrm{pW}/\sqrt{\mathrm{Hz}}$ at 300 K.

3.1.1. Golay detectors

The mechanism of a modern version of this detector, with its main features, is shown in **Figure 6**.

The main component is a sealed cell where a gas having a low thermal conductivity, usually xenon, is inserted. One side of the cell ends with a window that allows the transmission of the THz radiation in the adopted frequency interval. The other side instead is closed with a flexible mirror. The cell is completed with a thin absorbing metallic film whose impedance nearly matches that exhibited by free space. The metallic film absorbs the THz radiation thus heating the surrounding gas and causing a slight movement of the mirror. This displacement is subsequently converted into an electrical signal. In particular, a lens system is exploited to concentrate the light emitted by the source through a grid towards the mirror; then, the mirror reflects the light back to the detector through the grid. If a displacement of the mirror occurs, the reflected image of the grid is distorted, and thus, the amount of light reaching the light detector changes. For THz frequencies, the most useful window materials are high-density polyethylene (HDPE), high-resistivity Si, crystalline quartz, and diamond [22].

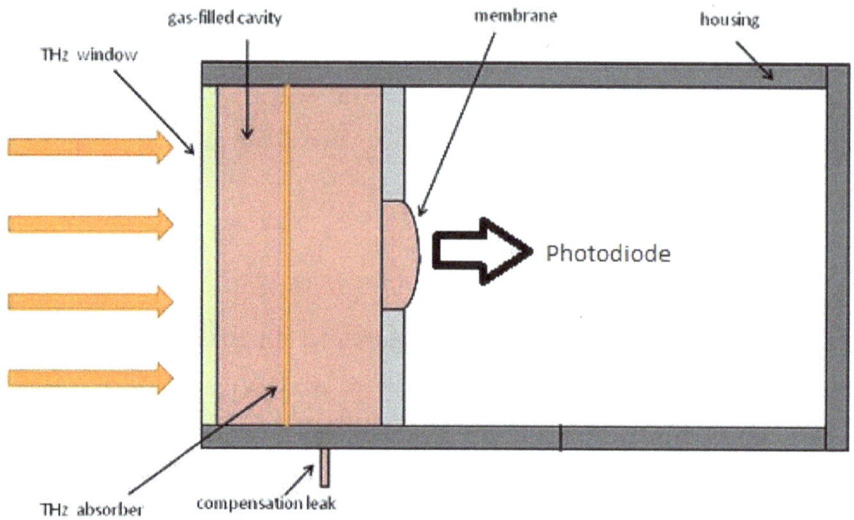

Figure 6. Structure of a Golay cell.

3.1.2. Pyroelectric detectors

A pyroelectric detector (or pyrometer) is an ac thermal sensor characterized by a frequency response spanning a large frequency range, which includes the THz region. It is based on the pyroelectric effect exhibited by a thin, permanently poled, ferroelectric crystal (i.e., $LiTaO_3$), in which the instantaneous polarization is dependent on the rate of change of the crystal temperature. Pyrometers are commercially available either as single devices or as arrays for the entire IR and THz spectral regions. Besides being very sensitive, they have many advantages, including being relatively cheap and rugged, with room temperature operation. Their most useful property is that, with appropriate design of an associated amplifier, they can exhibit response times ranging from milliseconds to less than a nanoseconds [23]. In **Figure 7**, the scheme of a pyroelectric detector and its equivalent circuital model is presented.

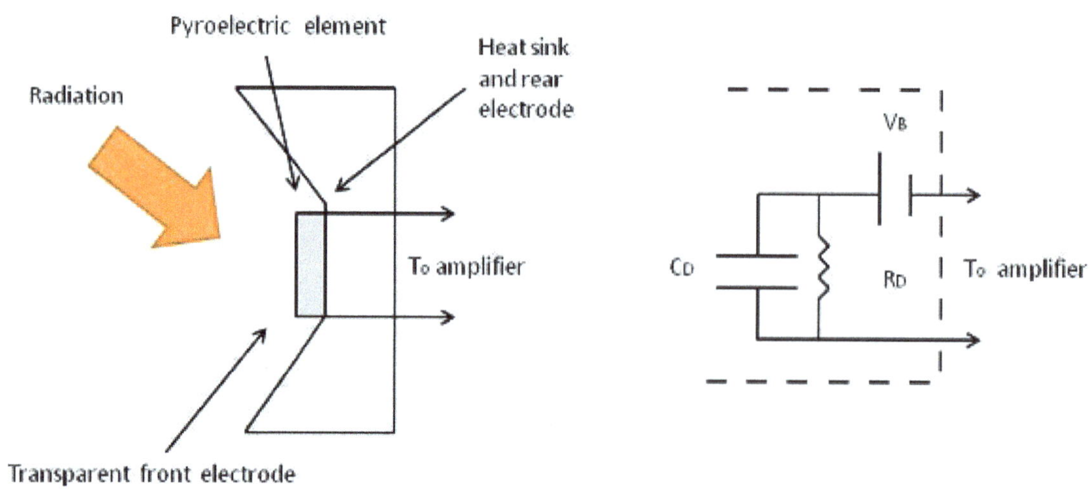

Figure 7. Pyroelectric detector and equivalent representation of electric circuit diagram.

3.1.3. Bolometer detectors

Semiconducting bolometers are among the most important of THz detectors. A design for a bolometer is shown in **Figure 8**. It consists of a small chip of doped semiconductor, Si or Ge. Typical doping concentrations are 10^{16} cm^{-3} for Ge and 10^{18} cm^{-3} for Si. The detector element is suspended in vacuum by two thin lead wires between the electrical contacts, which provide the electrical connections as well as the thermal link to the heat sink. Two aspects determine the optimum level of doping: (i) the temperature coefficient of the resistance should be large and (ii) the bolometer should have a resistance that allows for an efficient coupling to a low noise amplifier [24].

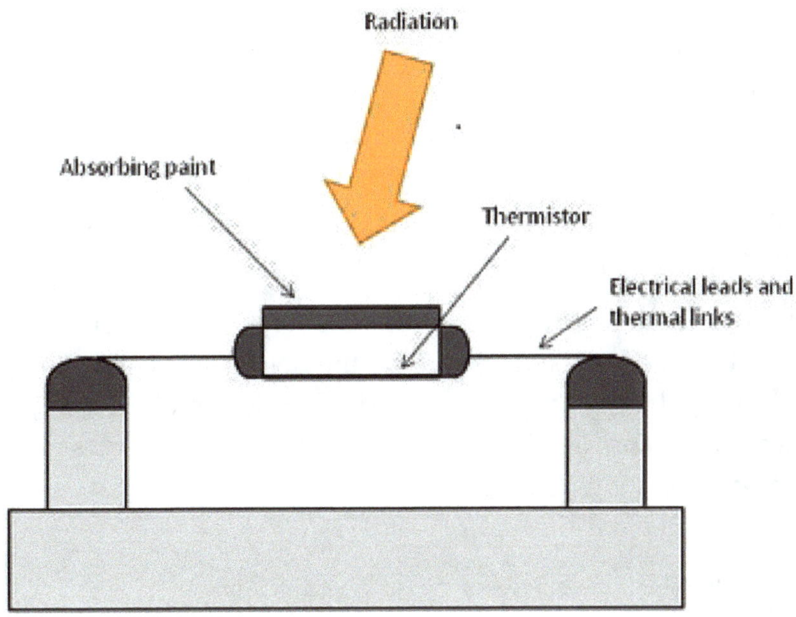

Figure 8. Simple design of a semiconducting bolometer.

A specific bolometer is the superconducting hot-spot air bolometer (SHAB), where the detector element consists of a microscopic narrow Nb or NbN strip creating a free-standing bridge structure on top of a substrate. When a voltage bias is applied to the structure, this produces the formation of a hot spot in the middle of the strip where the incoming radiation energy is dissipated, thereby switching from the superconducting to the normal state. The overall effect is a modulation of the suspended strip resistance, with a consequent modulation of the current flowing because of the voltage bias. From the recorded current, one can extract a measure of the THz radiation [25].

3.2. Photoconductive antenna

A photoconductive antenna could be considered also to detect THz waves. The structure is very similar to the structure of PCA for emission. In this case, PCA as detector measures the photocurrent, which is collected by the electrons generated by the probe beam across the antenna gap and biased by the THz electric field. When no electric field is present, the photocarriers produced by the laser pulse move randomly and no current is observed. On the contrary, when the THz wave irradiates the gap, it generates an electric field separating electrons from holes, and therefore, a current, proportional to the amplitude of the electric field, is observed. It is important to emphasize that PCA for detection and PCA for emission are differently designed. The narrower is the gap, the lower is the electric field required for obtaining an appreciable current; therefore, PCA for detection exhibits narrower gaps (~10 μm) when compared with typical gaps of PCA for emission (> 50 μm) [14, 26, 27]. Factors affecting the performance of a PCA are similar to those for the emitter: semiconductor bandgap, carrier lifetime and mobility, and antenna gap [14].

3.3. Electro-optical sampling (EO)

Electro-optical (EO) sampling is based on the Pockels effect, in which the application of an electric field, on a material, induces or modifies the birefringence properties of it. In other words, the Pockels effect is a change in the refractive index or birefringence that depends linearly on the electric field. It is important to say that the Pockels effect can be observed only in crystals characterized by no inversion symmetry, like those belonging to the zinc-blende group such as the ZnTe [19, 27].

Using this detection method, the THz electric field is sensed by measuring the change of the birefringence properties of the crystal, caused by the field itself. These changes can be measured by analyzing the polarization properties of an optical probe beam going through the crystal. The most common setup to measure the THz waveform with EO sampling is a balanced measurement approach, as shown in **Figure 9**.

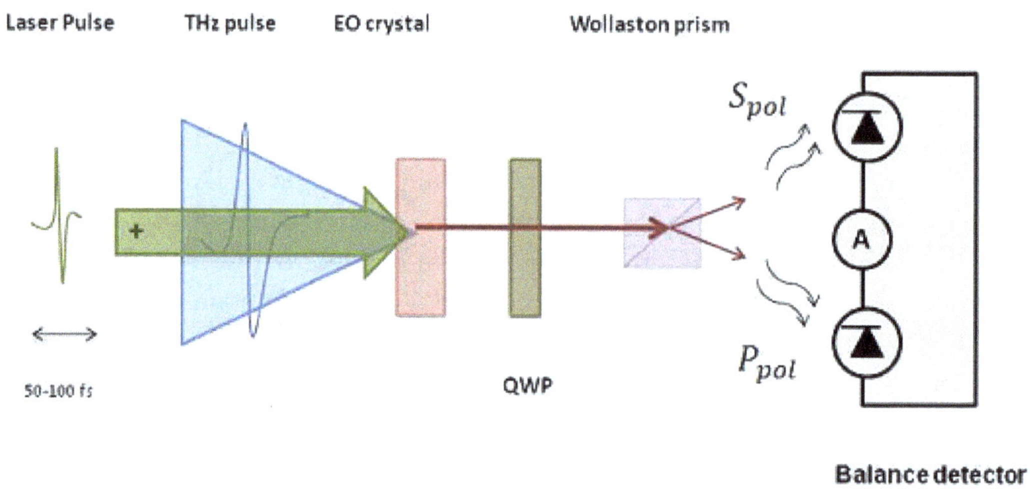

Figure 9. Sketch of the electro-optical sampling.

The operating principle is as follows. An optical probe beam characterized by linear polarization is first passed through a polarizer and then propagates within the EO crystal. A quarter-wave plate (QWP) is positioned just after the EO crystal in order to modify the probe beam ellipticity; moreover, a suitable Wollaston prism is exploited to split the elliptical polarization into its two perpendicular components. A proper photodiode assembly mounted in differential configuration detects the diverse polarization intensity. In the absence of THz radiation impinging on the EO crystal, the probe beam ellipticity can be regulated in such a way that both polarization intensities are equal; as a consequence, the net current flowing from the differential photodiodes is equal to zero. On the contrary, when the THz wave is present, the birefringence of the EO crystal is modified by the electric field, thus changing accordingly the ellipticity of the probe beam. As a result, the balance between the two polarizations is broken and the photodiodes assembly can generated a net current whose intensity is proportional to the amplitude of the electric field associated with the impinging THz wave.

Table 2 summarizes the major strengths and weaknesses for the operation of the detectors reported above in the THz region.

Detector	Advantages	Drawbacks
Golay cell	✓ Broad spectral response	✗ Limited dynamic range
	✓ High sensitivity	✗ Slow response
	✓ Large sensing area	✗ Bulky
		✗ Fragile
		✗ Expensive
Pyrometer	✓ Broad spectral response	✗ Large NEP
	✓ High dynamic range	✗ Microphonic response
	✓ Fast response	
	✓ Compact	
	✓ Inexpensive	
Bolometer	✓ Broad spectral response	✗ Slow response
	✓ Highest sensitivity	✗ Low operation temperature
	✓ Lowest NEP	✗ Small sensing area
	✓ Compact	
Photoconductive antenna	✓ Fast response	✗ Low detection current
	✓ Moderate sensitivity	✗ Relatively expensive
Electro-optical crystal	✓ Broad spectral response	✗ Complex readout
	✓ Fast response	✗ Relatively expensive
	✓ Moderate sensitivity	

Table 2. List of pros and cons for different THz detectors.

4. THz systems

As already described in Section 2, the operation of THz systems can be schematically divided into continuous wave and pulsed mode.

Frequency domain (THz-CW) systems work at a fixed frequency, which depends on the type of THz emitter. As an example, in **Figure 10**, the case of a QCL source coupled to a thermal detector (a pyrometer or a bolometer) is shown. Between THz emitter and detector, there is an ellipsoidal mirror that allows to collimate the laser beam. In this configuration scheme, the reference signal produced by the source and the output signal recorded by the detector is sent to a lock-in amplifier in order to provide a coherent detection [28].

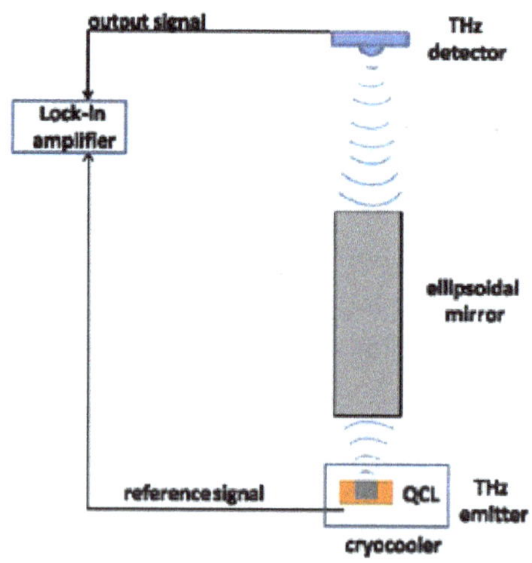

Figure 10. Sketch of a THz-CW system. A QCL is used as THz source and a bolometer or a pyrometer as THz detector.

In the following, the attention will be focused primarily on the characteristics and performance of THz measurement system working in the time domain (THz-TD). A typical architecture is shown in **Figure 11**.

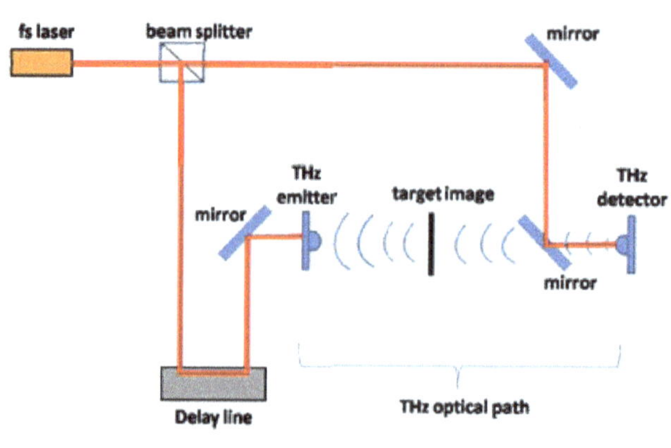

Figure 11. Typical scheme of a THz-TD system in transmission mode.

A beam splitter is adopted to divide the ultrafast optical pulse generated by an ultrafast laser into two beams, referred to as probe and pump, respectively. T-ray pulsed radiation is stimulated at the emitter by the optical pump beam via either charge transport or optical rectification effect, according to the specific type of exploited emitter. A suitable set of lenses and a pair of parabolic mirrors is adopted in this configuration to collimate and focus the diverging T-ray beam on the sample of interest. A similar combination of lenses and mirrors is then needed to recollimate and focus onto the receiver the T-ray beam passed through the sample. At receiver side, the originally split probe beam acts as an optical gate for the T-ray receiver; the optical gate signal is characterized by a shorter time duration compared with that of the arriving T-ray pulse. It is worth noting that a proper synchronization between the gating and T-ray pulse is mandatory to assure T-ray coherent detection at a time instance. By carefully controlling the optical delay line by means of the proper micromotion of a mechanical stage, it is possible to achieve a complete temporal scan of the T-ray signal [29].

Reflection measurements can be also used for practical applications, when bulky samples are considered, that are impossible to measure in a transmission mode (see **Figure 12**) [30].

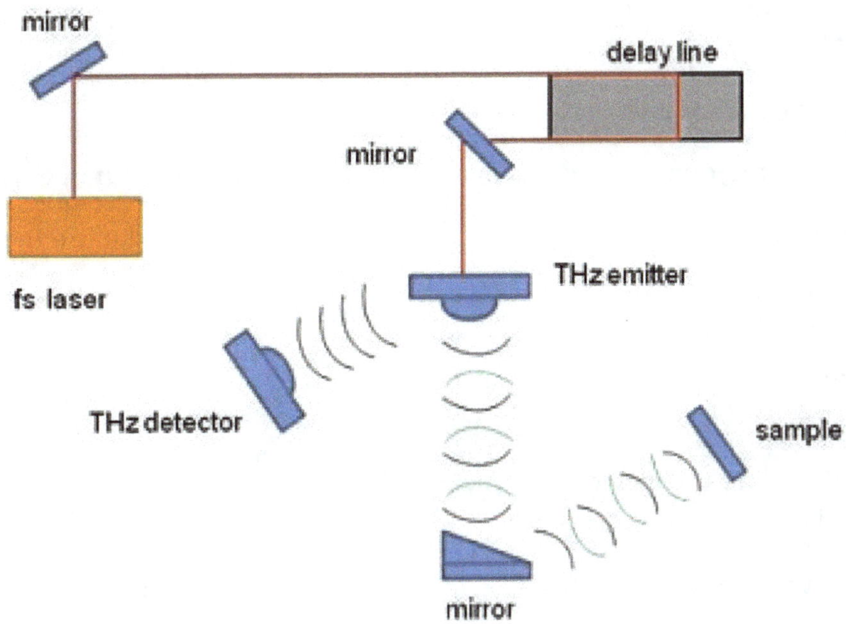

Figure 12. Typical scheme of a THz-TD system in reflection mode.

4.1. Performance of THz systems

In this paragraph, the various characteristics and limitations associated with a THz system, in particular when defining its performance, are discussed. The parameters commonly used are the dynamic range (DR) and the signal-to-noise ratio (SNR), which mainly affect the accuracy in THz measurements. They should always be evaluated during the measurements, to avoid false interpretation of the results; therefore, some recommendations for the best practice are presented [31].

4.1.1. Dynamic range and signal-to-noise ratio

The dynamic range (DR) is defined as the ratio between the highest and lowest measurable signal and therefore describes the maximum signal change that can be quantified. As a matter of practice, it is calculated as the ratio between the maximum signal amplitude and the root-mean-square of noise floor. The signal-to-noise ratio (SNR) is defined as the ratio between the mean peak amplitude and the standard deviation (SD) of the signal amplitude. It indicates the minimum detectable signal change and is a complementary system parameter with respect to the dynamic range. Therefore, the DR determines the measurement bandwidth, whereas the SNR reflects the amplitude resolution or sensitivity.

DR and SNR may be evaluated either with respect to the time-domain waveform or to the spectrum obtained through Fourier transform. Two specific aspects are involved when these parameters are considered. First, data are acquired as time-domain waveforms, whereas measured optical parameters are derived from the Fourier transform (FT) spectra. In particular, the DR and SNR of time-domain signals may result different from that of the corresponding spectra, and there is not a straightforward analytical relationship between the parameters estimated according to the two methods. Moreover, both DR and SNR of spectral data are strongly frequency dependent and typically decrease steeply with frequency.

The recommended procedure for estimating the DR and SNR of THz through time-domain data is based on the following steps:

1. Acquiring the time-domain waveform and measuring the maximum peak value.
2. Acquiring the noise signal in the absence of THz, for example, before the arrival of the main pulse.
3. The mean signal in the absence of THz should be constant (zero for electro-optic detection, nonzero for a photoconductive antenna). However, its standard deviation (SD) has to be measured.
4. *DR* is given by the ratio between the mean value of the measured peaks and the SD of the noisy signal.
5. *SNR* is given by the ratio between the mean value of the measured peaks and their SD.

The recommended procedure for estimating the DR and SNR of THz through the amplitude spectrum is based on the following steps:

1. Calculating the FTs of a defined number of time-domain records, their mean and SD, and estimating the noise floor of the mean FT. If only the DR is required, one FT spectrum is sufficient.
2. *DR* is given by the ratio between the mean FT amplitude and the noise floor.
3. *SNR* is given by the ratio between the mean FT amplitude and the SD.
4. The sampling frequency and observation interval of the time-domain acquisition have to be constant throughout the procedure, because both affect the SNR and the DR.

5. It is desirable to test the performance of the system by varying the scan parameters in order to identify the ranges characterized by the best SNR and DR values.

If DR and SNR are evaluated through the FT, the signal averaging is recommended, in order to reduce the noise effects. It is worth to notice that if a jitter affects the peak position, which often is due to errors in the initial position of the delay stage, the time-domain average will be distorted; the FT amplitude spectra and their average value, on the contrary, will remain correct. As expected, the approach for the evaluation of DR and SNR of a THz TDS system strictly depends on the specific domain adopted for the measurements. In other words, if the measurements are carried out in time domain, then the DR and SNR must be directly calculated from the available time-domain data. On the contrary, DR and SNR must be evaluated from FT amplitude spectra when measurement-based spectroscopic data are taken into account.

Another fundamental parameter to describe characteristics of a THz system is the spectral resolution. It is determined from the span of the time delay sweep, and it is given by the ratio between the light velocity c and the effective delay line length (multiplied by 2).

In principle, the pump laser pulse repetition rate is the only limitation of resolution if the ideal conditions of noise-free system and unlimited delay line are met. On the contrary, the actual frequency resolution is hardly reduced in presence of noise and mainly influenced by the time-domain SNR of the system. For increasing length of the delay line, the signal amplitude is reduced because of the increasing delay from the main pulse, and the SNR approaches unity.

As expected, THz pulses in the train must be identical with one another to make optical sampling works successfully. If this condition is not satisfied (i.e., the evolution of THz pulse shape occurs on time scales comparable to (or shorter than) the measurement time), no reliable samples of the waveform can be acquired. Besides this main drawback, another minor disadvantage affects the performance of optical sampling. As for the other sampling techniques, optical sampling also takes long time to acquire the desired data. The lower bound for the acquisition time is given by $N * \partial t$, (N and ∂t being, respectively, the number of measured electric fields and the train pulse-to-pulse distance). Since it is possible to take advantage of signal averaging, the acquisition time is usually much longer than this minimum value. Another problem inherent to sampling measurements is that they require a method for varying the delay of the sampling gate relative to the THz pulse. This requirement is often accomplished by means of a mechanical delay line consisting of a mirror that is moved to vary the optical path length [1].

4.1.2. TDS calibration

A typical device used for calibrating the linearity of amplitude/power measurements of THz system has to exhibit constant loss within the THz bandwidth. The most convenient and preferred solution is the employment of optically flat silicon plates as loss elements. Fresnel reflections are solely responsible for transmission losses in such a plate. Using a stack of plates parallel to each other, orthogonal to the incident THz beam and separated by air gaps, one can manage the loss level since it is dependent on the number of plates in the stack.

When the device is placed in the beam path, particular attention has to be paid to avoid distorting the THz beam or altering its focusing; in fact, it is desirable to position the device in a collimated beam. The device can be used in both THz CW and THz TDS systems, where the plates must be angled to the incident beam so as to destroy the etalon interference. There are two methods to verify the linearity and either time domain or frequency domain data can be processed:

a. In the time-domain method, the signal is processed in order to identify the peak value of the amplitude signal. Then, the obtained peaks are plotted in a semilog graph versus the number of Si plates in the beam path. The system is supposed to be linear if the plotted curve exhibits a linear behavior characterized by a slope equal to 0.7.

b. The second method of testing a TDS involves the calculation of THz spectra. The amplitude at chosen frequencies is plotted against the number of Si plates in the beam path. As in the time-domain method, the semilog plots are expected to be linear with a slope of 0.7.

It is worth noticing that the linearity of a TDS should not be assumed, but it should be experimentally verified.

4.2. Errors and uncertainty in THz-TDS

Many sources of error can affect a THz-TDS measurement and data extraction. As for example, laser intensity fluctuations, optical and electronic noise, delay line stage jitter, registration noise are common error sources. Moreover, contributions to the error in the estimated optical

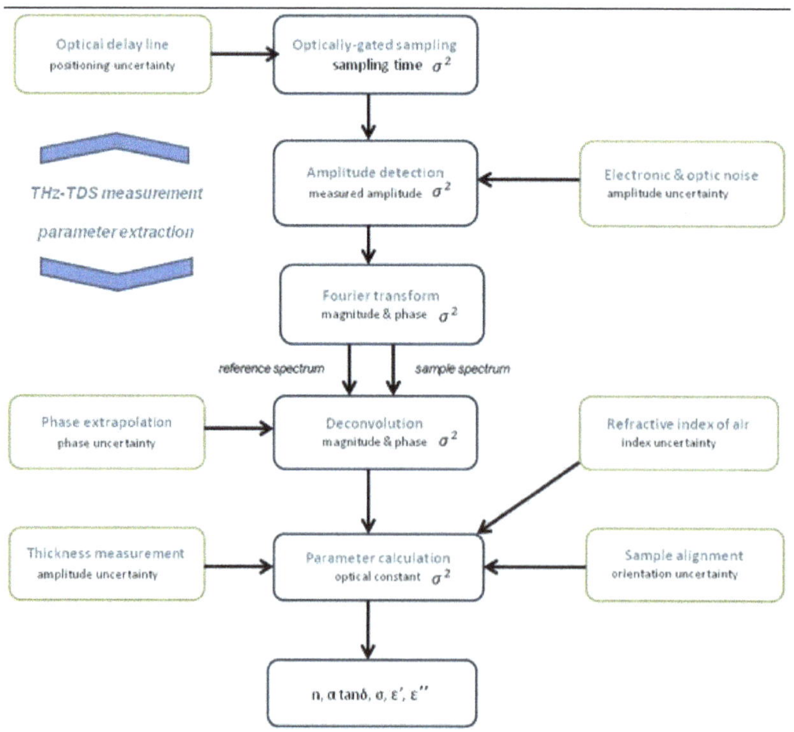

Figure 13. Scheme showing the propagation of uncertainties in THz-TDS measurements (taken from Ref. [33]).

constants are not only from the randomness in the signal, but also from imperfections in the physical setup and parameter extraction process. Examples are the sample thickness measurement, the sample alignment, and so on. Significant sources of error are shown in the scheme of **Figure 13**. The uncertainty sources (green lines) can influence both the THz-TDS measurements and the parameter extraction process. Each of them determines a variance that can be propagate along the process and determine a variance on optical constants [32, 33].

Two major sources of thermal noise can be singled out in a THz-TD system:

1. Johnson-Nyquist noise is generated when charge carriers fluctuate in a substrate. It results in an artifact voltage measured with no T-ray incident electrical field, whatever is the presence of optical gating pulses.
2. Background noise gives rise to a random voltage across the receiving antenna.

Other relevant noise sources are quantum fluctuations and laser and shot noise. One of the most efficient methods to remove noise in a T-ray signal is by means of digital signal processing technique such as wavelet de-noising that can actually improve the signal SNR, particularly when intensities are strongly reduced in biological samples. Ultimately, the noise in a THz-TDS system limits the spectral resolution; the best achievable resolution in a defined frequency interval depends, in fact, on the maximum time duration, which is, in turn, directly related to the actual SNR within that frequency range. In this way, improvement of the system dynamic range (obtained by either increasing the power of the transmitted THz waveform or decreasing the system noise floor) turns out to be mandatory to assure suitably high-frequency resolution.

Another uncertainty source is the positioning of the stage for optical delay line (ODL); thanks to the exploitation of a couple of moving mirrors, ODL mechanically induces a delay either on the probe or, equivalently, pumping pulse. As a consequence, the sampling time of the optically-gated detector is characterized by uncertainty; due to its combination with the other sources (electronic and optical noise), a final uncertainty on the amplitude measurements of the sampled T-ray pulse arises. The uncertainty associated with the amplitude of the acquired T-ray pulse affects also the spectral components obtained through the Fourier transform and the deconvolution process. Another uncertainty source that cannot be neglected involves the procedure for unwrapping the phase. Moreover, the thickness and the alignment of the samples have to be known in order to extract the model parameters; as a consequence, the uncertainty associated with these inputs affects the whole parameter detection process. Finally, the overall uncertainty is affected by the uncertainty associated with the estimation of the air refractive index. Each uncertainty source, however, can be taken into account by means of a proper model describing the uncertainty propagation in the measurement process.

4.3. THz spectroscopy

The term spectroscopy refers to a series of experiments aiming to investigate the excited states of a specimen, exploiting the interaction of a proper electromagnetic perturbation with a sample. Reflected and/or transmitted waves release specific information on the electromagnetic properties of the sample as function of the frequency. Therefore, according to the spectral content of the electromagnetic signal-probe, different excitations can be investigated ranging from the quantum properties (energy levels of atomic bonds, roto-vibrational states, etc.) of molecules to the impedance of a macroscopic samples or transmission lines. The THz band is ideal to study electrodynamic properties of materials from metals to insulators, since the frequency is lower than the typical plasma frequency of metals (about 10^{15} Hz) that defines the frequency above which the metal becomes transparent to radiation. Coherent THz radiation can provide valuable information on the complete set of the complex electrodynamic parameters [34] (refractive index ($\tilde{n}$) permittivity ($\tilde{\varepsilon}$) and /or conductivity ($\tilde{\sigma}$) characterizing a material whatever it is an insulator or metallic like. The complex refractive index ($\tilde{n} = n + i\,k$) furnishes information on both the delay (n) and the absorption (k). Once n and k are obtained (see below), the permittivity $\tilde{\varepsilon} = \varepsilon_1 + i\varepsilon_2$ can be reached exploiting the following relationship $\tilde{n} = \sqrt{\tilde{\varepsilon}\tilde{\mu} / \varepsilon_0\mu_0}$ where $\tilde{\mu}$ is the complex magnetic permeability, ε_0 is the vacuum permittivity, and μ_0 is the vacuum permeability. Since most of materials have $\tilde{\mu} = 1$, a direct relation between the refractive index and permittivity can be extracted $\tilde{n} \cong \sqrt{\tilde{\varepsilon}}$ which deals $n = \sqrt{\left(\sqrt{\varepsilon_1^2 + \varepsilon_2^2} + \varepsilon_1\right)/2}$ and $k = \sqrt{\left(\sqrt{\varepsilon_1^2 + \varepsilon_2^2} - \varepsilon_1\right)/2}$. Conductivity ($\tilde{\sigma} = \sigma_1 + i\sigma_2$) and permittivity are also reciprocally related through the formulas $\sigma_1 = \varepsilon_0\omega\varepsilon_2$ and $\sigma_2 = \varepsilon_0\omega(\varepsilon_\infty - \varepsilon_1)$, where $\omega = 2\pi\nu$ and $\varepsilon_\infty = \varepsilon\ (\omega \to \infty)$ can be obtained through a fitting procedure. From the practical point of view, the most important parameter to obtain for a sample characterization is $\tilde{n}$, since the other parameters are just a combination of n and k.

The peculiar characteristic of using a TDS consists into directly manipulating the time-dependent electric field E(x, t) transmitted through the sample. The ratio between the Fourier transforms of the transmitted signal and the reference signal is directly function of the refractive index. The sketch of the measurement on a generic sample L thick is reported in **Figure 14.** E(x,t) propagating from the transmitter (Tx) to the receiver (Rx) is linearly polarized along y. Since the signal is generated and detected in air, the proposed scheme allows to generalize the measurements in multilayer samples.

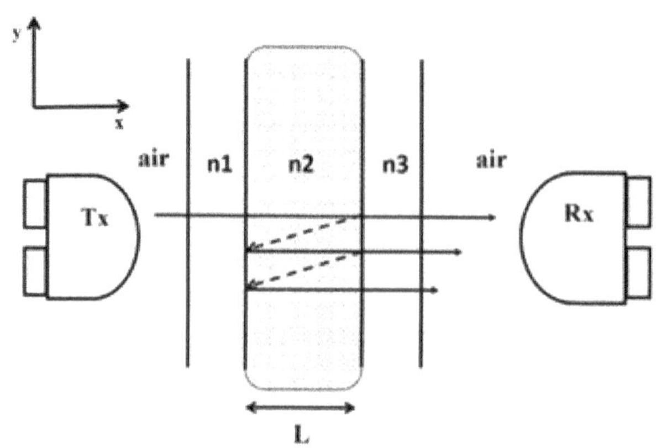

Figure 14. Scheme of the measurements through the THz-TDs system. Tx and Rx stand for transmitter and receiver, respectively. L represents the mean size of the sample, while n_i with i = 1, 2, 3 are the refractive index of different through THz pulse.

The transmitted signal through the sample $S(\omega)$, can be expressed as function of the Fresnel coefficients $T_{a,b}(\omega) = 2\tilde{n}_a/(\tilde{n}_a + \tilde{n}_b)$ and $R_{a,b}(\omega) = (\tilde{n}_a - \tilde{n}_b)/(\tilde{n}_a + \tilde{n}_b)$ and the propagation factor $P_a(\omega, d) = exp\{-i\ \tilde{n}_a\ \omega\ d\}$, where the labels refer to the material [35]. The complete signal can be expressed as:

$$S(\omega) = \eta(\omega)T_{1,2}(\omega)P_2(\omega,d)T_{2,3}(\omega)\sum_{k=0}^{\infty}\{R_{2,3}(\omega)P_2^2(\omega,L)R_{2,1}(\omega)\}^k \cdot E(\omega) \tag{5}$$

where $E(\omega)$ is the generated THz pulse, and $\eta(\omega)$ accounts for all the reflected and transmitted signals which do not reach Rx. In Eq. (1), the factors $T_{1,2}(\omega)P_2(\omega, d)T_{2,3}(\omega)$ take into account of the fraction of signal reaching Rx in one path, whereas the term $\sum_{k=0}^{K}\{R_{2,3}(\omega)P_2^2(\omega, L)R_{2,1}(\omega)\}^k$ accounts for the delayed K-pulses originated by the reflections of the primary pulse between the sample boundaries (usually $K \leq 3$). This phenomenon known as Fabry–Perot (FP) effect is depicted in **Figure 14** through the dashed arrows displaying the reflected signals. In the time domain, the FP effects show up in the appearance of copies of the primary transmitted signal. In **Figure 15**, a comparison is proposed between the reference signal (air) and the signal (Si) through a silicon slab 500 μm thick. Black arrows point out the copies of the primary signal. The delay between copies is due to the roundtrip walk in the sample and is about $\Delta t \cong 2Ln/c \sim 11.4ps$ for $n(Si) = 3.46$.

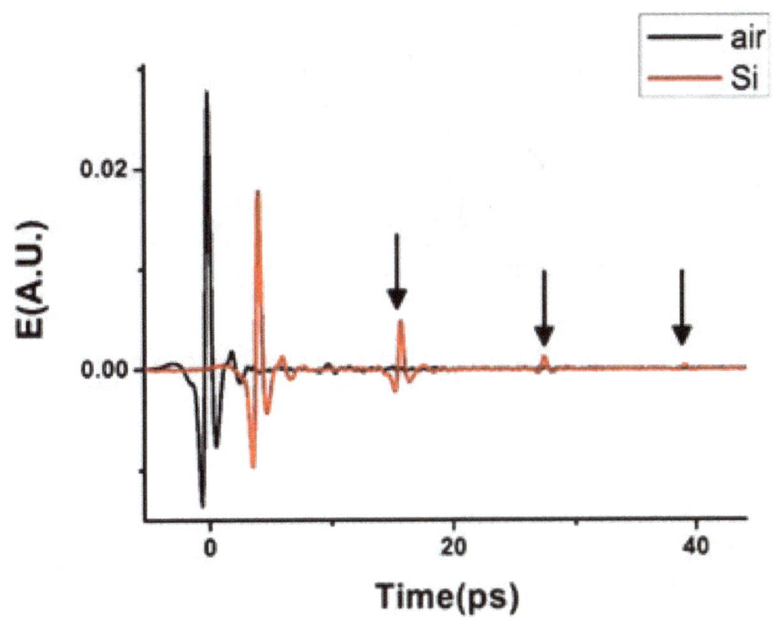

Figure 15. Time-dependent signal measured through THz-TDS system. The black curve is the reference signal acquired in free space, whereas the red curve is the signal through a Si sample of 500 μm thick. The arrows indicate the primary signal copies generated by the Fabry–Perot effect.

Eq. (1) also defines the transmittance

$$T(\omega) = {E_{sample}} \big/ {E_{ref}} \tag{6}$$

that, according to Eq. (5), in the case of a simple slab can be expressed as

$$H(\omega) = \frac{2\tilde{n}_2(\tilde{n}_1 + \tilde{n}_3)}{(\tilde{n}_2 + \tilde{n}_1)(\tilde{n}_2 + \tilde{n}_3)} exp\{-\frac{i(\tilde{n}_2 - \tilde{n}_{air})\omega L}{c}\} FP(\omega) \tag{7}$$

where

$$FP(\omega) = \frac{1}{1 - (\frac{\tilde{n}_2 - \tilde{n}_1}{\tilde{n}_2 + \tilde{n}_1})(\frac{\tilde{n}_2 - \tilde{n}_3}{\tilde{n}_2 + \tilde{n}_3}) \exp\{-i\tilde{n}_2\omega 2L / c\}} \tag{8}$$

is the explicit form of the Fabry-Perot term when the echos in media 1 and 3 are negligible [35].

The transfer function $H(w)$ in Eq. (7) is used as theoretical reference for $T(\omega)$ to calculate optical parameters of samples. Eqs. (6) and (7) describe the transmission through a homogeneous slab with refractive index $\tilde{n}_2$ when the equivalence $\tilde{n}_1 = \tilde{n}_3\ \tilde{n}_{air}$ is verified. Instead by putting $\tilde{n}_1 = \tilde{n}_{air}$, Eqs. (6) and (7) are able to describe a system composed by two layers as a thin metallic film on a dielectric substrate [36, 37].

Several techniques [36–40] have been developed in order to extract $\tilde{n}$ by computing the minimum difference between the moduli and the phases of $H(\omega)$ and $T(\omega)$:

$$\delta\rho(\omega) = | H(\omega) | - | T(\omega) |$$

$$\delta\varphi(\omega) = \arg[H(\omega)] - \arg[T(\omega)] \tag{9}$$

Eq. (9) allows to define an error function, the total variation (TV) [38], defined by the sum of differences $\delta 1$ and $\delta\psi$ for each frequencial point

$$ER = \sum_{\omega} | \delta\rho(\omega) | + | \delta\varphi(\omega) | \tag{10}$$

This is a tridimensional paraboloid as function of the frequency and the sample thickness. The computational search of the minimum of $ER(L, \omega)$ implies the contemporary knowledge of the main quantities describing the system: the sample thickness L the refractive index n, and the extinction coefficent k [39]. A fast resolution of the TV approach is affected by the noise in the measured spectrum of $T(\omega)$. The most relevant noise source in thin samples is the Fabry–Perot oscillations which show a frequency inversely proportional to L. This problem can be overcome imposing the quasi-space (QS) optimization [39], where the periodicity of the FP effect is employed to achieve the effective optical thickness of the sample. The quasi-space is defined by the Fourier transform of an electro dynamical parameter $y(\omega_n)$ which could be the refractive index or the extinction coefficient. Therefore, a new set of variables can be defined as follows

$$Q_{S_k} = \sum_{n=1}^{N-1}[y(\omega_n)\exp(-i\frac{2\pi}{N}k\,n)], k = 0,1,2\ldots, N-1 \quad (11)$$

This function can be displayed in terms of the variable $L_{Q_S} = x_{Q_S} c_0 / 2$, where c_0 is the speed of light in vacuum and $x_{Q_S} = 2\pi/\omega$, showing a pronounced peak at the effective optical length. Alternatively, the sample thickness L can be accounted by locating the minimum of Q_{S_k} (at a fixed frequency) for different L values [40]. The QS approach is limited just by the performances of the TDS system. In particular, the maximum and minimum detectable thickness can be expressed as $L_{max} = c_0/4n\ df$ and $L_{min} = c_0/2n\ \Delta f$, where df is the minimum detectable frequency of the system while Δf is its bandwidth. The former thickness is based on the application of Nyquist theorem, whereas the latter is based on the resolution of neighbor QS's peaks [40]. For instance, assuming as parameters $df \cong 3$, and, as effective refractive index, $n \cong 2$, the maximum and minimum length are $L_{max} \cong 12.5$ mm and $L_{min} \cong 12.5$ μm, respectively. Whenever the best optical length is found, the quality of the retrieved electrodynamic parameter depends also by the choice of a good thickness of the sample. Thinner samples become transparent to THz radiation, whereas thick samples do not release much information because the transmitted signal is low. In Ref. [41], authors aim to find the best thickness in order to minimize the standard deviations (s_n^2, s_n^2) of electrodynamical parameter as (ω) and (ω) inherited by the standard deviations of signals $E_{sample}(t)$ and $E_{ref}(t)$ acquired in the time domain. The functions s_n^2, s_n^2 can be minimized with respect L leading to get the optimal thickness as function of the absorption coefficient:

$$L_{opt} = c_0 / \omega k(\omega) = \frac{2}{\alpha(\omega)} \quad (12)$$

Eq. (12) shows that $n(\omega)$ and $k(\omega)$ are affected by some indetermination as consequence of the fixed extension of the sample. On the other hand, Eq. (12) enables the possibility to get reliable results also on very thin samples provided that the absorption coefficient $\alpha(\omega)$ is enough high. Indeed, full two-dimensional systems like single graphene layers have been extensively investigated through THz-TDS systems thanks to the robust absorption coefficient $\alpha_{graphene} \sim 5$ μm^{-1} [42–44].

5. Conclusions

Without claiming to be exhaustive, we presented a short overview of THz measurement systems presently under development for scientific and industrial applications. We first described the most common sources and detectors that are routinely in use for the manipulation of T-waves. We then focused on the typical architectures that are presently employed in time-domain spectroscopy and imaging. The importance of metrological aspects in THz systems performance and measurements and most of their related issues and solutions were discussed. Since each material has its own "identity card" in this band of the spectrum, a THz-TDS waveform transmitted through a sample is typically rich in information. We therefore presented what are the main parameters that can be measured from the material frequency response, namely optical or electrical complex quantities like the refractive index, the conductivity and the dielectric constant, and what are the data extraction methods and the related errors and uncertainty.

Author details

Leopoldo Angrisani[1*], Giovanni Cavallo[2], Annalisa Liccardo[2], Gian Paolo Papari[3] and Antonello Andreone[3]

*Address all correspondence to: angrisan@unina.it

1 DIETI – Department of Information Technology and Electrical Engineering, CeSMA – Center of Advanced Measurement Services, University of Napoli Federico II, Napoli, Italy

2 DIETI – Department of Information Technology and Electrical Engineering, University of Napoli Federico II, Napoli, Italy

3 DF – Department of Physics, University of Napoli Federico II, Napoli, Italy

References

[1] Siegel P. H. THz technology: an overview. Int. J. High Speed Electron. Syst. 2003; 13(2): 351–394.

[2] Shen Y. C., Lo T., Taday P. F., Cole B. E., Tribe W. R., Kemp M. C. Detection and identification of explosives using terahertz pulsed spectroscopic imaging. Appl. Phys. Lett. 2005; 86(24): 241116.

[3] Taday P. F. Applications of terahertz spectroscopy to pharmaceutical sciences. Phil. Trans. R. Soc. Lond. A. 2004; 362: 351–364.

[4] Wilmink G. J., Grundt J. E. Terahertz radiation sources, applications, and biological effects. In: Lin J.C, editors. Handbook of Electromagnetic Fields in Biological Systems. CRC Press Taylor & Francis Group. 2012. p.369–423.

[5] Aurele J. L. A. Review of near field terahertz measurement methods and their applications. Infrared Milli Terahz Waves. 2011; 32: 976–1019.

[6] Popovic Z., Grossman E. N. THz metrology and instrumentation. IEEE Trans. Terahertz Sci. Technol. 2011; 1(1): 133–144.

[7] Kozlov G., Volkov A. Coherent Source Submillimeter Wave Spectroscopy in Millimeter and Submillimeter Wave Spectroscopy of Solid. Springer Verlag; Berlin; 1998. p. 51–109.

[8] Faist J., Sirtori C. InP and GaAs-Based Quantum Cascade in Long-Wavelength Infrared Semiconductor Lasers. John Wiley and Sons; New York; 2004. p. 217–278.

[9] Tredicucci A., Kohler R. Terahertz Quantum Cascade Lasers in Intersubband Transitions in Quantum Structures. McGraw-Hill; New York; 2006. p. 45–105.

[10] Williams B. Terahertz quantum cascade lasers. Nat. Photonics. 2007; 1: 517.

[11] Kleiner R., Steinmeyer F., Kunkel G., Müller P. Intrinsic Josephson effects in $Bi_2Sr_2CaCu_2O_8$ single crystals. Phys. Rev. Lett. 1992; 68: 2394.

[12] Ozyuzer L., Koshelev A. E., Kurter C., Gopalsami N., Tachiki Q., Li, M., Kadowaki K. Emission of coherent THz radiation from superconductors. Science. 2007; 318: 1291.

[13] Tani M., Matsuura S., Sakai K., Nakashima S. Emission characteristics of photoconductive antennas based on low-temperature grown GaAs and semi-insulating GaAs. Appl. Opt. 1997; 36(30): 7853–7859.

[14] Shen Y. C., Upadhya P. C., Beere H. E., Linfield E. H., Davies A. G., Gregory I. S. Generation and detection of ultra-broadband terahertz radiation using photoconductive emitters and receivers. Appl. Phys. Lett. 2004; 85(2): 164–166.

[15] Shen Y. C., Upadhya P. C., Linfield E. H., Beere H. E., Davies A. G. Ultra-broadband terahertz radiation from low-temperature-grown GaAs photoconductive emitters. Appl. Phys. Lett. 2003; 83(15): 3117–3119.

[16] Leitenstorfer A., Hunsche S., Shah J., Nuss M. C., Knox W. H. Detectors and sources for ultra broadband electro-optic sampling: experiment and theory. Appl. Phys. Lett. 1999; 74(11): 1516–1518.

[17] Sinyukov A. M., Leahy M. R., Hayden M., Haller M., Luo J., Jen A. K-Y. Resonance enhanced THz generation in electro-optic polymers near the absorption maximum. Appl. Phys. Lett. 2004; 85(24): 5827–5829.

[18] Sinyukov A. M., Hayden L. M. Generation and detection of terahertz radiation with multilayered electro-optic polymer films. Opt. Lett. 2002; 27(1): 55–57.

[19] Wu Q., Zhang X. C. Ultrafast electro-optic field sensor. Appl. Phys. Lett. 1996; 68(12): 1604–1606.

[20] Sands D. Diode Lasers. Taylor & Francis; London; 2004.

[21] Auston D. H., Cheung K. P., Smith P. R. Picosecond photoconducting Hertzian dipoles. Appl. Phys. Lett. 1984; 45: 284–286.

[22] Golay M. J. E. A pneumatic infrared detector. Rev. Sci. Instr. 1947; 18: 347.

[23] Cooper J. A fast-response pyroelectric thermal detector. Rev. Sci. Instr. 1962; 39: 462.

[24] Kinch M. A. Compensated silicon-impurity conduction bolometer. J. Appl. Phys. 1971; 42: 5861.

[25] Andrews D. H., Milton R. M., DeSorbo W. J.A Fast Superconducting Bolometer Opt. Soc. Am. 1946; 36(9): 518.

[26] Liu T. A., Tani M., Nakajima M., Hangyo M., Pan C. L. Ultra-broadband terahertz field detection by photoconductive antennas based on multi-energy arsenic ion implanted GaAs and semi-insulating GaAs. Appl. Phys. Lett. 2003; 83(7): 1322.

[27] Kono S., Tani M., Sakai K. Ultra-broadband photoconductive detection: comparison with free-space electro-optical sampling. Appl. Phys. Lett. 2001; 79(7): 898–900.

[28] Angrisani L., Bonavolontà F., Schiano Lo Morello R., Andreone A., Casini R., Papari G.P. First steps towards an innovative compressive sampling based-THz imaging system for early crack detection on aerospace plates. In: IEEE International Workshop on Metrology for AeroSpace (IEEE 2014); 29–30 May 2014.

[29] Withayachumnankul W., Lin H., Mickan S. P., Fischer B. M., Abbott D. Analysis of measurement uncertainty in THz-TDS. Phot. Mater. Dev. Appl. II. 2007; 6593:659326-1-18.

[30] Zhang C., Zhong H., Zhang L. A phase extraction technique for terahertz reflection spectroscopy. Proc. SPIE. 2009; 7158: 71580Q–71580Q–14.

[31] Naftaly M., Dudley R. Methodologies for determining the dynamic ranges and signal-to-noise ratios ofTerahertz time domain spectroscopy. Opt. Lett. 2009; 34(8): 1213–1215.

[32] Naftaly M. Metrology issues and solutions in THz time domain spectroscopy: noise, errors, calibration. IEEE Sens. J. 2013; 13(1): 8–17.

[33] Withayachumnankul W., Naftaly M. Fundamentals of measurement in terahertz. J. Infrared Milli. Terahz Waves. 2014; 35: 610–637.

[34] Hangyo M. Development and future prospects of terahertz technology. Jpn J. Appl. Phys. 2015; 54: 120101.

[35] Duvillaret L., Garet F., Coutaz J. L. A reliable method for extraction of material parameters in terahertz time-domain spectroscopy. IEEE J. Selected Top. Quantum Electron. 1996; 2(3): 739-746.

[36] Yasuda H., Hosako I. Measurement of terahertz refractive index of metal with terahertz time-domain spectroscopy. Jpn J. Appl. Phys. 2008; 47: 1632–1634.

[37] Zhou D. X., Parrott E. P. J., Douglas J. P., Zeitler J. A. Determination of complex refractive index of thin metal films from terahertz time-domain spectroscopy. J. Appl. Phys. 2008; 104: 053110.

[38] Dorney T. D., Baraniuk R. G., Mittleman D. M. Material parameter estimation with terahertz time-domain spectroscopy. J. Opt. Soc. Am. A. 2001; 18(7): 1562-1571.

[39] Pupeza I., Wilk R., Koch M. Highly accurate optical material parameter determination with THz time-domain spectroscopy. Optics Express 2007; 15: 4335.

[40] Scheller M., Jansen C., Koch M. Analyzing sub-100-lm samples with transmission terahertz time domain spectroscopy. Opt. Commun. 2009; 282: 1304–1306.

[41] Withayachumnankul W., Fischer B. M., Abbott D. Material thickness optimization for transmission-mode terahertz time-domain spectroscopy. Opt. Express 2008; 16(10): 7382.

[42] Kuzel P., Nemec H. Terahertz conductivity in nanoscaled systems: effective medium theory aspects. J. Phys. D: Appl. Phys. 2014; 47: 374005.

[43] Ivanov I., Bonn M., Mics Z., Turchinovich D. Perspective on terahertz spectroscopy of graphene. Eur. Phys. Lett. 2015; 111: 67001.

[44] Denisultanov A. K., Azbite S. E., Balbekin N. S., Gusev S. I., Khodzitsky M. K. Optical properties of graphene on quartz and polyethylene substrates in terahertz frequency range. PIERS Proceedings, Guangzhou, China, August 25–28, 2014.

4

Microwave Power Measurements: Standards and Transfer Techniques

Xiaohai Cui, Yu Song Meng, Yueyan Shan and Yong Li

Abstract

In this chapter, precision power measurement, which is probably the most important area in RF and microwave metrology, will be discussed. Firstly, the background of RF and microwave power measurements and standards will be introduced. Secondly, the working principle of primary power standard (i.e., microcalorimeter) will be described, followed by the discussions of direct comparison transfer technique. Finally, there will be some discussions about the performance evaluation and uncertainty estimation for microwave power measurements.

Keywords: Direct comparison transfer, Microcalorimeter, Primary standard, RF and microwave power, Thermistor mount

1. Introduction

Recently, there are growing interests in higher frequency such as microwave and millimeter-wave applications, which is becoming a promising solution for satellite communications [1, 2] and millimeter-wave mobile backhauling [3]. For proper deployments of these applications and services, accurate and reliable signal power measurements are essential and important for system designers. Normally for the end users (i.e., system designers), microwave and millimeter-wave power measurements are highly relied on a conventional power detector and power meter combination or a spectrum analyzer. These measuring instruments have to be properly calibrated with traceability to the International System of Units (SI) for assuring the quality of measurement results as required by the industry.

As stated in ref. [4], the traceability of measuring instruments shall be achieved by means of an unbroken chain of calibrations or comparisons linking to relevant primary standards of the SI units of measurement as illustrated in **Figure 1**. The link to SI unit could be realized by a primary standard developed and maintained by a national metrology institute (NMI) such as the National Institute of Metrology (NIM) of China and the National Metrology Centre (NMC), A*STAR of Singapore. For RF, microwave, and millimeter-wave measurements and standards, power measurement has been recognized as one of the primary areas [5] and probably the most important research area by the NMIs. For simplicity in the rest of this chapter, microwave measurement will be synonymous to "RF, microwave, and millimeter-wave measurement."

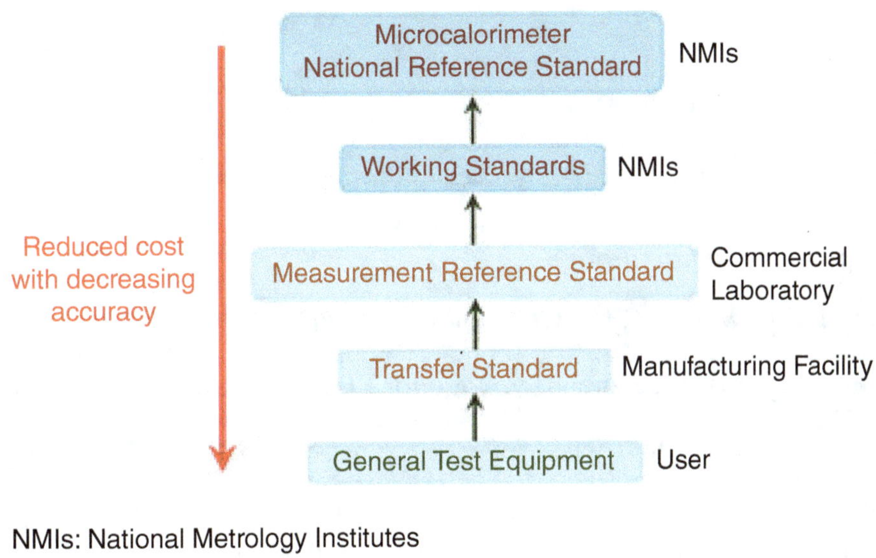

Figure 1. Typical traceability chain of RF, microwave, and millimeter-wave power measurements.

In the following, we will firstly give a background of microwave power measurements and standards. Secondly, a primary power standard (e.g., a microcalorimeter) will be discussed with recent developments at NIM, China. This will be followed by the discussions of the working principle of the microcalorimeter measurement system. The direct comparison transfer technique will be then introduced, together with some improvements at NMC, A*STAR of Singapore. Finally, performance evaluation and uncertainty estimation for microwave power measurements will be discussed.

2. Background of Power Measurements and Standards

Basically, microwave power can be measured by the combination of a power detector and a power meter, as shown in **Figure 2**. The power detector is a key instrument for power measurements, and its function is to convert high-frequency (i.e., RF, microwave, and millimeter-wave or higher) power to a direct current (DC) or low-frequency signal that a power meter can measure with a display. Different working principles and fabrication techniques have led to several power detectors that have been widely used in commercial applications.

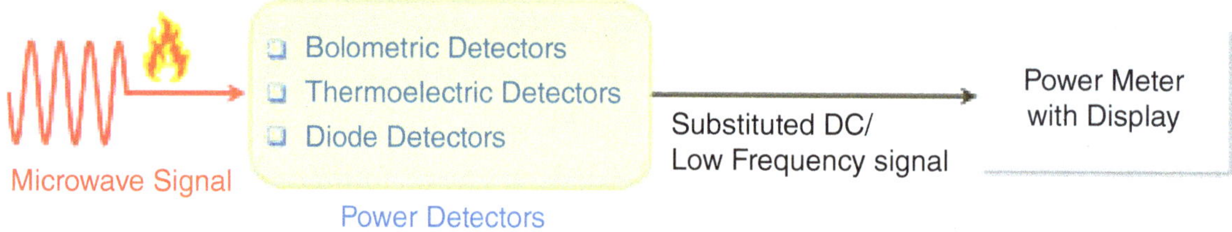

Figure 2. Microwave power measurements in the combination of a power detector and a power meter.

2.1. Commercial Power Detectors

Three main types of power detectors have been commercially available, which are designed and based on the bolometric element, thermoelectric element, and diode. The respective working principles behind are through

1. substituting DC power for RF power (bolometric type),
2. representing a thermally generated voltage for RF power (thermoelectric type), or
3. using the rectification property to convert RF power to DC voltage (diode type).

It is noted that each type of power detector indicated above has its own strengths and weaknesses for its application. In the early days [5], the diode detector was very sensitive to ambient temperature and also with a poor linearity, and therefore, it was rarely adopted as a transfer standard. Most of the NMIs have been continuously using bolometric detectors (i.e., bolometers) as transfer standards, since they are very nearly linear when used with a primary power standard (e.g., a microcalorimeter [6]) through the DC substitution technique. Bolometric detectors can also offer an extremely high long-term stability with a very low measurement uncertainty [7, 8].

However, a bolometric detector normally has a narrow dynamic range and limited power capability (e.g., with a power range from 10 μW to 10 mW [7]). Additionally, its productions have been discontinued, accompanying a new industry production trend toward other types of power sensors (e.g., diodes and thermocouples). Some NMIs therefore have attempted to use thermoelectric detectors (e.g., thermocouples) as the transfer standards [9], which are linear with a better sensitivity and dynamic range. Performance comparison between bolometric and thermoelectric sensors has recently been reported in ref. [10] using the same microcalorimeter. The results revealed that the two power standards (bolometric and thermoelectric sensors) in the comparison can be considered equivalent.

In the following, the bolometric detector as the transfer standard will be introduced, since it has been widely used by most of the NMIs including NIM of China and NMC, A*STAR of Singapore. Its calibration using a microcalorimeter will be focused upon and described.

2.2. Reference Power Standard: Bolometer

Bolometers have a very high reliability and have been used as the reference power standards in most of the NMIs, together with a microcalorimeter. A bolometer consists of a small temperature-sensitive resistor. It is operated by changing its own resistance following a change in its temperature resulted from the incident microwave power being dissipated in the bolometric element.

Two types of bolometers have been commonly used, namely, barretter and thermistor. The barretter is a thin metal wire with a positive temperature coefficient of resistance, and the thermistor is a small bead of semiconductor material with a negative temperature coefficient of resistance [11]. It is noted that the thermistor is more sensitive than the barretter due to a much greater temperature coefficient, but it has a slower response time due to its larger thermal time constant [12].

The thermistor is therefore more popularly used. Typically, a thermistor bead has a diameter of around 0.05–0.5 mm with a small-size (diameter of 15–100 μm) metal wire embedded inside. A waveguide or coaxial termination that houses a thermistor with an internal matching circuit to obtain specified impedance conditions (e.g., 100 Ω or 200 Ω) with appropriate DC bias power applied [11] is called a thermistor mount. The schematic diagram of a popular type of waveguide thermistor mount is shown in **Figure 3**, and some samples of waveguide thermistor mounts used at NIM, China, are shown in **Figure 4**.

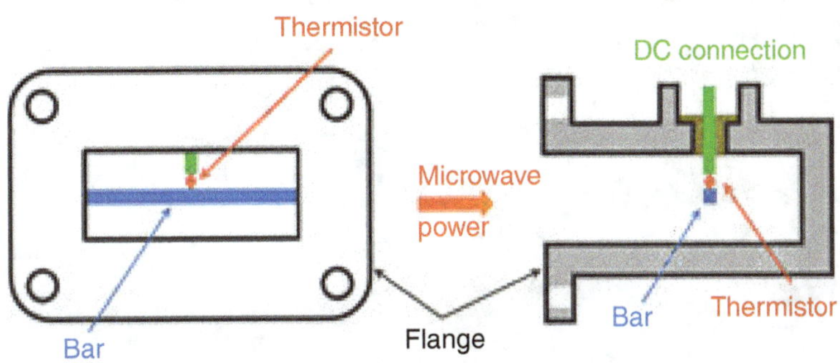

Figure 3. One popular type of waveguide thermistor mount.

Figure 4. Waveguide thermistor mounts.

2.3. Primary Power Standard: Microcalorimeter

At present, calorimeters have been accepted as the basis of primary standards for microwave power measurements and calibrations within the NMIs or the standards laboratories [5]. Among several different types of calorimeters (e.g., dry load calorimeters and flow calorimeters [8]), microcalorimeters [13, 14] have been popularly used. The microcalorimeter technique is based on the DC substitution method, and its traceability is established through the principle of "thermal effect equivalence." It allows the experimental determination of effective efficiency of thermal power sensors (i.e., bolometric and thermoelectric sensors).

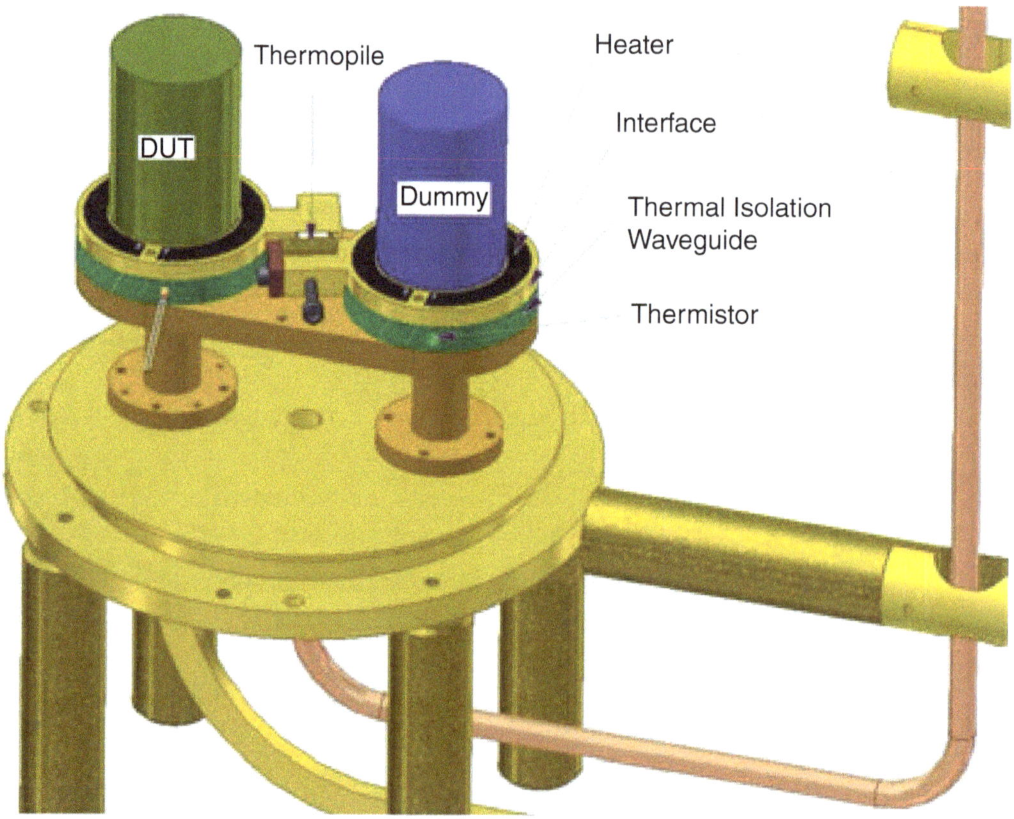

Figure 5. Prototype of a waveguide microcalorimeter with twin-line structure.

Figure 5 presents the design and configuration of a waveguide microcalorimeter, which is China's national primary power standard developed and maintained at NIM, China, for the thermistor mount [device under test (DUT) in this case] measurements [15–18] through the DC substitution technique (i.e., applied microwave power is compensated by an appropriate reduction of DC power). As shown in **Figure 5**, the design of a waveguide microcalorimeter at NIM, China, is based on a twin-line structure with a symmetrically located inactive mount (i.e., a "dummy" thermistor mount) as the temperature reference. A thermopile is attached in between to monitor the temperature difference, when the microcalorimeter is in operation with a thermistor mount (DUT). With the same DUT and dummy mounts, nearly the same thermal transmission paths could be achieved and therefore produce almost identical response to the ambient temperature at both the terminals of the thermopile. This twin-line design makes the microcalorimeter less affected by the ambient temperature. More specifically, it could effectively reduce the influence of a long-term ambient temperature drift on the measurement results.

The core part of the microcalorimeter as shown in **Figure 5** consists of a base extension, two thermal isolation sections (TIS), and two interface plates. The DUT is attached to a standard waveguide flange on the interface plate with screws that pass through all the three core components. The TIS is about 6 mm thick and is made of gold-coated ABS plastic so that the waveguide section has little loss. A thermistor has been embedded into the TIS for monitoring its temperature rise due to unexpected power consumption within the thermal isolation waveguide. A sample of fabricated WR-22 waveguide microcalorimeter at NIM, China, is shown in **Figure 6**.

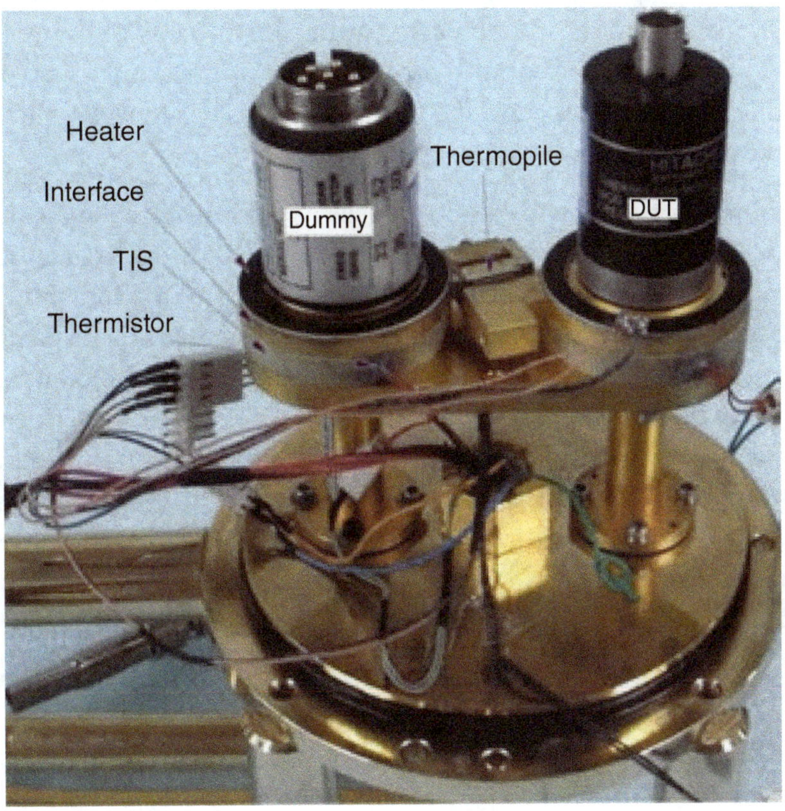

Figure 6. Fabricated WR-22 waveguide microcalorimeter at NIM, China.

3. The Working Principle of Microcalorimeter

The microcalorimeter is used to measure the effective efficiency η_e of a thermistor mount which serves as the reference standard for power measurements. The effective efficiency η_e of a bolometer unit (e.g., a thermistor mount) is defined as the ratio of the changes in the DC-substituted power P_{sub} to the total microwave power P_{rf} dissipated within the bolometer unit, as specified in ref. [11]. That is,

$$\eta_e = \frac{P_{sub}}{P_{rf}}. \quad (1)$$

It is noted that, in practice, the effective efficiency η_e of a thermistor mount is determined using the DC substitution technique with a microcalorimeter, in conjunction with a self-balancing bridge circuit.

3.1. DC Substitution Technique

The DC substitution technique has been implemented through automatically reducing the DC bias power to keep the operating resistance of a thermistor constant, when the microwave power is applied to the thermistor mount. Ideally, if the applied microwave power is totally absorbed by the thermistor element and the thermistor also has the same thermal reaction for DC and microwave power, $P_{sub} = P_{rf}$. Thereby, $\eta_e = 1$.

However, practically, there are some existing losses in the input transmission line, the mount structure, and others. For example, as shown in **Figure 7**, some unexpected power consumptions could be on the wall of the thermistor mount (P_w) and within the thermal isolation waveguide (P_i), besides the net power P_{rft} dissipated on the thermistor bead. It is noted that practically the net power P_{rft} is very difficult/impossible to be measured and is represented by the compensated DC power (i.e., DC-substituted power P_{sub}). These effects can result in a measurement error, which is normally characterized as the mount efficiency. Moreover, the thermistor bead has a different thermal reaction and power distribution for DC and microwave powers and can cause a microwave-to-DC (or RF-to-DC) substitution error. Both the mount efficiency and the microwave-to-DC substitution error shall be considered in the correction factor g of a microcalorimeter for accurate effective efficiency determination.

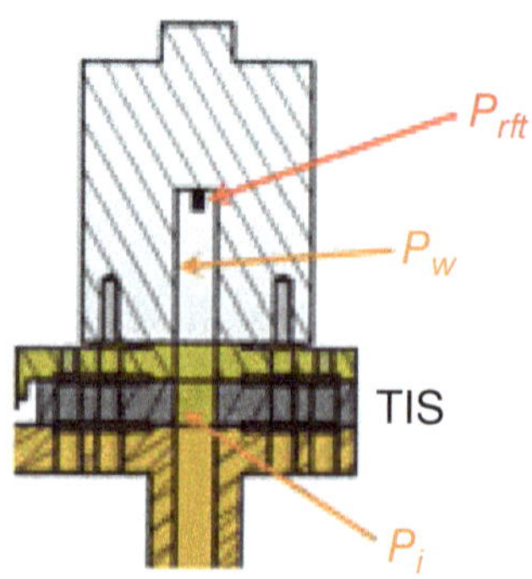

Figure 7. Main power absorptions when calibrating a thermistor mount using a microcalorimeter.

The DC substitution requires a self-balancing bridge circuit to work with the thermistor mount for keeping its operating resistance R_T constant, when the microwave power is applied.

Figure 8 shows a typical self-balancing bridge circuit for maintaining R_T. The initial resistance R_T of a thermistor is normally at 200 Ω (or 100 Ω for different types of thermistor mounts) with a DC bias. When the microwave power is fed to the thermistor, R_T will change due to the temperature rising of the thermistor. The DC bias power has to be reduced to balance the bridge circuit. It is noted that the reduced amount of DC-biased power is proportional to the microwave power P_{rft} applied onto the thermistor bead.

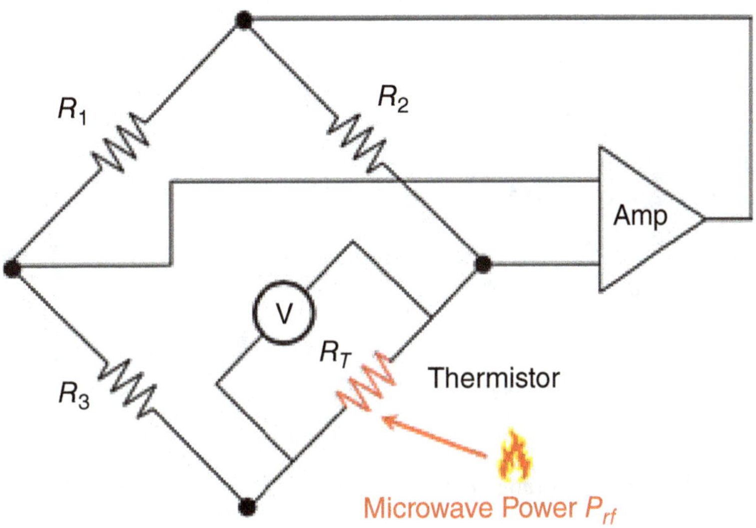

Figure 8. An example of a self-balancing bridge circuit for monitoring the resistance change in a thermistor mount.

3.2. Operation of a Microcalorimeter

A Type IV power meter has been fabricated at NIM, China, for realizing the DC substitution technique with a self-balancing bridge circuit inside. It works with the thermistor mount in a closed loop to keep R_T constant. The internal circuit of the NIM-manufactured Type IV power meter for calibrating the thermistor mount is shown in **Figure 9**.

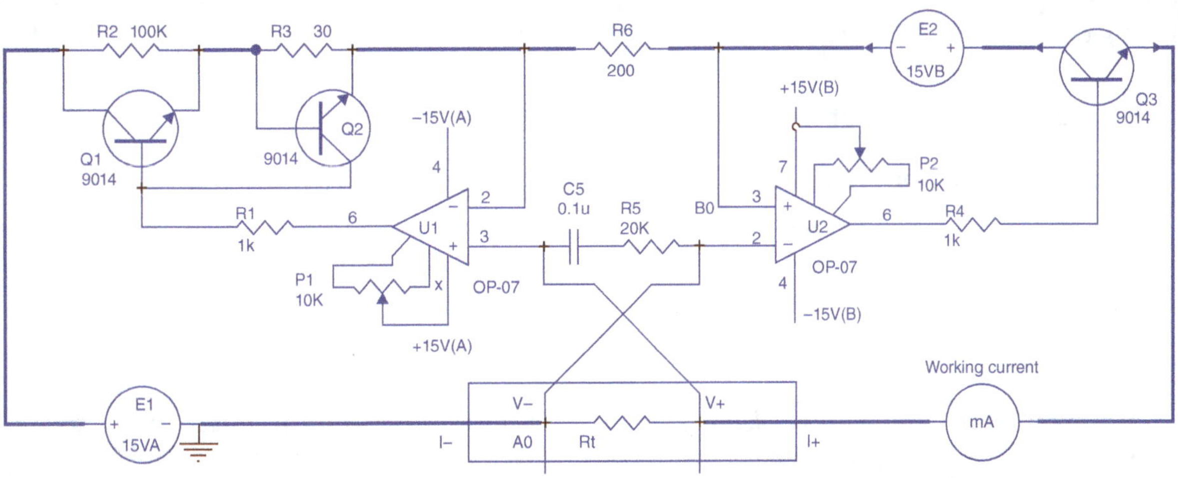

Figure 9. Internal circuit diagram of the NIM-manufactured Type IV power meter for measuring the thermistor mount.

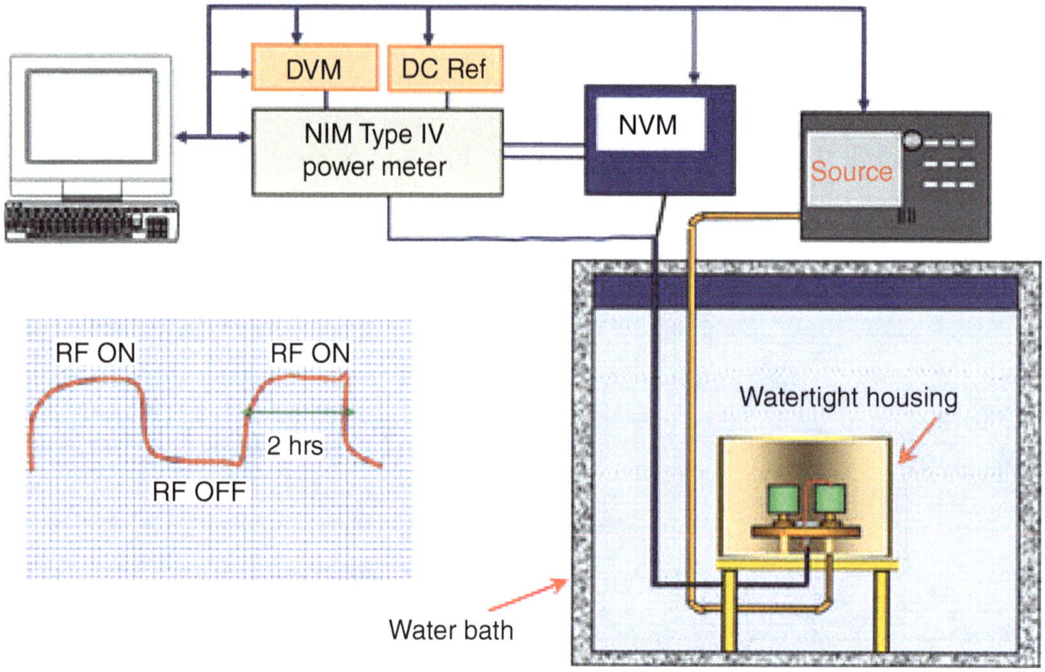

Figure 10. Schematic diagram for microcalorimeter measurements inside a water bath.

Figure 10 presents a complete operation setup for thermistor mount measurements and calibrations using a microcalorimeter at NIM, China. The microcalorimeter is sealed within a watertight housing and then is placed inside a water bath. The water bath has a very good thermal stability with a temperature fluctuation of less than 1 mK. During the measurements, signal source, digital voltmeter (DVM), nanovoltmeter (NVM), DC reference (DC Ref), and Type IV power meter were controlled by a computer for automation.

The measurement system is used to determine the DC bias voltages (V_1 and V_2) and thermopile outputs (e_1 and e_2) when the applied microwave power is off/on and the system reaches the thermal equilibrium. A typical output curve from the thermopile is also shown in **Figure 10** as a reference when the applied microwave power is off/on. With V_1, V_2, e_1, e_2, and correction factor g of a microcalorimeter, the effective efficiency η_e can be determined.

3.3. Measurement and Calibration Model

From the definition, the effective efficiency η_e of the thermistor mount is

$$\eta_e = \frac{P_{sub}}{P_{rf}} = \frac{P_{sub}}{P_{sub} + P_{dw}}. \tag{2}$$

Here, the total microwave power P_{rf} dissipated within the thermistor mount includes the DC-substituted power P_{sub} on the thermistor bead and the total loss P_{dw} (including the loss P_w on the wall of the thermistor mount and some portion of unsubstituted power). The DC-substituted power P_{sub} can be estimated through the DC bias voltages V_1 and V_2, as

$$P_{Sub} = \frac{V_1^2 - V_2^2}{R}. \tag{3}$$

The loss P_{dw} contributes to the portion of thermopile output change ($\Delta e = e_2 - e_1$) due to the temperature rising. The thermopile output change Δe has the following relationship:

$$\Delta e \propto \left(P_{dw} + cP_i\right). \tag{4}$$

The coefficient c is the weighted thermal factor due to the loss P_i within the thermal isolation waveguide onto the thermopile output. If the loss is uniformly distributed along the axial direction of the thermal isolation waveguide (typical case), $c = 0.5$.

Taking into consideration all the contributions from heat dissipated in different locations (such as the mount wall, TIS) of a microcalorimeter into its correction factor g, the effective efficiency η_e of thermistor mount at each frequency of interest can be calculated using the following recommendation [14]:

$$\eta_e = g\frac{1-\left(\frac{V_2}{V_1}\right)^2}{\frac{e_2}{e_1}-\left(\frac{V_2}{V_1}\right)^2} = g\eta_{e,unc}. \tag{5}$$

Here, $\eta_{e,unc}$ is the uncorrected effective efficiency, and g is the correction factor, which is the most important characteristic of a microcalorimeter. Several different techniques have been proposed in refs. [17–20] for evaluating the correction factor g, in order to determine the effective efficiency of the reference standard, thermistor mount, accurately. It is noted that sometimes the calibration factor K of the thermistor mount is of interest for applications. The calibration factor K can be derived from the effective efficiency η_e, as

$$K = \eta_e\left(1 - |\Gamma|^2\right). \tag{6}$$

Here, Γ is the input reflection coefficient of the thermistor mount.

4. Transfer Technique: Direct Comparison

The parameter of a reference power standard such as a thermistor mount can be transferred to the DUT sensor by means of the direct comparison transfer technique, which was proposed and summarized by the National Institute of Standards and Technology (NIST) of USA [21,

22]. **Figure 11** presents a basic idea of the direct comparison transfer for waveguide microwave power sensor calibration.

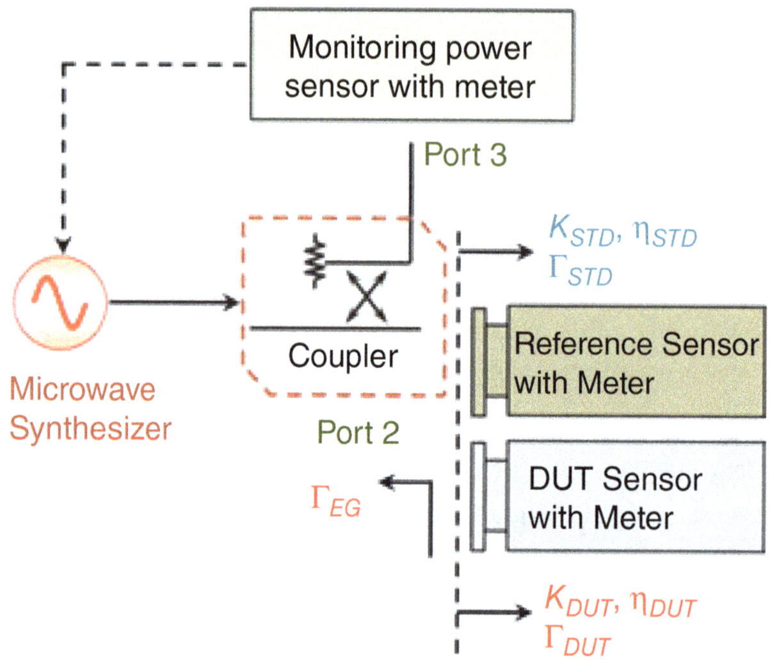

Figure 11. Calibration of a waveguide power sensor by means of direct comparison transfer using a coupler.

The system consists of a microwave synthesizer and a three-port directional coupler which is used to minimize the source mismatch [23]. As shown in **Figure 11**, a monitoring power sensor with a meter is connected to Port 3 of the coupler. The effective efficiency η_{DUT} and the calibration factor K_{DUT} of a DUT sensor are measured by alternately connecting a reference power standard (e.g., a thermistor mount with the effective efficiency η_{STD} and the calibration factor K_{STD}) and the DUT to Port 2 of the coupler. For the setup shown in **Figure 11**, the connectors of the DUT and the reference standard are kept the same. It is noted that for coaxial application, the coupler shall be replaced using a three-port power splitter.

The calibration factor K_{DUT} of the DUT sensor can be determined through

$$K_{DUT} = K_{STD} \times \frac{P_{DUT}}{P_{STD}} \times \frac{P_{3STD}}{P_{3DUT}} \times \frac{|1 - \Gamma_{DUT}\Gamma_{EG}|^2}{|1 - \Gamma_{STD}\Gamma_{EG}|^2}. \tag{7}$$

Here, P_{DUT} and P_{3DUT} are the powers measured at Port 2 using the DUT sensor and that at Port 3 using a monitoring power sensor, respectively. P_{STD} and P_{3STD} are the powers measured at Port 2 using the reference standard (e.g., a thermistor mount) and that at Port 3 using the same monitoring power sensor. Γ_{STD} is the input reflection coefficient of the reference standard, and Γ_{DUT} is the input reflection coefficient of the DUT. Γ_{EG} is the equivalent source match term of Port 2 [8] and equal to

$$\Gamma_{EG} = S_{22} - \frac{S_{21}S_{32}}{S_{31}}, \tag{8}$$

where S_{ij} ($i, j = 1, 2$, or 3) are the scattering parameters (S-parameter) of the directional coupler. A more detailed description of eq. (7) can be obtained in refs. [24, 25], and the derivation of eq. (8) can be found in ref. [6].

However, sometimes, a DUT sensor has an unmatched connector with the reference standards, and then an adaptor has to be used. Some measurement models with an adaptor at DUT/ reference standards have been proposed in refs. [25–27] and will be briefly introduced below.

4.1. Calibration Scenario with an Adaptor before Reference Standard

This is the application scenario where an adaptor has been connected between the reference standard and Port 2 of the coupler, while the DUT sensor is alternatively connected to Port 2 directly. The calibration factor K_{DUT} of the DUT sensor can be calculated with

$$K_{DUT} = K_{STD} \times \frac{P_{DUT}}{P_{STD}} \times \frac{P_{3STD}}{P_{3DUT}} \times \left| \frac{S_{21A}\left(1 - \Gamma_{DUT}\Gamma_{EG}\right)}{1 - \Gamma_{STD}S_{22A} - \Gamma_{EG}\Gamma_{A-STD}} \right|^2 . \tag{9}$$

Here, $\Gamma_{A-STD} = S_{11A} + \Gamma_{STD}S_{21A}S_{12A} - \Gamma_{STD}S_{22A}S_{11A}$, and S_{lmA} is the S-parameter of adaptor A, and $l, m = 1$ or 2. **Figure 12** presents a typical measurement setup when a coaxial-to-waveguide adaptor has been used before a waveguide reference standard (a thermistor mount as shown in **Figure 12** (a)).

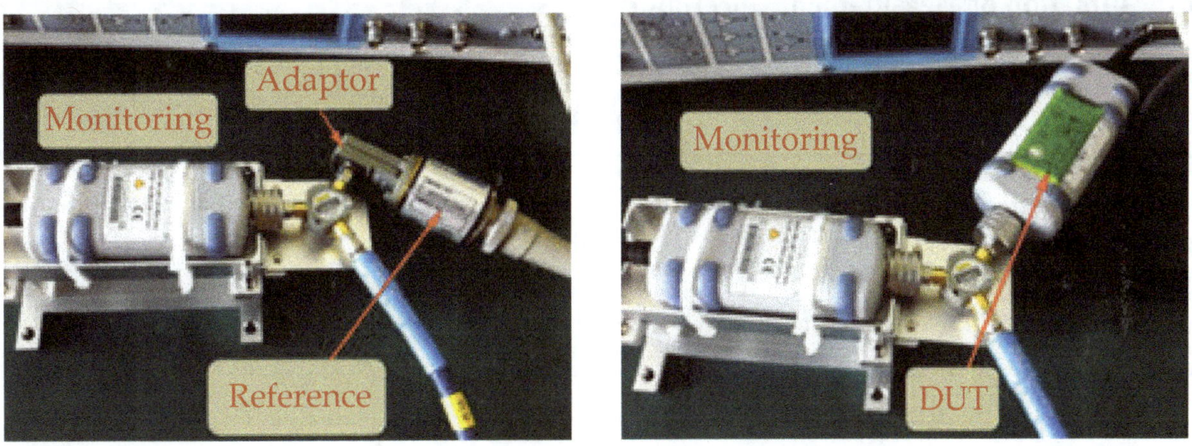

(a) (b)

Figure 12. Calibration of a DUT sensor by means of direct comparison transfer with an adaptor before the reference standard.

4.2. Calibration Scenario with an Adaptor before DUT Sensor

This is the application scenario where an adaptor has been connected between the DUT sensor and Port 2 of the coupler, while the reference standard is alternatively connected to Port 2 directly. The calibration factor K_{DUT} of the DUT sensor can be calculated with

$$K_{DUT} = K_{STD} \times \frac{P_{DUT}}{P_{STD}} \times \frac{P_{3STD}}{P_{3DUT}} \times \left| \frac{1 - \Gamma_{DUT} S_{22A} - \Gamma_{EG} \Gamma_{A-DUT}}{S_{21A}\left(1 - \Gamma_{STD} \Gamma_{EG}\right)} \right|^2 . \tag{10}$$

Here, $\Gamma_{A-DUT} = S_{11A} + \Gamma_{DUT} S_{21A} S_{12A} - \Gamma_{DUT} S_{22A} S_{11A}$. The mathematical model [eq. (10)] was previously derived using signal flow graphs together with the non-touching loop rule analysis in ref. [25]. It was later comparatively investigated in ref. [27] through the analysis of physical measurement processes. A consistent mathematical model has been observed.

The proposed model was successfully used to calibrate a high-sensitivity (lower power range) power sensor with an attenuator (the attenuator can be treated as a two-port adaptor with high loss) between the DUT and Port 2 of the coupler in ref. [27]. Good performance has been achieved referring to the data from the manufacturer.

5. Performance Evaluation and Uncertainty Estimation

In this section, the evaluation of measurement uncertainty is briefly introduced with the *Guide to the Expression of Uncertainty in Measurement* (GUM) [28]. The GUM method has been accepted and used in most of the current routine calibration works at NMIs or the standards laboratories. This is followed by an example of calibrating waveguide thermistor mounts with uncertainty evaluation in an international comparison.

5.1. Estimation of Measurement Uncertainty

To evaluate the measurement uncertainty, the GUM shall be followed. According to the GUM, there are two methods to evaluate the standard uncertainty $u(x_i)$ associated with the physical quantity x_i in the measurements, namely, *Type A Evaluation* and *Type B Evaluation*.

Type A Evaluation is a method of evaluating the standard uncertainty through the statistical analysis of a series of observations. It is normally referred to as "repeatable" measurement uncertainty. *Type B Evaluation* is a method of evaluating the standard uncertainty from other information including previous measurement data, specifications from manufacturers, data provided in calibration and other certificates, and uncertainties assigned to reference data taken from handbooks.

For evaluating the uncertainty of a measurand y from the standard uncertainty information of other physical quantities $(x_1, x_2, \ldots, x_N)$ with $y = f(x_1, x_2, \ldots, x_N)$, combined standard uncertainty $u_c(y)$ associated with y is adopted. According to the *Law of Propagation of Uncertainty* (LPU) in the GUM [28], $u_c(y)$ can be estimated from the standard uncertainties of $x_1, x_2, \ldots, x_N$, as

$$u_c(y)=\sqrt{\sum_{i=1}^{N}\left[\frac{\partial f}{\partial x_i}\right]^2 u^2(x_i)+2\sum_{i=1}^{N-1}\sum_{j=i+1}^{N}\frac{\partial f}{\partial x_i}\frac{\partial f}{\partial x_j}u(x_i,x_j)}, \tag{11}$$

where $u(x_i, x_j)$ is the covariance between x_i and x_j.

The expanded uncertainty U, which defines an interval about the result of a measurement that may be expected, can be estimated through $U = ku_c$. Here, k is the coverage factor and equal to 2 for a one-dimensional physical quantity at a level of confidence of approximately 95% assuming a Gaussian distribution. However, for a complex-valued physical quantity (e.g., S-parameter), the coverage factor k for 95% coverage probability is around 2.45 [29, 30].

5.2. Performance Evaluation in an International Comparison

CCEM.RF-K25.W Key Comparison: The precision measurement capabilities of NIM-fabricated WR-22 microcalorimeter have been validated and demonstrated in a key comparison (CCEM.RF-K25.W) on high-frequency power in the frequency range 33–50 GHz. The comparison is an exercise to establish the metrological equivalence of signatory NMIs' standards as stated in the Mutual Recognition Arrangement (MRA) of the Bureau International des Poids et Mesures (BIPM).

In the CCEM.RF-K25.W comparison, the effective efficiency and calibration factor of the travelling standards (Hughes Model 45772H-1100) as shown in **Figure 13** were compared. The effective efficiency of the travelling standard at each frequency of interest was determined by measuring the heating of mount in the microcalorimeter during the DC substitution. As introduced previously, a Type IV power meter was used as a bolometer bridge. The correction factor g of NIM's microcalorimeter was characterized through the measurements where a foil short was inserted between the test port and the DUT in the microcalorimeter [16].

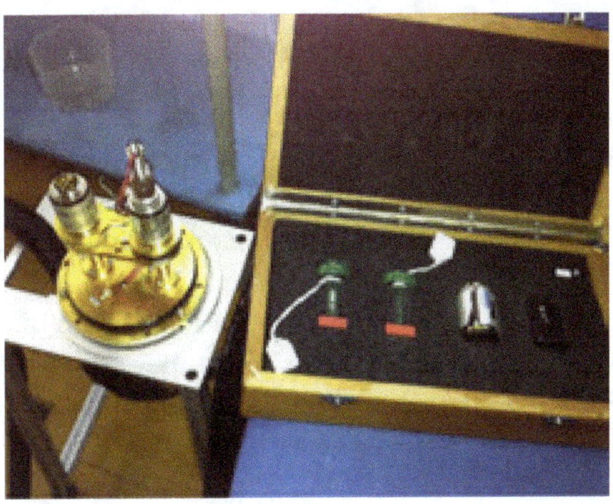

Figure 13. Participation in the international key comparison, CCEM.RF-K25.W.

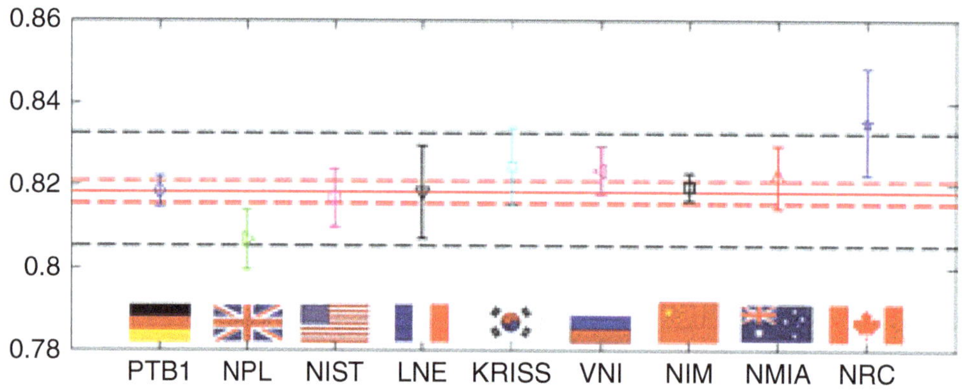

Figure 14. Calibration factor for the travelling standard Hughes Model 45772H-1100 (SN 216) at 33.0 GHz [31].

Figure 14 presents the measured calibration factor *K* for the travelling standard Hughes Model 45772H-1100 (SN 216) at 33.0 GHz. From the results shown in **Figure 14**, it can be observed that the NIM's microcalorimeter has a good measurement capability, and very good equivalence has been achieved referring to the results reported by other NMIs. An example of NIM uncertainty budget at 33.0 GHz is shown in Table 1 as a reference.

Quantity X_i	Estimate x_i	Standard uncertainty $u(x_i)$	Probability distribution/Type	Sensitivity coefficient c_i	Uncertainty contribution $c_i u(X_i)$	DOF υ_i
DC voltage ratio	0.8689	0.00002	Gaussian/Type B	1.5	0.00003	50
Thermal voltage ratio	1.0172	0.0004	Gaussian/Type B	0.93	0.0004	50
Correction factor *g*	1.0035	0.0036	Gaussian/Type B	0.85	0.003	50
Repeatability	0.8850	0.0009	Gaussian/Type A	0.96	0.0009	3
$\|\Gamma\|$	0.2778	0.0025	Gaussian/Type B	0.46	0.001	50
K	0.8196	0.0033				$\upsilon_{eff} = 27$

Table 1. An example of NIM uncertainty budget at 33.0 GHz [31].

6. Summary

In this chapter, we mainly focused on the introduction of microwave power measurements and standards. Primary power standards (e.g., microcalorimeter) and reference standards (e.g., thermistor mounts) have been discussed. Some recent developments of the waveguide microcalorimeter at NIM, China, and further applications of the direct comparison transfer technique at NMC, A*STAR of Singapore, have been reported. This is followed by an introduction of uncertainty evaluation for calibrating a WR-22 waveguide thermistor power sensor during an international comparison.

Furthermore, we have attempted to calibrate a WR-15 (50–75 GHz) waveguide thermistor mount using the direct comparison transfer technique [32]. Good performance has been observed preliminarily. Further improvement works have been planned and will be carried out in the near future.

Acknowledgements

This work was supported in part by the National Science and Technology Supporting Program "Wireless Communication Power Measurement Standard and Traceability Technology Research" of China under Grant No. 2014BAK02B02 and the Agency for Science, Technology and Research (A*STAR) of Singapore under Grant No. 0920170078.

Author details

Xiaohai Cui[1*], Yu Song Meng[2*], Yueyan Shan[2*] and Yong Li[1]

*Address all correspondence to: cuixh@nim.ac.cn, meng_yusong@nmc.a-star.edu.sg, and shan_yueyan@nmc.a-star.edu.sg

1 National Institute of Metrology, Beijing, China

2 National Metrology Centre, Agency for Science, Technology and Research (A*STAR), Singapore

References

[1] Panagopoulos AD, Arapoglou PM, Cottis PG. Satellite Communications at Ku, Ka, and V Bands: Propagation Impairments and Mitigation Techniques. IEEE Communications Surveys & Tutorials 2004; 6(3) 2–14.

[2] Lee YH, Natarajan V, Meng YS, Yeo JX, Ong JT. Cloud attenuation on Ka-band satellite link in the tropical region: Preliminary results and analysis. In: 2014 IEEE Antennas and Propagation Society International Symposium, 6–11 July 2014, Memphis, TN, USA.

[3] Dehos C, González JL, De Domenico A, Kténas D, Dussopt L. Millimeter-Wave Access and Backhauling: The Solution to the Exponential Data Traffic Increase in 5G Mobile Communications Systems? IEEE Communications Magazine 2014; 52(9) 88–95.

[4] ISO/IEC 17025:2005. General Requirements for the Competence of Testing and Calibration Laboratories. Geneva, Switzerland; 2005.

[5] Estin AJ, Juroshek JR, Marks RB, Clague FR, Allen JW. Basic RF and Microwave Measurements: A Review of Selected Programs. Metrologia 1992; 29(2) 125–151.

[6] Fantom A. Radio Frequency and Microwave Power Measurement. London, UK: Peter Peregrinus Ltd.; 1990.

[7] Rohde & Schwarz. Voltage and Power Measurements — Fundamentals, Definitions, Products. In: PD 757.0835.23, 1999.

[8] Agilent Technologies. Fundamentals of RF and Microwave Power Measurements (Part 2). In: Application Note 1449-2, 5988-9214EN; 2006.

[9] Brunetti L, Oberto L, Vremera ET. Thermoelectric Sensors as Microcalorimeter Load. IEEE Transactions on Instrumentation and Measurement 2007; 56(6) 2220–2224.

[10] Brunetti L, Oberto L, Sellone M, Vremera ET. Comparison Between Thermoelectric and Bolometric Microwave Power Standards. IEEE Transactions on Instrumentation and Measurement 2013; 62(6) 1710–1715.

[11] IEEE Standards 470-1972. IEEE Standard Application Guide for Bolometric Power Meters. New York; 1972.

[12] Ham RE. Microwave detectors. In: Chang K. (ed.) Encyclopedia of RF and Microwave Engineering. John Wiley & Sons, Inc.; 2005. P2736–2742.

[13] Clark RF. The Microcalorimeter as a National Microwave Power Standard. Proceedings of the IEEE 1986; 74(1) 102–104.

[14] Allen JW, Clague FR, Larsen NT, Weidman MP. NIST Microwave Power Standards in Waveguide. NIST Technical Note 1511; 1999.

[15] Crowley TP, Cui X. Design and evaluation of a WR-15 (50 to 75 GHz) microcalorimeter. In: Conference on Precision Electromagnetic Measurements Digest, CPEM2008, 8–13 June 2008, Broomfield, CO, USA.

[16] Cui X, Li Y, Gao X. A new design and evaluation technique of a microcalorimeter. In: Conference on Precision Electromagnetic Measurements Digest, CPEM2012, 1–6 July 2012, Washington, DC, USA.

[17] Cui X, Li Y, Gao X, Song Z, Sun W. Correction factor evaluation technique for a rectangular waveguide microcalorimeter. In: Digest of the 82nd ARFTG Microwave Measurement Conference, 18–21 November 2013, Columbus, OH, USA.

[18] Cui X, Crowley TP. Comparison of Experimental Techniques for Evaluating the Correction Factor of a Rectangular Waveguide Microcalorimeter. IEEE Transactions on Instrumentation and Measurement 2011; 60(7) 2690–2695.

[19] Clague FR. A Method to Determine the Calorimetric Equivalence Correction for a Coaxial Microwave Microcalorimeter. IEEE Transactions on Instrumentation and Measurement 1994; 43(3) 421–425.

[20] Judaschke R, Ruhaak J. Determination of the Correction Factor of Waveguide Microcalorimeters in the Millimeter-Wave Range. IEEE Transactions on Instrumentation and Measurement 2009; 58(4) 1104–1108.

[21] Weidman MP. Direct Comparison Transfer of Microwave Power Sensor Calibration. NIST Technical Note 1379; 1996.

[22] Ginley R. A Direct Comparison System for Measuring Radio Frequency Power (100 kHz to 18 GHz). Measure 2006; 1(4) 46–49.

[23] Engen GF. Amplitude Stabilization of a Microwave Signal Source. IRE Transactions on Microwave Theory and Techniques 1958, 6(2) 202–206.

[24] Shan Y, Cui X. RF and microwave power sensor calibration by direct comparison transfer. In: Cocco L. (ed.) Modern Metrology Concerns. Rijeka: InTech; 2012. P175–200.

[25] Shan Y, Meng YS, Lin Z. Generic Model and Case Studies of Microwave Power Sensor Calibration Using Direct Comparison Transfer. IEEE Transactions on Instrumentation and Measurement 2013; 62(6) 1834–1839.

[26] Kang TW, Kim JH, Kwon JY, Park JI, Lee DJ. Direct comparison technique using a transfer power standard with an adapter and its uncertainty. In: Conference on Precision Electromagnetic Measurements Digest, CPEM2012, 1–6 July 2012, Washington, DC, USA.

[27] Meng YS, Shan Y. Measurement and Calibration of a High-Sensitivity Microwave Power Sensor with an Attenuator. Radioengineering 2014; 23(4) 1055–1060.

[28] BIPM, IEC, IFCC, ILAC, ISO, et al. Evaluation of measurement data – guide to the expression of uncertainty in measurement. In: JCGM 100:2008 (GUM 1995 with Minor Corrections). Joint Committee for Guides in Metrology; 2008.

[29] Ridler NM, Salter MJ. An Approach to the Treatment of Uncertainty in Complex S-Parameter Measurements. Metrologia 2002; 39(3) 295–302.

[30] Meng YS, Shan Y. Measurement Uncertainty of Complex-Valued Microwave Quantities. Progress In Electromagnetics Research 2013; 136 421–433.

[31] Judaschke R. CCEM.RF-K25.W — RF power from 33 GHz to 50 GHz in waveguide. Final Report (Draft B); 2014.

[32] Cui X, Meng YS, Shan Y, Yuan W, Ma C, Li Y. Evaluation and validation of a national WR-15 (50 to 75 GHz) power measurement system. In: Digest of the 84th ARFTG Microwave Measurement Conference, 2–5 December 2014, Boulder, Colorado.

5

Innovative Theoretical Approaches Used for RF Power Amplifiers in Modern HDTV Systems

Daniel Discini Silveira, Marcos Paulo de Souza Silva,
Marcel Veloso Campos and Maurício Silveira

Abstract

The essential purpose of this chapter is to introduce theoretical and numerical approaches that can be used for modeling nonlinear effects that appear intrinsically in the design of power amplifiers that have been used widely in many modern high-density television (HDTV) architectures. Important effects like the pre-distortion using adaptive techniques, with distinct characteristics like amplitude, phase, and frequency, as well as, their specific nature such as *AM/AM*, *AM/PM*, *PM/AM*, and *PM/PM*, and constitute one of the main directions of this research. All theoretical and technological approaches have been supported by a consistent set of numerical data performed with one of the most important platform of simulations used in the great area of Radio Frequency (RF) and Microwave structures. As a direct application, we are introducing some efficient processes that can be used for the characterization of RF systems with a set of consistent laboratorial measures that permit us to visualize the effective cost and a complete architecture for the characterization of high-power amplifiers. With the continuous and innovative technological demand that is imposed by the international marketing has a great importance to find versatile systems that are capable of measuring several amplifier characteristics, as gain, output power, inter-modulation distortion of different signals, efficiency, current, and temperature that constitute another direction of research that has been demanded strongly for news advanced technologies used widely in modern HDTV systems.

Keywords: innovative theoretical modeling, digital pre-distortion, nonlinear power amplifiers, characterization of RF systems, numerical approach, measurement systems, HDTV architectures

1. Introduction

Nowadays, the research and development of the modern *RF* power amplifiers have been demanded some continuous and accurate studies concentrated over the linearity of the equipment. Besides the distortions caused by the nonlinear devices, we have been having a constant improvement in the implementation of new modulation techniques by exploring the spectrum properties [1]. The nonlinear performance caused by the devices is a fundamental effect, which have invoked the attention of many researchers in many famous research centers around the entire world [2, 3]. In Section 2, we will focus our research in the development on an adaptive prototype for the linearization of the power amplifier with resources of the digital pre-distortion, in a such way that it is possible to generate a feedback distortion through the power amplifiers [4, 5]. This will provide a much more linear final answer, being the main application applied for modern high-density television (*HDTV*) transmitter equipment. In the current literature, we can classify this type of pre-distortion as being adaptive, since that we will must to compare the input signal with that transmitted one, which it includes the distortion generated by the power amplifier [6]. The theoretical as well as experimental approach starts by set an open-loop prototype until to reach the closed loop by adopting an adaptive system for analog signals, such that the digital treatment uses the adaptive techniques over the entire transmission system [7–14].

Actually, for many modern implementations of broadcast TV transmitters, the association of power amplifiers (PAs) is performed in order to obtain the desired output power, usually higher than 1 kW. The filtered output signal is expected to present low levels of distortion, so it does not interfere in adjacent broadcast channels. A signal applied to a power amplifier (PA) will suffer from distortions, independently of its input level and the technology (BJT and LDMOS) [1, 2, 15]. Basically, three types of distortions are caused in the input signal: amplitude, frequency, and phase. Amplitude distortions are caused by the amplifier gain compression, P_{1dB}, a device-dependent characteristic. By clipping, the signal harmonics are generated, also classified as frequency distortion. The magnitude of these harmonics is increased with the amount of signal clipping, consequently increasing the overall distortion. Phase distortions are mainly caused by memory effects present in amplifiers due to the different reasons, as

cant *back-off* can avoid some of these undesirable distortions, but the efficiency of the amplifier will be very poor. A powerful way to decrease distortion in the output signal is to use linearization techniques, thus operating the PA in a more efficient way [16–18]. Some initial research for the *AM/AM* analysis can be obtained, by imposing that the transfer function of the output voltage of the power amplifier had third-order nonlinear dependence with respect to the input signal. Although the numerical approach adopted in that occasion was efficient for this particular case, the nonlinear effect for the *AM/PM* analysis was not compensated [19–22]. Example of models which can compensate these and also other distortions can be found with a study over several practical models. An accurate model is the first step for the implementation of an efficient pre-distortion system using a parallel model has obtained good results, compensating for both *AM/AM*, *AM/PM* distortions, and also for memory effects, the main causes of distortions in the PA output signal. These main theoretical and experimental approaches constitute the most important contributions presented in the Section 3 [23, 24–26].

During these last years, a wide class of theoretical approaches have been presented for the current literature that involves the project of the power amplifiers modeled with field-effect transistor (*FET*) structures, where the main focus is concentrate over the linear region of the device [2–5, 27–29]. Recently, important studies were done for nonlinear microwave circuits as well as the new design for wireless digital structures. Also some innovative studies using linearization techniques with high gain of the power amplifiers, as well as the great importance it acquires the presence of an active pre-distortion circuit for monolithic microwave integrated circuit (*MMIC*) components implemented with field-programmable gate array (*FPGA*) architectures [30–32]. More recently, a set of innovative research appears in the current literature, with the introduction of a new theoretical modeling for the nonlinear analysis of the power amplifiers, where the electronic structure modeling was strongly motivated by the involvement of the authors with the project of modern HDTV equipments. With respect to industrial implementations, this research has an expressive contribution related with polarization of the digital structures using the common source configuration in the design of all hybrid RF circuits [33]. Therefore, the desired great stability can be reached independent of the particular pre-distortion that must be corrected in the real time, like: *AM*/*AM*, *AM*/*PM*, and all the others cases. The theoretical approach presented in the Section 4 can be used for a wide class of modern power amplifiers independent of the particular modulation system that will be adopted for the design of the HDTV systems. Some experimental measurements are performed, and we make a comparison with all numerical simulations using a very powerful numerical platform, that is, the Advance Design Systems (*ADS*) that confirm all ours theoretical approaches [34, 35].

The characterization of RF power systems or products over the entire Ultra High Frequency (*UHF*) band is very difficult and sometimes expensive task that demands time and accuracy. This means expensive and specialized setups capable of performing the necessary measurements [36–45]. In Section 5, we present a high performance and low-cost setup designed for test and characterization of RF high-power amplifiers, capable of generate reports that will help the operator to decide about the conformity of the equipment with other ones tested, as we can see in the characterization of a power drawer [46–52]. The drawer is composed by four power amplifiers, capable of delivering a maximal output power of 860 Watts peak sync (Wps), 430 (W) for the USA digital TV broadcast system (*ATSC*), or 220 (W) for the standard adopted in the Japanese/Brazilian systems (*ISDB-T*). The architecture presented in this section can naturally also be used to characterize other RF power amplifiers, where the power drawer was only a particular case.

2. Efficient experimental techniques to control the adaptive linearization: digital signals *I* and *Q*

In this section, we intend to focus the implementation of an efficient FPGA electronic architecture than can serve for those systems that have a very sensible characteristic, with respect to the nonlinear distortions [1]. In principal, this research will be directed for the analysis of both I and Q complex signals in order to minimize both *Phase* (*AM-PM*) and *Amplitude* (*AM-AM*) distortions that are present intrinsically in the functioning of the architecture. One of the most important steps to be conquered with this design is to combine the power amplifier with a digital pre-distortion system, in order to get an acceptable inter-modulation level at the output.

With respect to the numerical implementations, all programs are made using classic computational methods, such as *linear correlation* and *linear regression*, which are very convenient for this kind of distortion as well as permits for the designer to work close to the linear operation region of the amplifier [2–5].

2.1. Fundamental relationships adopted for the linear numerical approach

Recently, the implementation of innovative electronic architectures that involve numerical implementation has an important requirement to investigate the relationship existent between two or more variables presented in the system. The verification of the degree of relationship between the variables is object of the correlation studies, with the essential purpose to measure and to evaluate the relationship existent between two random variables [6–8].

Typically, the linear correlation seeks to measure the relationship between the variables and through the disposition of the points around a straight line and is used exhaustively for the determination of the correlation coefficient. For a more general computational modeling, we must to detect the correlation that exists between two arbitraries variables. For that case, it is possible to allocate all points over a straight line, which corresponds the known classic method of the linear correlation, while the points in the two-dimensional real domain must be placed close of some curve and establish a nonlinear correlation.

If both variables x and y have the same characteristics such that the increase of the first variable must impose the same performance for the second one, we are in front of the positive linear correlation, where the alternative case is corresponding to the negative linear correlation. The perfect linear correlation happens when the points x and y are perfectly aligned. The null correlation occurs when there is not a relationship among the variables, which implies that all variations can happen independently [9–12].

The linear coefficient, α_r, which in the international literature is known as the Pearson correlation can be described by Eq. (1) in the analytic form:

$$|\alpha_r| = \frac{\sum_{i=1}^{n} |(x_i - \bar{x})(y_i - \bar{y})|}{\sqrt{\sum_{i=1}^{n}(x_i - \bar{x})^2 \sum_{i=1}^{n}(y_i - \bar{y})^2}} \tag{1}$$

In the previous equation, we have: n = number of pairs (x, y); $\bar{x}$ and $\bar{y}$ = averages of both x and y variables. The coefficient α_r has values in the interval I = [−1, 1] that corresponds the correlation level, as is illustrated in **Table 1**.

Amplitude	Correlation level
$\lvert\alpha_r\rvert = 0$	Zero
$0 < \lvert\alpha_r\rvert \leq 0.30$	Weak
$0.30 < \lvert\alpha_r\rvert \leq 0.60$	Medium
$0.60 < \lvert\alpha_r\rvert \leq 0.90$	Strong
$0.90 < \lvert\alpha_r\rvert < 1$	Very strong
$\lvert\alpha_r\rvert = 1$	Perfect

Table 1. Amplitude of the parameter α_r.

The proximity of the parameter α_r close to zero permits us to identify that there does not exists a significant linear correlation between the variables x and y, while if this coefficient is close to ±1, we can conclude that there exists a strong correlation. Typically, for the general case, we have: $0 \leq \lvert \alpha_r \rvert \leq 1$.

The straight line of the minimum square that adjusts the set of points $\{(x_1, y_1), (x_2, y_2), \ldots (x_n, y_n)\}$ that has a better adjustment performed through the linear equation:

$$y = a.x + b \tag{2}$$

For this numerical approach, the constants a and b can be determined by solving the normalized system composed by the equations:

$$a = \frac{\sum_{i=1}^{n}[(x_i - \bar{x}).(y_i - \bar{y})]}{\sum_{i=1}^{n}(x_i - \bar{x})^2} \qquad b = \bar{y} - a.\bar{x} \tag{3}$$

2.2. An initial hardware implementation

In order to reach the demanded linearization, we must to take into account some very important parameters, such as operation frequency, the temperature, and other ones. It is great to emphasize that those parameters are not constant as a function of the time, and thus we must to set their control under an adaptive digital system.

If our project has the technological demand to work with a fast control and the system must to operate in real time, we looked for to substitute the used memory for a model implemented experimentally by a *Programmable Logic Device* (*PLD*) just because its present some countless advantages. In our case, we have opted for a *Look-Up Table* (*LUT*) digital architecture.

The analog video signal is transformed in a digital one that proceeds for the *LUT* and for the retard circuit, aiming to establish the best relationship among the samples. In other words, we must to get a better synchronism, and at that point, the adjustment algorithm compares the samples of the correlator and the *up-date-algorithm* will make the updating at the *LUT*.

An initial architecture for the pre-distortion system can be visualized through the block diagram shown in **Figure 1**. Despite of we have many parameters to adjust, has a great importance the fact that we still must to be sure that we have comparing the samples in the same instant of time. A program can be developed in C++ language to control the whole system, and this initial idea proved to us a great effectiveness in the linearization as well as in the correction of the amplitude.

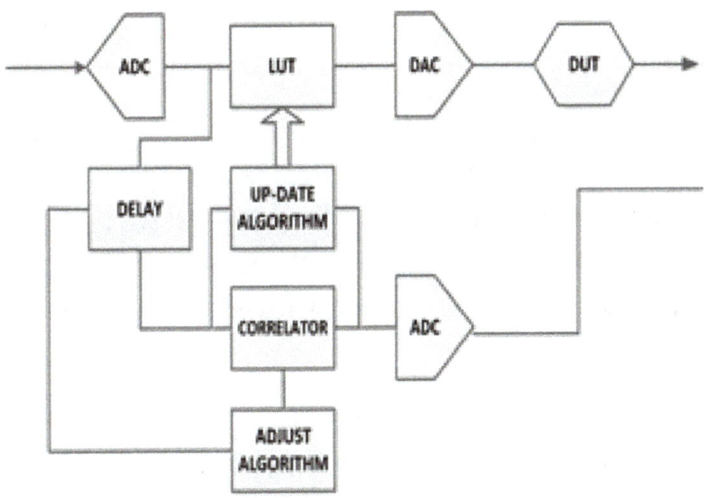

Figure 1. Block diagram: digital pre-distortion system.

Figure 2. An initial hardware architecture: pre-distortion system.

At the same time, our hardware implementation must be projected as well as operate with distinct input digital signals: *I* (*Phase*) and *Q* (*Quadrature*), such that it is possible to correct both amplitude and the phase of each one. An initial whole architecture for the Module of Programmable Logic (*MLP*), the *PLD* plate and the converters, can be visualized in **Figure 2**. At first, the control is done through the C++ software; but with the progress of this project, it is possible to set the control through a device that make the digital signal processing (*DSP*), as well as making the use of a microcontroller, by placing him in the same architecture besides both *AD* and *DA* converters.

2.3. An improvement of the hardware architecture

The linearization of the signals has a great importance, because independently of the distinct *TV* digital patterns that have been adopted in many countries, the systems have a very sensible characteristic with respect to the nonlinear distortions. These effects always have been demanding the need of to implement circuits that can provide an efficient linearization [9–15].

With respect to the analysis of both *I* and *Q* complex signals, the new linearizer circuit must allow to get a better performance for the whole architecture. In order to minimize both *Phase* (*AM-PM*) and *Amplitude* (*AM-AM*) distortions, the ideal solution would be to combine the power amplifier with a digital pre-distortion system, in order to get an acceptable inter-modulation level at the output. In this project, we concentrate our choice over the *FPGA* electronic architectures.

2.4. Experimental measurements in accordance with a set of compatible simulations

At first, it is important to emphasize that the *FPGA* architecture works like a *LUT* that stores the input data and permits to establish a comparison between all of them with the coming set of feedback data. At the same time, this comparison is performed; the numerical codes inserted inside the digital architecture shall supply to the *LUT* a new set of values that can be stored, being this process a constant and continuous routine.

For the flow of the data inside the entire architecture, we can use the *Matlab* platform, for the generation of the input signal with *two tones* [10–12]. This function can be accomplished by the first *FPGA* structure, and in the sequence, we can pass the signal through the pre-distortion processor. The output signal will enter in the second *FPGA* architecture that has the essential function of to simulate the power amplifier. After the amplification is performed, we can remove the feedback signal that will be compared with that one to be transmitted, and as one direct consequence, it is possible to realize all corrections that are demanded by the project in both signals.

An update of the first architecture exhibited previously can be visualized in **Figure 3**.

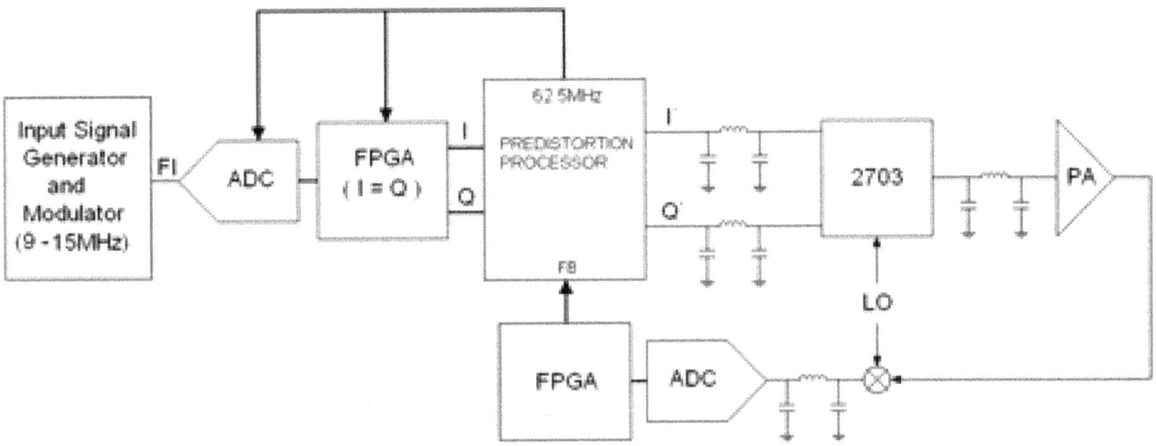

Figure 3. An improvement of the first hardware architecture.

In **Figure 4(a–c)**, we can visualize the input signal, the distorted signal by the amplifier and the error vector, respectively, and the error vector corresponds to the subtraction of the two previous signals at the same time t.

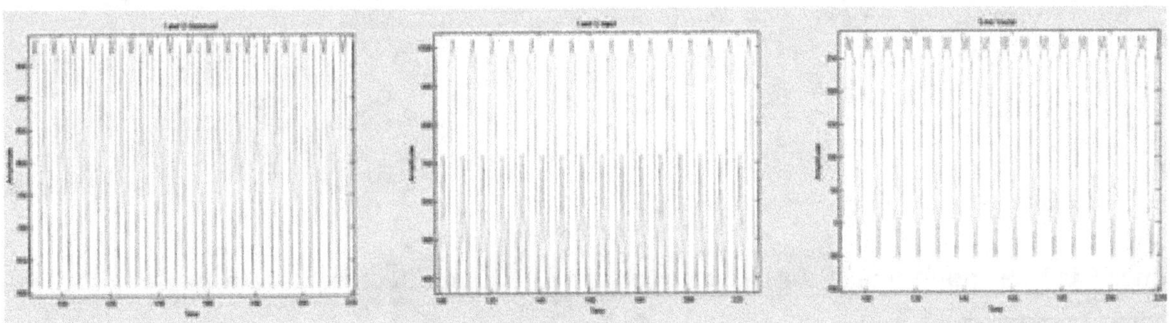

(a) *Input* Signal. (b) *Distorted* Signal. (c) *Error* Vector.

Figure 4. Some of the most important signals detected in some points of the architecture, plotted with the *MatLab* as a function of the time t (s). (a) *Input* Signal. (b) *Distorted* Signal. (c) *Error* Vector.

The *linear regression method* combined with *linear correlation* using the *FPGA* technology permits to adjust the linear function $y = f(x) = ax + b$. A reverse effect for the distortion is shown in **Table 2**, using as a numerical approach the inverse linear function $x = g(y) = (1/a).[y - b]$, where the parameters a and b are defined as in Section 2.1, where evidently we are presenting all numerical data placed in distinct lines.

For this case, if we consider the entire set of ideal values, all coefficients defined in Section 2.1 acquire the values: $\bar{x}$=48.5, $\bar{y}$=48.39, a = 1.1837 and b = −9.02. In addition, in **Figure 5**, it is possible to visualize the transition of the most important curves with respect to the ideal linear performance of the amplifier, by adopting this numerical approach for the adaptive pre-distortion linearization.

y ideal	40	41	42	43	44	45	46	47	48
x up-date	41.4	42.2	43.1	43.9	44.7	45.6	46.4	47.3	48.1
y ideal	49	50	51	52	53	54	55	56	57
x up-date	49.0	49.8	50.7	51.5	52.4	53.2	54.0	55.0	55.8

Table 2. *Up-Date* of the input signals.

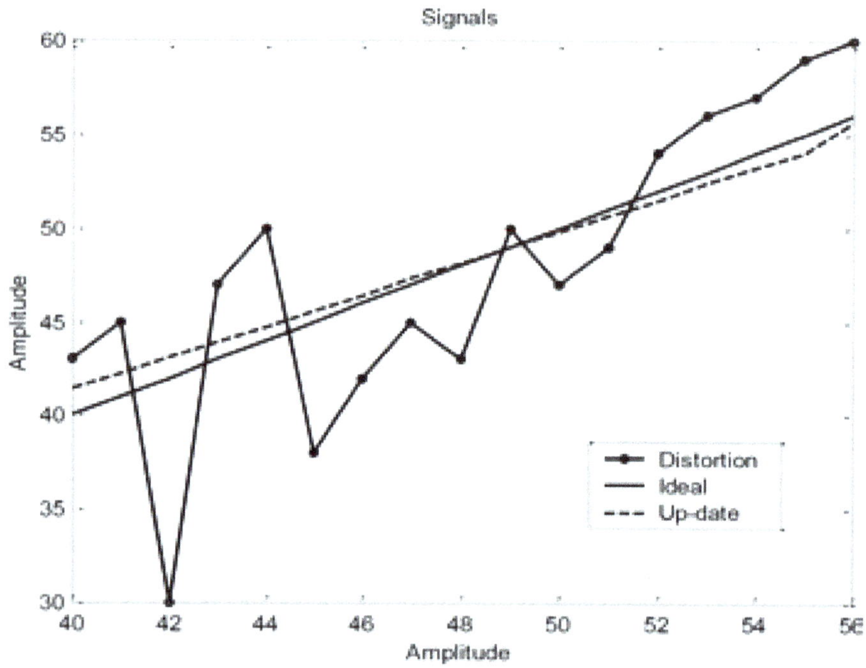

Figure 5. An improvement of the hardware architecture.

Figure 6 is presenting one of the most important *FPGA* architecture used for all experimental measurements.

Figure 6. One FPGA programmable logic module architecture.

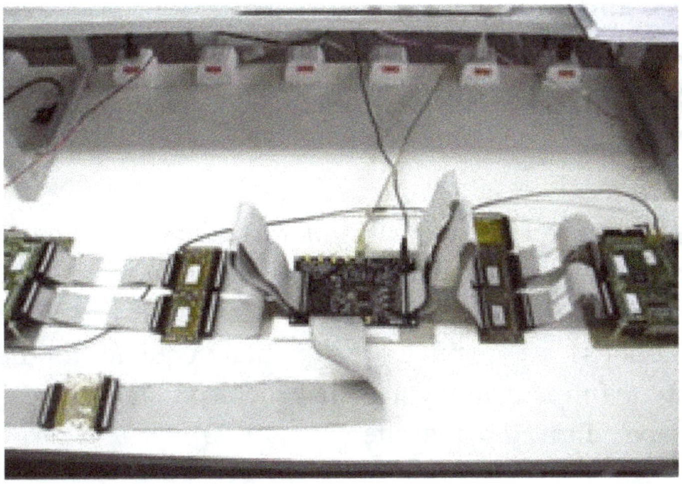

Figure 7. The entire pre-distortion digital architecture.

Finally, in **Figure 7**, we are showing the entire architecture implemented for experimental tests of a modern HDTV transmission equipment. In this Section, we present a programmable logic module architecture that have as the main function to control the linearization process, which permits for us to acquire more knowledge about how to set the programming codes more easily [13–15].

It is easy to confirm the nonlinear performance of the power amplifiers provokes the harmonic distortion as well as the inter-modulation products as a function of how many carriers we have used for setup the system. This apprenticeship permits for the designer to get more background with the linearization approaches and uses an appropriate software for the adaptive digital pre-distortion techniques.

3. Distortions for *ISDB*-T OFDM signals for RF power amplifiers architectures

The essential purpose of this Section is to present some recent development for many types of digital modulations, capable of transmitting large amount of data, which occurs for the standard *ISDB-T*. For this case, the carrier combination can generate a reasonable number of voltage spikes in the time domain, depending of its phase.

Experimentally, some important parameters like the figure of merit, the peak-to-average ratio (PAR), assume an important role with respect to improve our background of how to determine the level of the compression point of 1 (dB) for a wide class of AB power amplifiers. Depends on how efficient is the architecture will be implemented, it is possible to confirm that a large back-off of all amplifiers used in the *ISDB-T* transmitter chain needs to be performed, in comparison with the case of an analog broadcast transmission.

Normally, the gain compression is caused by inter-modulation characteristics of the transistors, with major influence of the third harmonic component. The inter-modulation distortion

(*IMD*) can be easily measured when more than one carrier is used as a PA input signal, as two tones with equal amplitude. For the case that a two-tone test signal is applied to an amplifier, it is possible to observe several inter-modulation components at the output spectrum generated due to the transistor nonlinear behavior.

For an *ISDB-T* standard Orthogonal Frequency Division Multiplexing (OFDM) signal with thousands of carriers, the distortion at the fundamental zone appears as a set of multiple distortions, and the bandwidth of the distortion is related to the signal bandwidth. Therefore, it is possible to establish some relationships between the signal bandwidth with the second and higher-order distortions for a specified fundamental frequency for the system. In addition, it is possible to detect the fundamental zones that corresponding to odd or even distortions, and some of them can be removed using filters [17, 19, 23].

Another set of important parameters and effects must be controlled for a system with memory, such as low (high) frequency, thermal effects, trapping effects, biasing circuits, Automatic Gain Control (*AGC*) loops; transistor (transit time and parasitic reactance), matching networks (group delay). The amount of memory is defined as the time between the origin of the kernel until the point where it passes through zero and is directly dependent of the transmitted signal bandwidth.

Some particular cases can be analyzed for RF power amplifiers in order to get more information about general classification of memory effects, in linear and nonlinear ones, as well as the *memoryless* systems. Thus, it is possible to confirm the origin of linear memory and that the nonlinear memories are generated in the nonlinear dynamic interactions of the input signal. Another important experimental procedure confirms that a practical PA output signal has both linear and nonlinear memory, and an *ISDB-T* OFDM signal that occupies a fixed bandwidth with up to thousand of carriers will cause more memory effects in a PA than for an analog signal, which has a small number of carriers [20–22, 24].

3.1. Fundamental principles over the clipping and inter-modulation distortion effects

Generally, there are many types of digital modulations, capable of transmitting large amount of data. If the OFDM technique is used, the signal can contain thousands of carriers. Depending of the phase of these carriers, the combination can result in a signal with high *Crest Factor*, demanding very high output power from the system PA. The system will be then inefficient.

It is possible to quantify the magnitude of this *Crest Factor* by the following equation, named here peak-to-average ratio (PAR), defined by:

$$PAR = 20.Log(|X_{PEAK}|/X_{Av}) \tag{4}$$

where X_{PEAK} and X_{Av} are the maximum and the average values (V) of the signal, respectively. The X_{PEAK} value can be quantified by an oscilloscope with enough bandwidth to measure the signal in time domain. The X_{Av} can be measured using a Wattmeter, and later converting this value to voltage, assuming a 50 (Ohms) load. The PAR from some common used signals is: 3

(dB) for a continuous wave (CW), 3.5–4 (dB) for a *QPSK*, and approximately 12 (dB) for the *ISDB-T* standard OFDM.

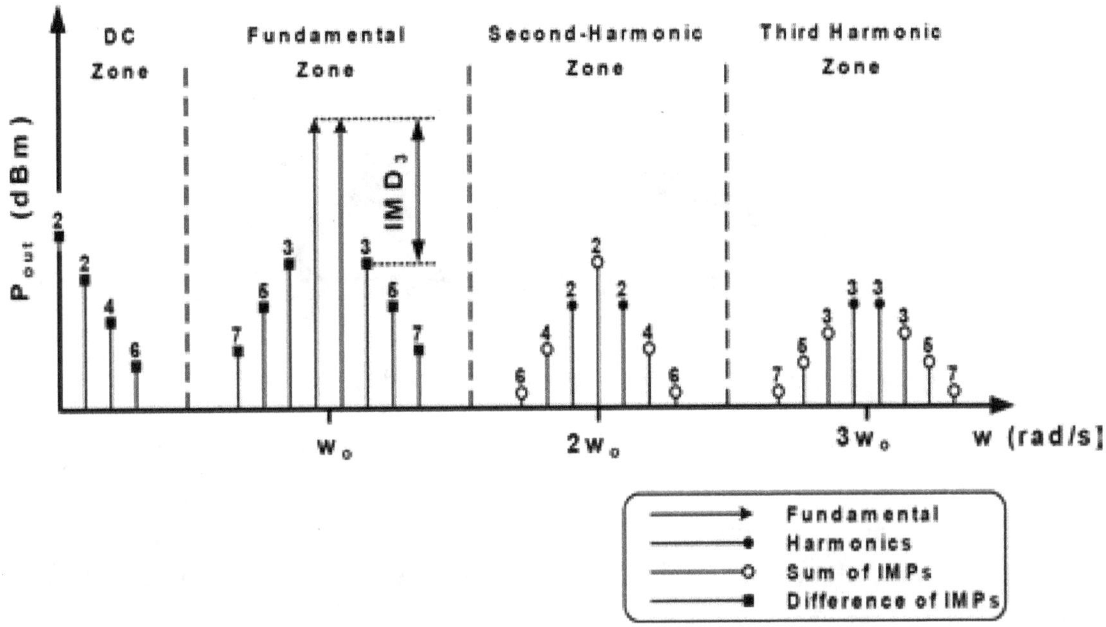

Figure 8. Several components of inter-modulation generated by a *two-tone* test signal applied to a PA.

Inter-modulation can be easily measured when more than one carrier is used as a PA input signal, as two tones with equal amplitude. When a *two-tone* test signal is applied to an amplifier, it is possible to observe several *IMD* components at the output spectrum generated due to the transistor nonlinear behavior, as seen in **Figure 8**.

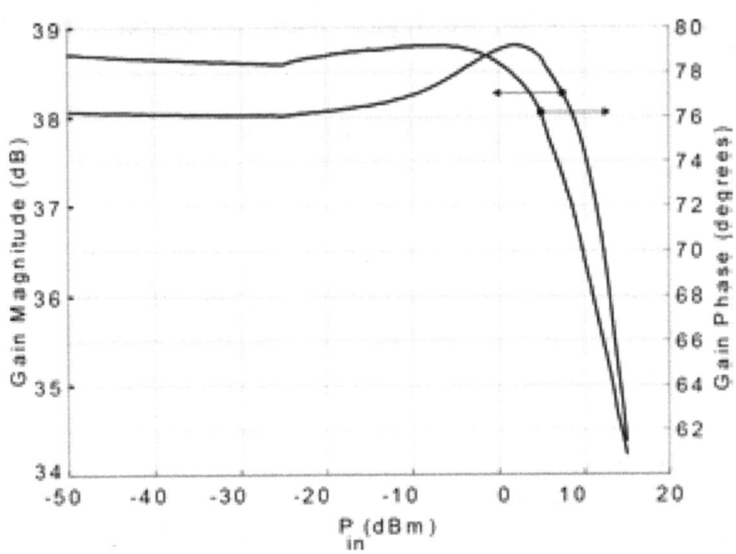

Figure 9. Typical gain curve of a class AB power amplifier.

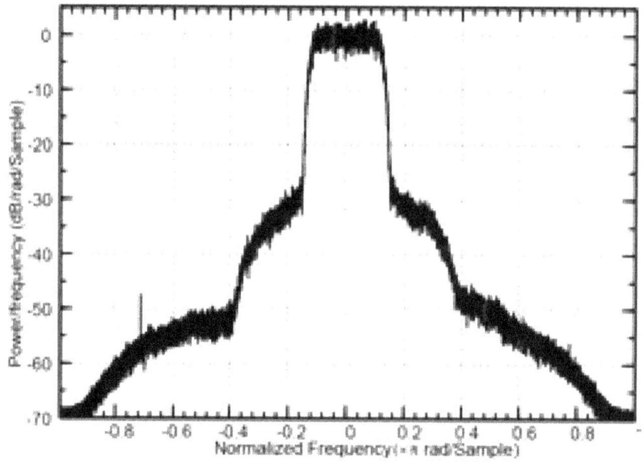

Figure 10. Distortion generated by a PA when excited by an *ISDB-T* standard signal.

By adopting this procedure, if a class AB PA has a compression point of 1 (dB) at 10 (*dBm*), as shown in **Figure 9**, an OFDM signal in the *ISDB-T* standard can be amplified until a maximum input power of −2 (*dBm*), using this PA, without compression of the signal peaks. Therefore, a large *back-off* of all amplifiers used in the *ISDB-T* transmitter chain needs to be performed, in comparison with the case of an analog broadcast transmission. The gain compression is caused by inter-modulation (IMD) characteristics of the transistors, with major influence of the third harmonic component.

In an *ISDB-T* standard OFDM signal with thousands of carriers, the distortion at the fundamental zone appears as a set of multiple distortions, as in **Figure 10**. The bandwidth of the distortion is related to the signal bandwidth. Thus, if a signal has a bandwidth of 5.6 (MHz) at the fundamental frequency, the bandwidth of the second-order distortion will be of (2)*(5.6) = 11.2 (MHz), the third-order distortion will have bandwidth of (3)*(5.6) (MHz), and so on. The odd-order distortions lie inside the fundamental zone, requiring a different treatment than the even order ones. These lie outside the fundamental and at the DC zone, and can be removed using filters.

A possibility to avoid IMD occurrence is to perform *back-off*, as said before. This procedure obviously increases costs in the transmission system, due to an increased number of PAs, but will guarantee a very high quality of the transmitted *ISDB-T* signal, measured by the MER in (dB), which excellent levels are over 40 (dB). On the other side, if high values of *back-off* are used, the PA efficiency is affected, as the amplifier will operate with less power than it was designed for. Therefore, a careful trade-off between *back-off*, or high values of MER, and PA efficiency has to be carefully established.

3.2. About the intrinsic dependence of memory effects

A system with memory is that one where the output signal has dependency of past states of the input signal. The memory of a system can be estimated through the observation of the first-order *Volterra* kernel. The amount of memory is defined as the time between the origin of the kernel until the point where it passes through zero and is directly dependent of the transmitted signal bandwidth.

Let us consider the particular case of a RF power amplifier, which has a complex structure and presents many kinds of memory. For our experimental procedure, we adopt the classification in accordance with the band of frequencies, in the form:

A. *Low frequency* (kHz to MHz): Thermal effects, trapping effects, biasing circuits, AGC loops.

B. *High frequency* (GHz): Transistor (transit time and reactance parasitics), matching networks (group delay).

These memories are mixed together in a nonlinear coupled power amplifier, and the problem of estimating behavioral models becomes very difficult. A general classification of memory effects, in linear and nonlinear ones, can be found, while the *memoryless* system is characterized if the output signal envelope follows the variations of the input signal envelope.

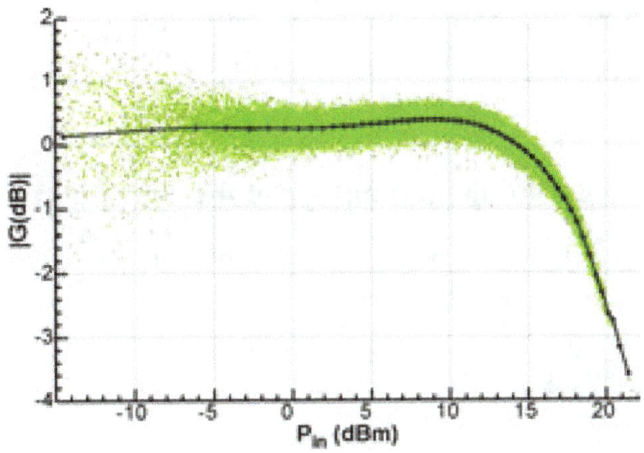

Figure 11. Plot of an *AM/AM* characteristic curve. The center line illustrates an estimate of the *memoryless* amplifier behavior.

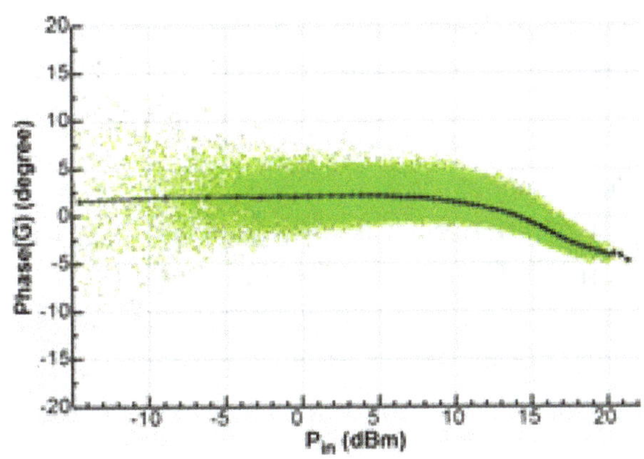

Figure 12. Plot of an *AM/PM* characteristic curve. The values close to the origin at the abscissas axis may include some errors of synchronization as well as measurement noise.

Matching networks are the origin of linear memory, and nonlinear memories are generated in the nonlinear dynamic interactions of the input signal. A practical RF PA output signal has both linear and nonlinear memory, and an *ISDB-T* OFDM signal that occupies a bandwidth of 5.7 (MHz) with up to 8.000 carries will cause more memory effects in a PA than an analog signal, which has only three carriers.

The effects caused by a system with memory in a signal can be easily viewed through *Amplitude/Amplitude* (*AM/AM*) in **Figure 11** and *Amplitude/Phase* (*AM/PM*) in **Figure 12**, where the central line shows a *memoryless* amplifier behavior. The spreading of points for a particular instantaneous power indicates more memory, although close to the origin at the abscissas axis may include some errors of synchronization as well as measurement noise also. These plots were obtained from time-domain measurements performed by a Vectorial Signal Analyzer (*VSA*). Another memory definition can be found in the literature that permits to include a "*time*" interval between "*AM/AM* and *AM/PM* curves".

The IMD effects mixed with memory effects of a PA can generate distortions that degrade seriously the output signal, which can make the task of achieving the required ISDB-T norm mask more difficult. As every amplifier presents distortion, the only way to produce cost-effective and efficient amplifiers is to implement methods to compensate or correct these distortions, as presented in next section.

3.3. Correction of distortions and the technology of the compensation methods

Distortions caused in the amplitude of a signal can be corrected basically with *back-off*, the simpler, and the less-efficient method. If linearization is used, many methods can be applied. The most frequently used ones are *feed forward* and *pre-distortion* with pros and cons listed in our initial references [16, 17, 19, 20, 23].

Choosing the right model is the key for the compensation experimental techniques to control a wide class of distortions generated intrinsically by the RF power amplifier. An efficient architecture for experimental measurements can be built using a basic platform that allows the simulation several models together with *ISDB-T* OFDM signal generation. The board architecture includes essential devices such as *A/D* and *D/A* converters, a FPGA architecture which realizes the *Mux* and *Demux* for *Trelis-Encoder* and *Reed-Solomon* digital communications, memories, and other circuits, as we are presenting in **Figure 13**.

A fundamental step surpassed for the hardware architecture setup implementation was the construction of a filter with excellent sideband attenuation, like surface acoustic Wave (*SAW*) filters that normally present problems regarding pass-band ripple, high insertion loss, and high cost. Using the FPGA and DSP techniques and digital filters, this problem was solved. In addition, it was also possible to compare two distinct theoretical and experimental approaches: *Weaver* and *Frequency Complex* methods combined with DSP techniques, which permit to generate *AM-VSB* analog TV and *8-VSB* digital TV signals through a simplified architecture [6, 7, 8, 23, 25, 26].

This *FPGA* structure was used to implement the nonlinear experimental approach discussed in this chapter and to achieve the best results for the pre-distortion. This architecture was designed, implemented, tested and was in industrial production, part of the first transmission HDTV equipment in Brazil for the *ISDB-T* standard and the electronic structure can serve also for other standards of modulation.

Figure 13. The implemented architecture used to check both theoretical and experimental approaches.

The approach discussed in this Section reveals different distortions that occur in an RF PA input signal, specifically the *ISDB-T* OFDM signal, and their main sources. Further on, common compensation methods were discussed: *back-off*, linearization using *feed forward*, and pre-distortion, including costs involved in each method. For the design of a good quality, efficient, and cost-effective transmitter, a careful study about the distortion generated by the RF amplifier has to be performed. The architecture was designed such that the pre-distortion methods were implemented in practice, and the correct compensation method can be chosen, in order to obtain the highest efficiency allied to high quality [7–9, 14, 17, 23, 25, 26].

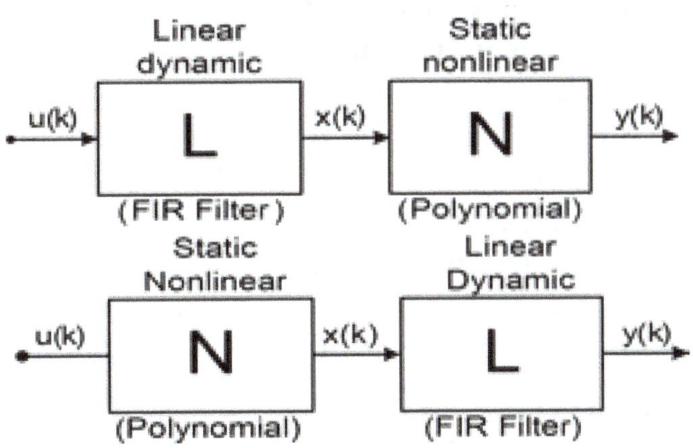

Figure 14. Examples of models including linear and nonlinear blocks. The *Wiener* model contains a filter before the nonlinearity block, and the *Hammerstein* model has the nonlinear block first.

By considering the fact that only those models that include memory in their architectures can be effective against the distortions generated by the modern RF PAs, **Figure 14** presents an example of a block diagram proposed by *Wiener* and *Hammerstein*. Normally, this approach can be adopted in accordance with the compensated distortion that varies as a function of the imposed bandwidth and also with the complexity of the adopted pre-distortion structure.

4. An innovative nonlinear technological modeling of RF power amplifiers for HDTV architectures

The essential purpose of this section will be to present an efficient theoretical approach for modeling with high precision power amplifiers that have quadratic characteristics for both input voltage and the output current. Although our approach will refer for the *FET* device, the same ideas can be adopted for the others structures such as *MOS*, *LDMOS*, and other similar ones. Through the entire section, we will assume that the signal is working in the active region of the transistor, which implies that the input voltage V_{gs} is bigger than the threshold V_t. Some particular cases can be analyzed, if we use the Fourier approach for modeling the drain current of the device [2–5, 15, 21, 28]. In order to compare our experimental results with our technological approach, all computational simulations will be performed using the most important numerical platform for this special kind of nonlinear analysis, which is the software *Advanced Design System* (*ADS*), that permits to show with high-resolution harmonic spectrum for any output signal.

In order to get a more complete idea of this theoretical approach, we must to consider the presence of signals of great amplitudes, such that it is possible to detect the occurrence of the saturation of the device as well as in some cases this effect can occurs abruptly. For that case, the more convenient modeling that can be adopted is to use a partial quadratic characteristic with multiple segments. Therefore, we must to extend our initial analysis for the Fourier series that represents a *pulse train* with distinct sinusoidal peaks for the amplitudes of the currents.

The last theoretical modeling permit us to make the calculations of the coefficients of the Fourier cosine series for all respective train of pulses, which turns out in the new series obtained with the coefficients we have found in the previous analysis. This study is strongly supported if we have enough involvements with new power architectures, just due to the fact that these equipments have some enormous requirements imposed over the project of all hybrid circuits in order to get more stability in the analysis for both *AM/AM* and *AM/PM* digital pre-distortion corrections on real time. The *ADS* platform permits to make to visualize the harmonic components of higher order, such that all numerical results can be compared with those ones that we can get with our hardware implementation.

Our technological approach will be tested with the experimental data that we can exhibit with the project of an electronic architecture that was built at the Laboratory of HDTV Systems of a Company that have a wide class of power amplifiers in this line of industrial production. This device was designed for validation our theoretical approach in order to get a complete set of experimental measurements, and the RF Amplifier was implemented using a *LDMOS MRF 9060* component.

4.1. Quadratic behavior of the *FET* transistor

We start by assuming that we have a quadratic characteristic between both input voltage and output current that can be approached with a good precision for a *FET* device, such that an input tension V_{GS} can control a current source I_d [15, 28–30]. In this case, we will have a quadratic behavior of the type:

$$I_d(t) = \frac{I_{DSS}}{V_T^2}(V_{GS} - V_T)^2, \quad V_{GS} = V_b + V_1.\cos\omega.t \tag{5}$$

where I_{DSS} = Intrinsic parameter of the transistor, V_T = Threshold voltage, V_{GS} = Gate to source voltage, V_b = Bias tension, and V_1 = Peak tension. We can assume always the signal is working in the active region of the transistor, which implies that the input signal V_{GS} is bigger than the threshold one V_T, with $V_x = V_T - V_b$. Therefore, we combine the previous definition, and we can derive in the equation:

$$I_d(t) = \left(\frac{I_{DSS}}{V_T^2}\right).\left(\frac{V_1^2}{2} + \frac{V_1^2.\cos 2.\omega.t}{2} - 2.V_1.V_x.\cos(\omega.t) + V_x^2\right) \tag{6}$$

It is easy to verify that an expansion in Fourier series for $I_d(t)$ has only three terms, and they are:

$$I_d(t) = I_0 + I_1 cos\omega.t + I_2 cos 2.\omega.t \tag{7}$$

If we compare the expression expanded in (6), with the *Fourier* series shown in (7), it is easy to check that:

$$I_0 = \frac{I_{DSS}}{V_T^2}\left(V_x^2 + \frac{V_1^2}{2}\right), \quad I_1 = -2.\frac{I_{DSS}}{V_T^2}V_x.V_1, \quad I_2 = \frac{I_{DSS}}{V_T^2}\frac{V_1^2}{2} \tag{8}$$

To check the efficiency of this approach, we consider an analytical model with the following set of parameters: $I_{DSS} = 16$ (mA), $V_t = -4$ (V), $V_b = -2$ (V), and $V_1 = 2.\cos(4.\pi.10^7 t)$ (V). Thus, if we make the calculations in accordance with the Eq. (6), the current I_d (t) in (mA) can be done by:

$$I_d(t) = [6 + 8\cos(4\pi.10^7 t) + 2\cos(8\pi.10^7 t)] \tag{9}$$

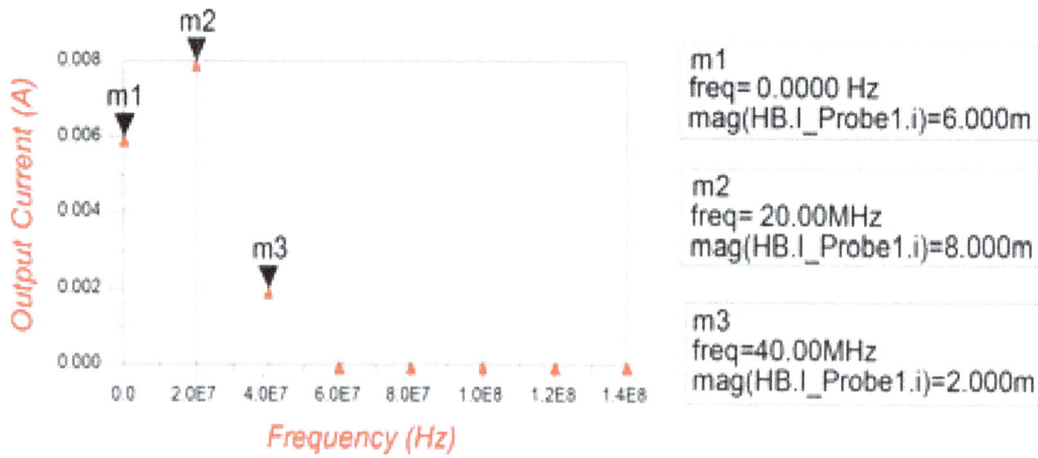

Figure 15. *Amplitude* versus *Frequency* of the output current signal.

Figure 15 shows the harmonic spectrum of the output signal using the software *ADS*. In this case, it is easy to detect that there is a very good agreement between the results obtained with the simulations and those ones obtained with the calculation for the analytical model proposed in this chapter.

In **Figure 16(a)**, we are showing all equipments used in the hardware implementation with the essential purpose of making the validation of the characteristic quadratic of the amplifier presented in this subsection. The graphic shown in **Figure 16(b)** depicts the output current signal, where we are showing into detail the spectral components of first and second order.

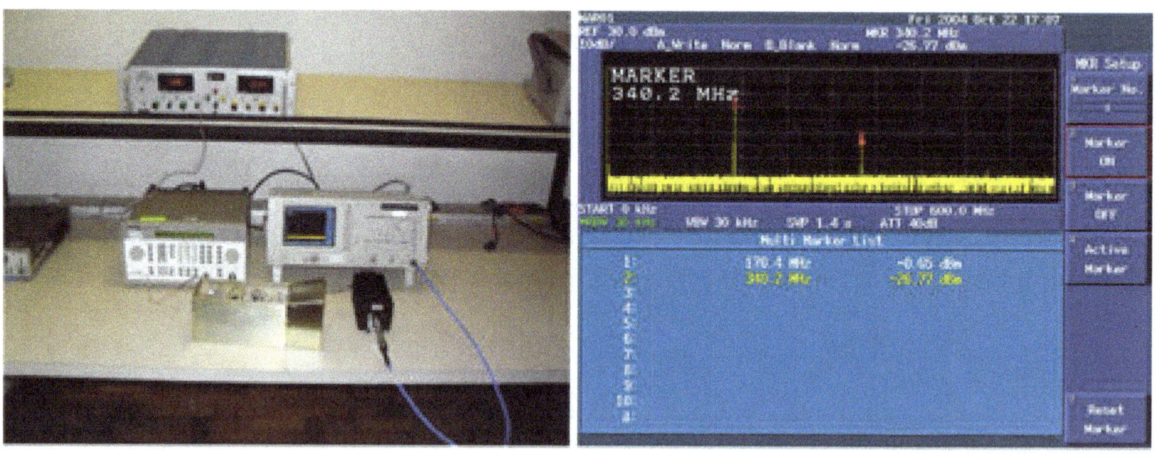

(a) Equipments used for experimental measures. (b) Output current signal of the amplifier.

Figure 16. The architecture designed for experimental measurements—power RF amplifier. (a) Equipments used for experimental measures. (b) Output current signal of the amplifier.

4.2. An innovative analytical approach in the presence of the saturation effect

Typically, in the presence of great signals, it is possible to detect the occurrence of the saturation, and in some cases, this can occurs abruptly. In this case, the more convenient modeling will be to use a partial quadratic characteristic with multiple segments, and we can adopt the same approach of Section 4.1, such that V_y and V_x are the maximum and minimum limits for the fluctuation of the signal. Starting on the Fourier cosine series, this modeling can be found by the direct subtraction of the series that represents a pulse train with sinusoidal peaks of amplitude I_{p1} and I_{p2}.

In addition, if the conduction angles are $2\varphi_1$ and $2\varphi_2$, we can get a new series where the coefficients are corresponding to the subtraction of those ones of the original Fourier series of each pulse train. Now, if we assume the quadratic behavior of the device and by imposing some limits of fluctuation over the signal, we can express the peak current and the drain currents of the device through the analytical expressions in the form:

$$I_{p1} = \frac{I_{DSS}}{V_T^2}(V_1 - V_x)^2, \quad I_{d1} = \frac{I_{DSS}}{V_T^2}(V_1 \cos \omega.t - V_x)^2, \quad \varphi_1 = \cos^{-1}\frac{V_x}{V_1} \tag{10}$$

$$I_{p2} = \frac{I_{DSS}}{V_T^2}(V_1 - V_y)^2, \quad I_{d2} = \frac{I_{DSS}}{V_T^2}(V_1 \cos \omega.t - V_y)^2, \quad \varphi_2 = \cos^{-1}\frac{V_y}{V_1} \tag{11}$$

In this case, the Fourier cosine series as a function of θ has the following representation:

$$f(t) = a_0 + \sum_{n\,1}^{\infty} a_n \cos n.\omega.t, \quad f(t) = I_d(t), \quad a_0 = \frac{1}{\pi}\int_0^{\pi} f\left(\frac{\theta}{\omega}\right).d\theta, \quad a_n = \frac{2}{\pi}\int_0^{\pi} f\left(\frac{\theta}{\omega}\right).\cos(n.\theta).d\theta \tag{12}$$

If we develop the Fourier series for both I_{d1} and I_{d2} currents, where the index j is associated with the current I_{dJ} that must to be determined. Therefore, for $j \in \{0, 1\}$, the iterative process will have the following steps:

$$I_{(J)0} = \frac{I_{pJ}}{\pi}\int_0^{\varphi_J}\left(\frac{\cos\theta - \cos\varphi_J}{1 - \cos\varphi_J}\right)^2 .d\theta \Rightarrow I_{(J)0} = \frac{I_{pJ}}{\pi}.\frac{\varphi_J - \frac{3}{4}\sin 2\varphi_J + \frac{\varphi_J}{2}\cos 2\varphi_J}{(1-\cos\varphi_J)^2} \tag{13}$$

$$I_{(J)1} = \frac{2.I_{pJ}}{\pi}\int_0^{\varphi_J}\left(\frac{\cos\theta - \cos\varphi_J}{1 - \cos\varphi_J}\right)^2 .\cos\theta.d\theta \Rightarrow I_{(J)1} = \frac{2.I_{pJ}}{\pi}.\frac{\frac{3}{4}\sin\varphi_J - \varphi_J\cos\varphi_J + \frac{1}{12}\sin 3\varphi_J}{(1-\cos\varphi_J)^2} \tag{14}$$

$$I_{(J)n} = \frac{2I_{pJ}}{\pi}.\frac{(4-n^2).\sin(n\varphi_j) + (3n).\sin[(n-2)\varphi_j] + (n-1).(n-2).[\sin(n\varphi_j).\cos 2\varphi_j]}{(n).(n^2-1).(n^2-4).(1-\cos\varphi_j)^2}, n \geq 2 \tag{15}$$

Finally, we can obtain the iterative current expressed analytically through the equation:

$$I_{d_j}(t) = I_{(j)0} + \sum_{n=1}^{\infty}[(I_{(j)n}).\cos(n.\omega.t)] \quad (16)$$

In addition, we can make the calculations for the coefficients of the Fourier series for both pulse trains that permit to obtain a new series with the subtraction of the coefficients found in the Eq. (16), and represented by:

$$I_d(t) = [(I_{(1)0}) - (I_{(2)0})] + \sum_{n=1}^{\infty}[(I_{(1)n}) - (I_{(2)n})].\cos(n.\omega.t) \quad (17)$$

In order to get more legitimacy of the analytical approach here proposed, we make a comparison similar to that one realized in the previous model. In this second model, we adopt the following characteristics for the device: $I_{DSS} = 16$ (mA), $V_x = -4$ (V), $V_y = 4$ (V), $V_b = 0$ (V), and $V_1 = 5.\cos(4.\pi.10^{7}t)$ (V). The output current for this case has the spectrum depicted in the **Figure 17(a)**, and all simulations done with the *ADS* software permit us to visualize the harmonic components of higher order. By making a comparison in **Figure 17(b)**, we present a Table where we can detect a very good agreement between the results obtained with the simulation and those ones established with the theoretical modeling that indicates this approach is very efficient and can serves as a tool in the designs of the RF amplifiers. Our professional experience confirms for us that the technological modeling proposed in this subsection can serve for a wide class of projects that involve many modern transmission equipments, with this continuous and rigorous demand imposed by the international market.

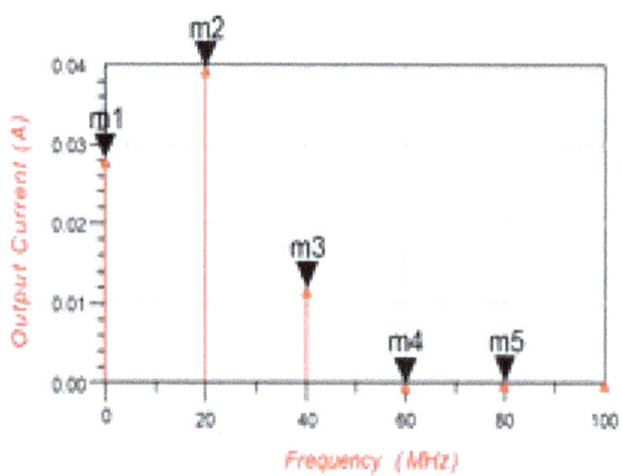

(a) *Amplitude* versus *Frequency* of the output current.

Harmonics	Analytical model	Simulation
DC	28.18 (mA)	28.32 (mA)
1st (Harmonic)	40 (mA)	39.94 (mA)
2nd (Harmonic)	12.1 (mA)	12.14 (mA)
3rd (Harmonic)	10 (μA)	64.71 (μA)
4th (Harmonic)	264 (μA)	361.3 (μA)

(b) Analytical Model and the Results of Simulation.

Figure 17. Analytical and numerical analysis of the saturation effect for the power RF amplifier. (a) *Amplitude* versus *Frequency* of the output current. (b) Analytical model and the results of simulation.

In conclusion, **Figure 18** presents some photos of the device implemented in the laboratory of the Company [7–14, 25–26, 34–35, 40]. In the item (a), we shows the zoom of the *RF* Amplifier built using the *LDMOS* MRF 9060 transistor, while that in the item (b), we present a complete view of the electronic architecture that was used for all experimental measurements that were presented into detail in **Figure 16(b)**.

(a) The *LDMOS 9060* Electronic Structure. (b) Power RF Amplifier (*10 W*) at the 3rd band.

Figure 18. Photographs of the hardware architecture for the power RF amplifier. (a) The *LDMOS 9060* electronic structure. (b) Power RF amplifier (*10 W*) at the 3rd band.

4.3. An appropriate theoretical approach for the amplifier with a source resistance

This approach starts by assuming that we will concentrate our analysis over the source resistance effect for the circuit depicted in the **Figure 19**. In this case, we must to get all analytical equations that describe the functioning of the model, and in consequence, it is possible to check that the V_{in} tension cannot be applied directly in the gate-source voltage V_{GS}. Therefore, only part of the voltage appears in the gate-source input of the FET and the complementary tension must be applied in the resistance *R*. Therefore, in this case, the equ. (18) presents the analytical dependence between these parameters, described analytically in the form:

$$V_{in} = V_r + V_{GS} = (I_d).R + V_{GS} \tag{18}$$

In accordance with the previous studies done by Cantrell and Davis, we can define the input incremental resistance expressed mathematically by:

$$r_{in(n)} = (\partial V_{in} / \partial I_d)_{I_d = I_{DC}} \Rightarrow r_{in(n)} = \{\frac{\partial}{\partial I_d}[(I_d).R]\} + [\frac{\partial V_{GS}}{\partial I_d}] \tag{19}$$

where I_d and I_{DC} are, respectively, the drain and the DC currents defined by the circuit polarization. Starting on the equation that defines the classic modeling of the FET structure, and after some algebraic calculations, it is possible to express the voltage V_{GS} in the form:

$$V_{GS} = \left(I_d / k \right)^{1/2} + V_T, k = \left(\frac{1}{2} \right).(k.n).\left(W / L \right) \tag{20}$$

Under these hypotheses, if we combine conveniently the last equations, both the incremental resistance and the new transconductance can be expressed by the equations:

$$r_{in(n)} = \left(\partial V_{GS} \Big/ \partial I_d \right)_{I_d = I_{DC}} = \left(\frac{1}{2.k} \right) \cdot \left(I_{DC} \Big/ k \right)^{-(1/2)}, g_{m(n)} = \left(\frac{1}{r_{in(n)}} \right) \left[\frac{g_{m(n)}}{1 + (g_{m(n)}).R} \right] \tag{21}$$

For a fixed value of the parameter g_m, Eq. (21) shows that the new transconductance decreases as a function of the resistance R. In particular, if the factor $(g_m).R >> 1$, in the limit condition, this expression will depends only of R. Therefore, the transconductance is independent of any change in the polarization current, that is: $g_{m(n)} = 1/R$.

4.4. Additional performance with respect to the effect of the feed-forward linearity

With the final purpose of to get a more knowledge about the effect of the feed-forward linearity, when the electronic structure has the presence of the passive component R, it acquires a great importance define a new curve of transconduction V_{in} versus I_d [3–5, 10–15, 18, 28–30]. This will demand a new modeling in the initial Eq. (18), using the voltage V_{GS} as in Eq. (20), which derives in the expression:

$$V_{in} - V_T + V_{GS} = (I_d).R + \left(I_d \Big/ k \right)^{1/2} \tag{22}$$

where V_T is the threshold voltage. This will demand a new definition of the voltage V_c, that indicates the dependence of V_{in} with respect to V_T, when the drain current vanishes, that is $I_d = 0$. Therefore, we can get:

$$V_c = (I_{DC}).(R + r_{in}) \Rightarrow V_c = \left(\frac{I_{DC}}{g_m} \right).[1 + (g_m).R] \tag{23}$$

In order to simplify the 2D plot analysis of the new transconductance, we normalize the Eq. (5) by the voltage V_c as a function of the parameter $(g_m).R$, which imply that:

$$\left(\frac{V_{in} - V_T}{V_c} \right) = \left(\frac{I_d}{I_{DC}} \right) \cdot \left(\frac{(g_m).R}{1 + (g_m).R} \right) + \left(I_d \Big/ I_{DC} \right)^{1/2} \cdot \left(\frac{2}{1 + (g_m).R} \right) \tag{24}$$

4.5. Analysis in the limit condition at the presence of the feed-forward effect

At first, in **Figure 20**, we are portraying a graphic that we obtain for three different values of the parameter $(g_m).R$. Therefore, it is possible to check the change in the quadratic curve with respect to the correspondent linear behavior of the amplifier, as well as the decreases of its gain. Some particular cases related with the real transition of the parameter $(g_m).R$ can be performed, and each one of them has a very simple mathematics derivation, if we look the previous equation at the limit condition.

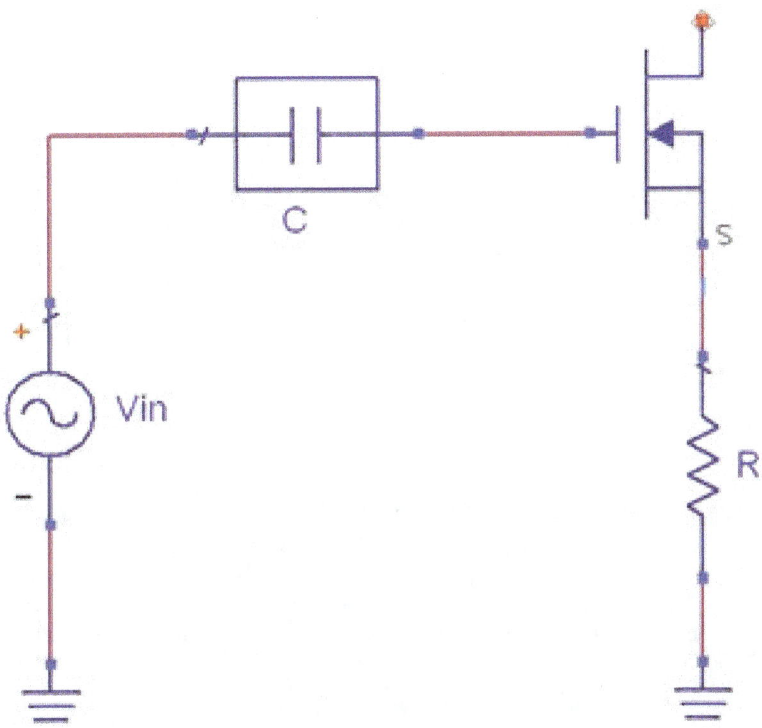

Figure 19. The FET device with the presence of the source resistance.

- At first, it is easy to check that when $(g_m).R \ll 1$, the behavior of the structure has a quadratic performance.
- For the case that $(g_m).R \gg 1$, we will have a linear dependence that involves all variables.
- It is important to emphasize that the limit condition, that is: $(g_m).R = 0$, characterizes the quadratic behavior too, which corresponds exactly the same one presented by the FET device.
- By the other hand, the 2nd limit condition, that is: $(g_m).R \rightarrow \infty$, can be reached with the linear behavior which will be analyzed in the next subsection, more specifically in the set of Eqs. (28) and (29), and is associated to the *threshold* condition of the amplifier.
- Generally, we must to concentrate our approach by making an appropriate choice on the cases discussed in the previous items. The almost linear dependence that corresponds to the limit condition $(g_m).R = 1$, is shown in **Figure 20**.
- Finally, it is important to put in evidence that the case $(g_m).R = 0$, can be solved with those analytical expressions that permits us to detect the quadratic behavior of the FET transistor.

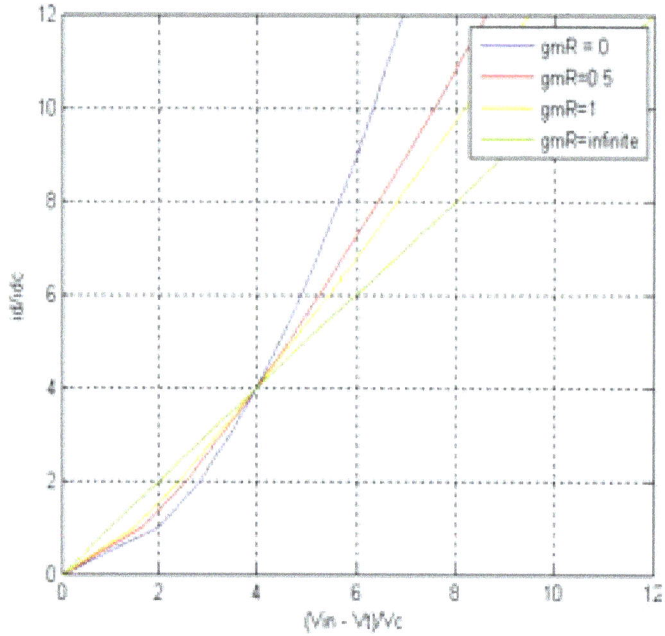

Figure 20. The 2d plot of the transconductance as a function of $(g_m).R$.

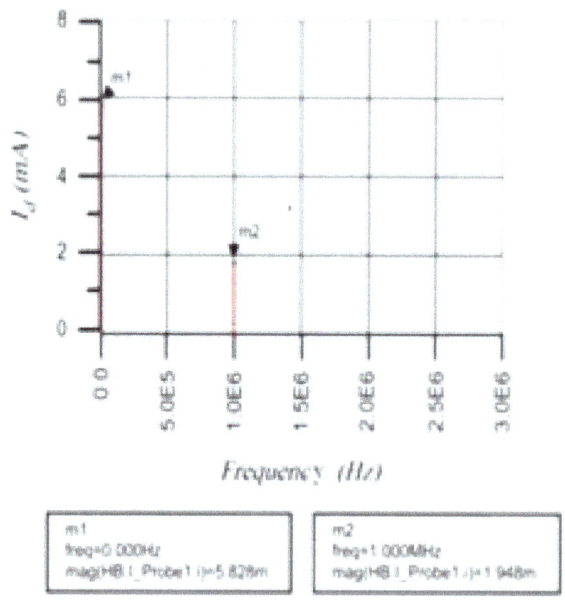

Figure 21. The I_d harmonic components plotted with the *ADS* platform, for $V_{in} < V_c$.

4.6. An important analysis corresponding to the presence of the threshold effect

The threshold condition, $(g_m.R) \to \infty$, imposes for us focus our analysis over the equation: $I_d = (V_{in} - V_T)/R$, for $V_{in} > 0$, and we will assume that the electronic structure has a voltage source $v_{in} = V_{in}.\cos(\omega.t)$ In this case, two different and important conditions must be analyzed that are as follows:

I. $V_{in} < V_c$—It is easy to check that, if we combine this inequality with the addition condition: $V_c = (I_{DC}).R$, we can assume the linear performance of the FET, which make possible to derive the analytical expression: $I_d = I_{DC} + [V_{in}/R].\cos(\omega.t)$.

In order to verify the higher dependence of the parameter R, we imposed in our simulations the following set of parameters for the FET structure: $g_m = 86$ (mS), $I_{DC} = 5{,}8$ (mA), $R = 500$ (Ω), and $v_{in} = 1.\cos.[(2\pi.10^6).t]$ (V). For this fixed value of R, we can obtain the transconductance g_m of the amplifier as well as its corresponding polarization, and with this particular choice of the parameters, the more convenient real dependence can be formalized with the following condition: $(g_m).R = 43$.

In addition, the current output I_d characterizes the linear behavior of the amplifier, that is: $(g_m).R >> 1$, and the drain current has the form: $I_d = (5.8) + 4.\cos(\omega.t)$, with $\omega = 2.(\pi).10.\exp(6)$. **Figure 21** depicts the harmonic composition of the output current I_d, simulated with the *"Premium Version"* of platform of simulations: *Advanced Design Systems—ADS (Agilent Tech. Co.)*. It is possible to confirm that the level of the fundamental component is reduced with the same rate that the transconductance increases. The same comparison can been performed in the case of the quadratic behavior of the FET, if we observe that the 2nd harmonic component does not exist practically, that characterizes the *feed-forward effect*, and this is one of the main reasons we can detect a small reduction in the fundamental component.

II. $V_{in} > V_c$—In this case, the current I_d takes the form of a periodic train of pulses with sine peaks, which can be visualized on **Figure 22**. Then, we must to modeling the device using the classical Fourier Cosine Series.

With respect to the drain current, we can write this signal in accordance with its series representation in the form:

$$I_d(t) = I_0 + I_1 \cos \omega.t + I_2 \cos 2.\omega.t + ... \quad (25)$$

which has an alternative analytical written as:

$$I_d(t) = I_0 + [1 + (I_1 / I_0)\cos \omega.t + (I_2 / I_0)\cos 2.\omega.t + ... \quad (26)$$

If we use this Fourier Cosine Series approximation and we make an accurate analysis over the graphic presented in the **Figure 22**, it is possible to derive the functional dependence like: $I_p = G[V_{in} - [V_{in} - V_T]$. Now, by assuming that $(g_m).R >>1$, if we follow the same line of ideas we have made previously, we can deduct that $G = g_{m(n)} = (1/R)$, which implies that we will obtain $I_p = (1/R).V_T$.

In the sequence, after some extensive calculations, we can derive the harmonics components for the output signal, using the Fourier series approach, and the set of equations that is possible to derive can be synthesized in the form:

$$\left(\frac{I_n}{I_0}\right)=\left(\frac{2.[\cos\varphi.sen(n.\varphi)-n.sen\varphi.\cos(n.\varphi)]}{n(n^2-1)(sen\varphi-\varphi.\cos\varphi)}\right),\left(\frac{I_1}{I_0}\right)=\left(\frac{\varphi-sen\varphi.\cos\varphi}{sen\varphi-\varphi.\cos\varphi}\right) \tag{27}$$

$$\left(\frac{V_{in}}{V_c}\right)=\left(\frac{\pi}{sen\phi-\varphi.\cos\varphi}\right),\ \varphi=\arccos\left(\frac{V_{in}-V_T}{V_{in}}\right) \tag{28}$$

For some application, we will adopt the same set of parameters used at the end of the last subsection. For this case, we use the input voltage: $V_{in} = 4\ (V)$, since that the condition $V_{in} > V_c$ must be verified. In addition, through the Eqs. (27) and (28), it is possible to determine all harmonic components; and for this next example, we intend to preserve the relation: $V_{in}/V_c =$ 1,372.

It is important to emphasize that, we will obtain a transcendental equation, which demands intrinsically a numerical approach to get the solution in the form: $I_d = [5{,}8 + 7{,}13.\cos.(2\pi.10^6.t) + 0{,}696.\cos.(4\pi.10^6.t) + 0{,}458.\cos.(6\pi.10^6.t) + 0{,}234.\cos.(8\pi.10^6.t)]$. For all numerical simulations, we can explore the resources of the ADS *software* to verify the veracity of the proposed analytical model presented in this section, and the results for the harmonic component of the drain current I_d are depicted in **Figure 23** [32–35].

A comparison between the theoretical and the simulated results is presented in **Table 3** and it is possible to see a small difference between the results determined for the current I_0. To justify these differences, we observe that we have a limitation with respect to the harmonic components involved in the simulation, which not only can provoke some alterations but also has some consequences due to the fact that these values are exceeding the voltage $V_{DC} = 15\ V$.

Harmonics	**Analytical model**	**Simulation**
DC	5.8 (mA)	6.13 (mA)
1st (harmonic)	7.13 (mA)	7.22 (mA)
2nd (harmonic)	696.0 (μA)	467.2 (μA)
3rd (harmonic)	458.0 (μA)	363.2 (μA)
4th (Harmonic)	234.0 (μA)	196.8 (μA)

Table 3. Comparison between theoretical and numerical results: harmonics components.

As a direct consequence, the output signal reaches the cutoff region. Furthermore, if the output signal operates outside the cutoff region, intrinsically it can be cut, due to the limits previously established for the excursion of the signal. **Figure 24** presents the output waveform that can reach at maximum the V_{DC} level.

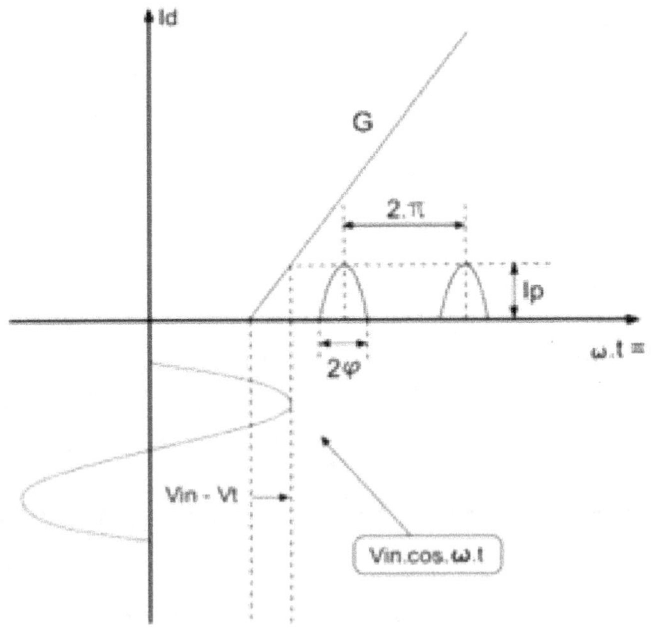

Figure 22. Graphic of the transconductance at the condition: $V_{in} > V_c$.

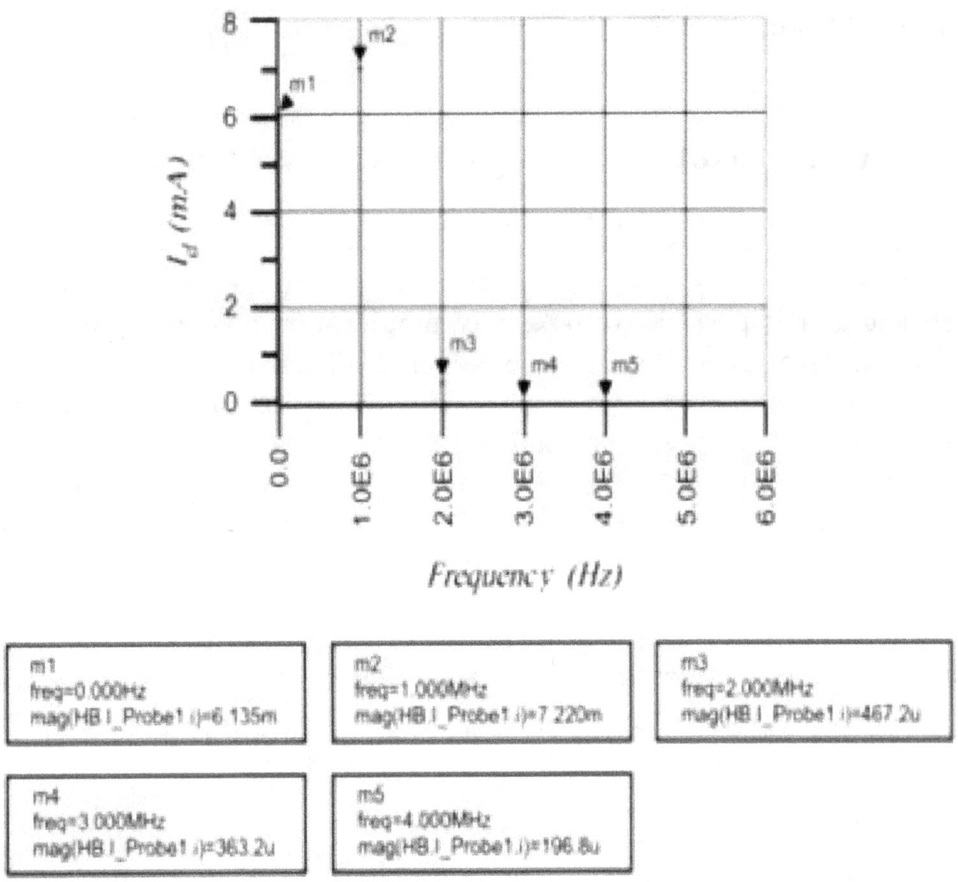

Figure 23. The I_d harmonic components plotted with the *ADS* platform, for $V_{in} > V_c$.

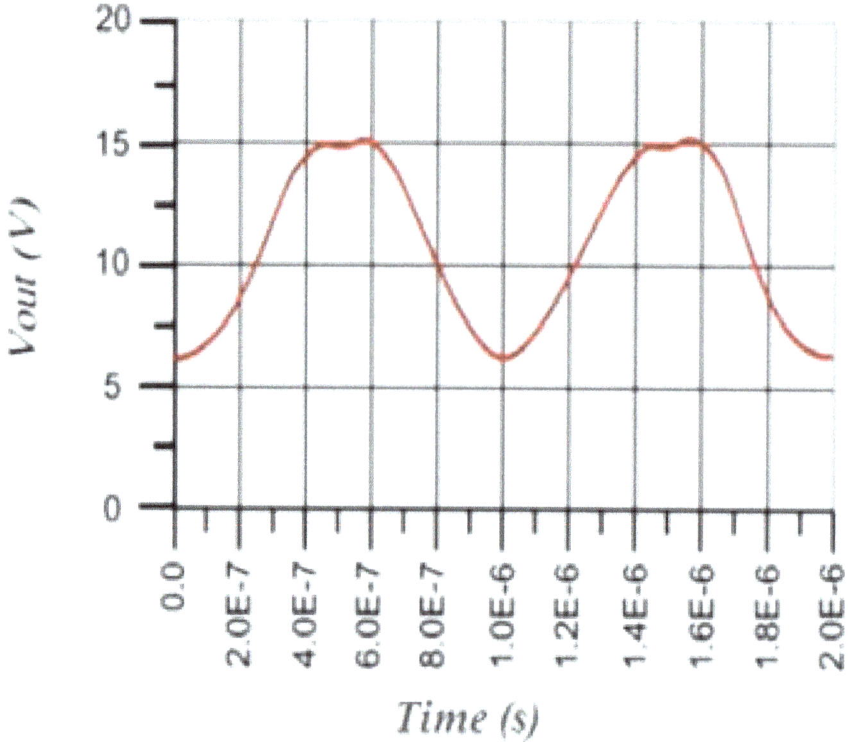

Figure 24. Temporal response the output voltage, for $V_{in} > V_c$.

5. An efficient architecture involving broadcast of RF power amplifiers

The main focus of this Section is to present an efficient in-house NTSC architecture that can works as an exciter with a central PC that controls various interfaces. With the setup of this electronic structure, it is possible to make a communication with different measurement equipment, and the levels generated by the exciter are sufficient to drive the power drawer to the nominal power 860 (Wps). All numerical codes of the central software that interfaced and controlled all equipment were written using both *Matlab* platform. For this design, the most important interfaces are the Serial *RS232*, *USB*, *GPIB*, and *LAN*, and the final software presents an easy-to-use interface that are strongly demanded currently in the production line of all modern RF architectures [7–10, 12–17, 23].

5.1. A wide class of important reports: principal measurements over the power amplifier

Figure 25 shows the amplifier we intend to make some experimental tests. This is a picture from a part of the power drawer that shows four RF power transistors, capable of delivering altogether 860 (Wps).

Typically, the operation of this device preserves the balanced configuration and has some similarity with other literatures as described previously. For the case that the setup has a correct operation, a wide class of different reports can be generated [20–22, 24, 36–42].

Hybrids structures were used to combine push–pull amplifiers, such that we should expect a difference in the current of the transistors, especially in the middle of the UHF band, when the coupled and the direct port present the largest dispersion from the ideal 3 *dB* value. The designer can use all reports to get enough background about the performance analysis of the amplifier under test, as well as to acquire a fine tuning of an amplifier that is being optimized using the same setup. At first, we shown in **Figure 26**, the current of each transistor recorded for each channel, in the entire UHF band.

Figure 25. The RF architecture implemented for experimental tests with four power amplifiers with 860 (Wps).

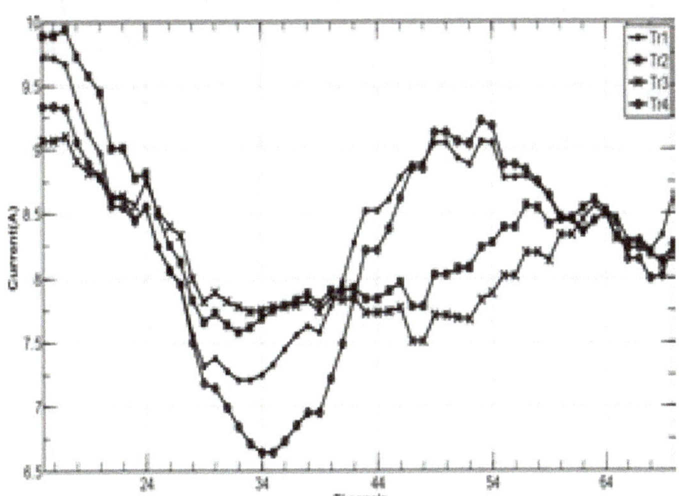

Figure 26. Currents of each transistor measured at nominal power for the UHF band (channels 14–69).

A 2D diagram in the entire UHF band for the temperature of each transistor is presented in **Figure 27**. Evidently, the transistors that were operating with less current due to the imperfect signal division of the hybrids presented lower temperatures, but the maximal ratings are important measures to verify the air cooling system of the drawer.

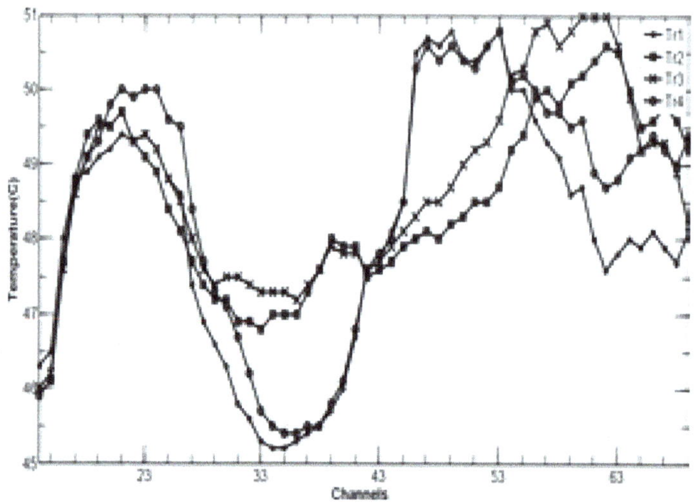

Figure 27. Temperatures of each transistor measured at nominal power for the UHF band (channels 14–69).

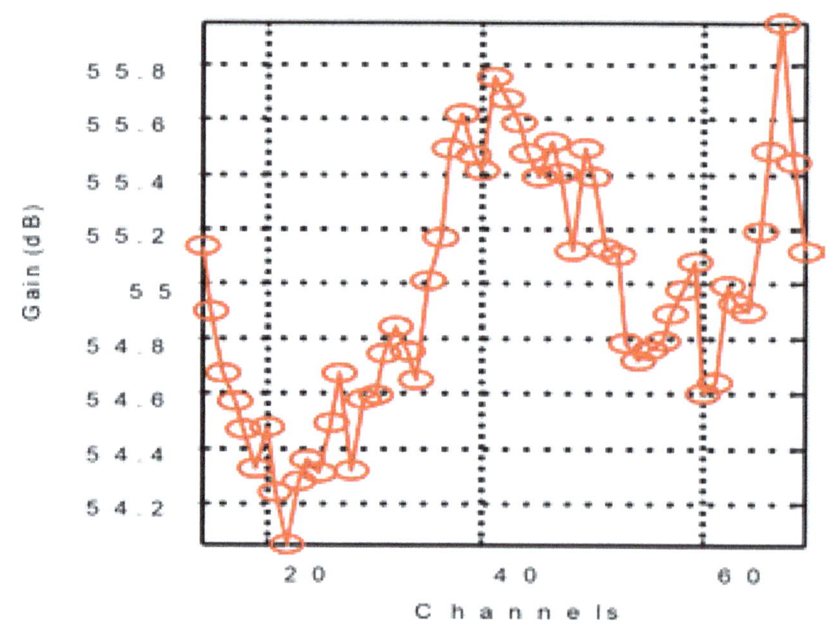

Figure 28. Plot of the gain (*dB*) for the UHF band (channels 14–69).

With this architecture, it is possible to perform a complete set of reports, as we can visualize in the graphic of **Figure 28** that shows the values of the Gain (*dB*) versus the UHF television channels (14–69), while **Figure 29** depicts the inter-modulation of 0% average power level (APL) black signal located at (−4.5) (MHz).

In order to consider the stress conditions generated by the amplifier by considering the signal that represents the highest output power generated by the device in an NTSC transmission, we must check the use of the APL 0% black signal. This experimental allows observe the *IMD* generated by the amplifier at the maximum output power in the entire band imposed for the operation of the device. Therefore, it is possible to plot both the lower side of the *IMD* (−4.5 MHz) and the upper side of the *IMD* (9 MHz), and we can detect a high asymmetry between the lower and the higher sides of the *IMD* frequencies.

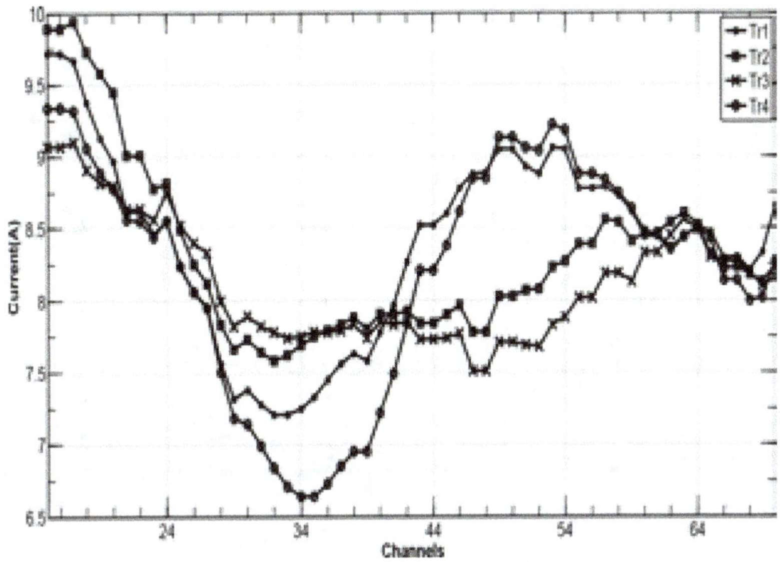

Figure 29. Plot of the inter-modulation (0%) APL black signal located at −4.5 (MHz), for the UHF band (channels 14–69).

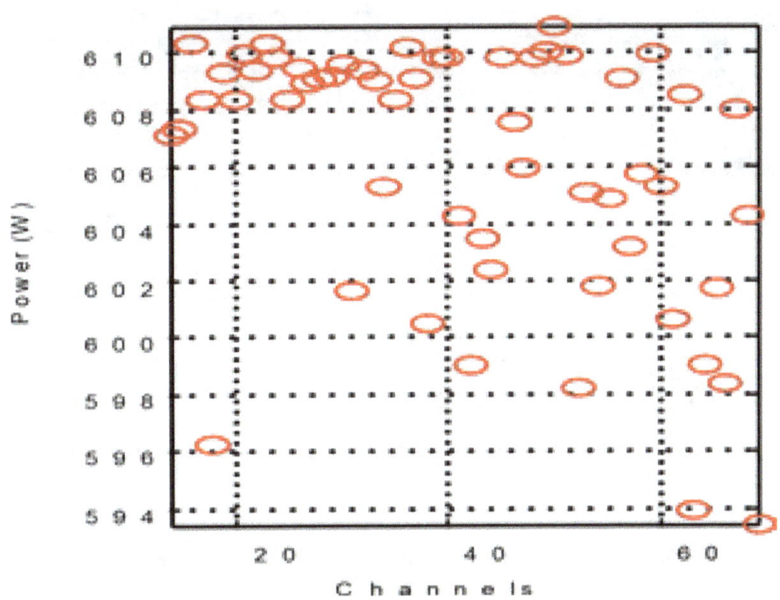

Figure 30. Plot of the (75%) red *Ramp* signal for the UHF band (channels 14–69).

If we taking into account that the human eye is sensitive to static noise intermixed in a red field, such that the *IMD* levels should be as low as possible in a good quality amplifier, it is important to get a 2D plot corresponding to the (75%) red *Ramp* test signal results for the same UHF television channels, which is represented in **Figure 30**.

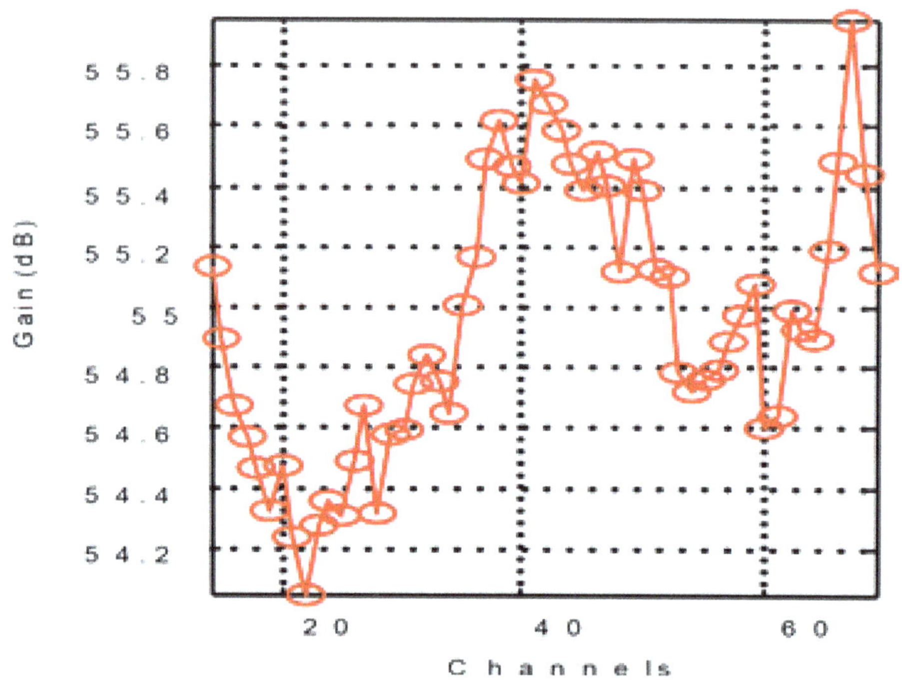

Figure 31. Plot of both local oscillator and the mirror signals for the UHF band (channels 7 14–69).

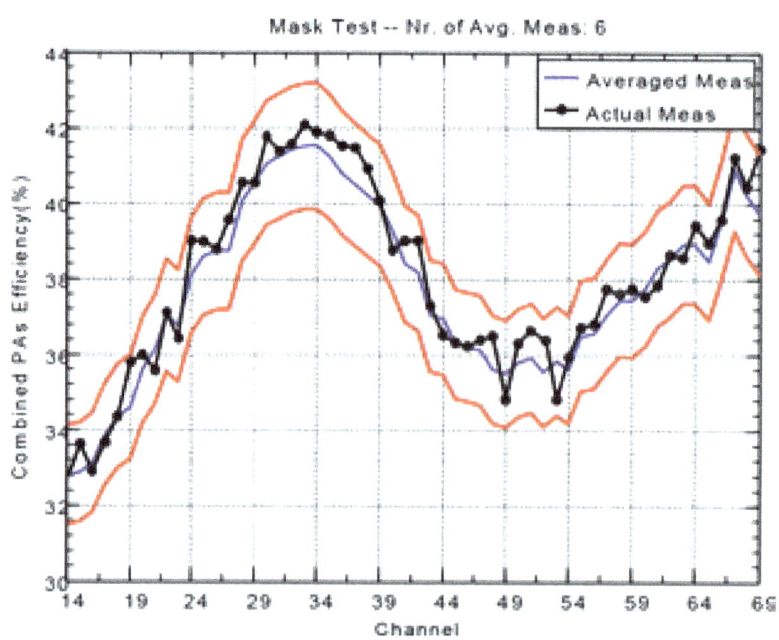

Figure 32. A statistical comparison of the combined PA's efficiency for validation of all measurement tests.

It is possible also to perform a graphic that corresponds the efficiency of the power drawer for the entire UHF band, since this test would indicate the RF power delivered over the total power consumed by the amplifier. Another important report is corresponding to the level of both local oscillator (LO) and the mirror signals as a function of the same channels of UHF television, which is plotted in **Figure 31**. This test ensures that the LO of the modulator and the mirror signals are low enough, which permit us the complete validation of this experimental test.

As a conclusion of all experimental measurements, it is really important to use all data available, which permit us not only to make some statistical comparisons as well as pass-fail tests were it is possible. This procedure presents a mean value of several results that were performed, such that it is possible to establish a comparison with the actual measurements and those ones we have made previously. If more tests are realized more precise, they become and the mean curve can be adjusted more conveniently, as it is being presented in **Figure 32** [22–26, 42–47].

Using this mean curve and upper and lower bounds, a defective piece can be easily identified and sent to repair if necessary. A good agreement can be observed for the amplifier tested in this graphic, and so the behavior of the amplifier was inside expectations.

5.2. Block diagram and the setup of the entire experimental hardware architecture

A block diagram corresponding to the flows of the signal for the implemented architecture is depicted in **Figure 33**. For both input and output signals, it will be necessary to check the system gain and to ensure that the output power is at the correct level, a set of instruments such as: Power Meters, USB2001, E4416B, Current Probe 34134A, and the 34401A Multimeter, in order to get the whole system that will be used to measure the drawer overall efficiency. The transistor currents were measured from the power drawer supply, and the temperatures were acquired from a sensor placed at the side of the transistor flange.

With the purpose of to maintain the exciter totally *"free of distortion"*; sometimes, we need to check and realize, the performance of both the local oscillator (LO) as well as the image suppression for each analyzed channel. An ESA spectrum analyzer can serve for this experimental procedure which permits also to detect inter-modulation levels of three different NTSC signals, namely 0% average power level (APL) black, 75% modulated red, and the modulated ramp.

In comparison with others commercial systems (e.g., Anritsu ME 7840A), this architecture is more flexible regarding the output power as well as all individual components. Besides of the RF structure being composed by several parts, these ones can be tuned for a single amplifier module of 220 (Wps) or a power drawer 860 (Wps), which permits that some parts can also be substituted by low-cost equivalent, as for example, power meters, current meters, attenuators, and couplers.

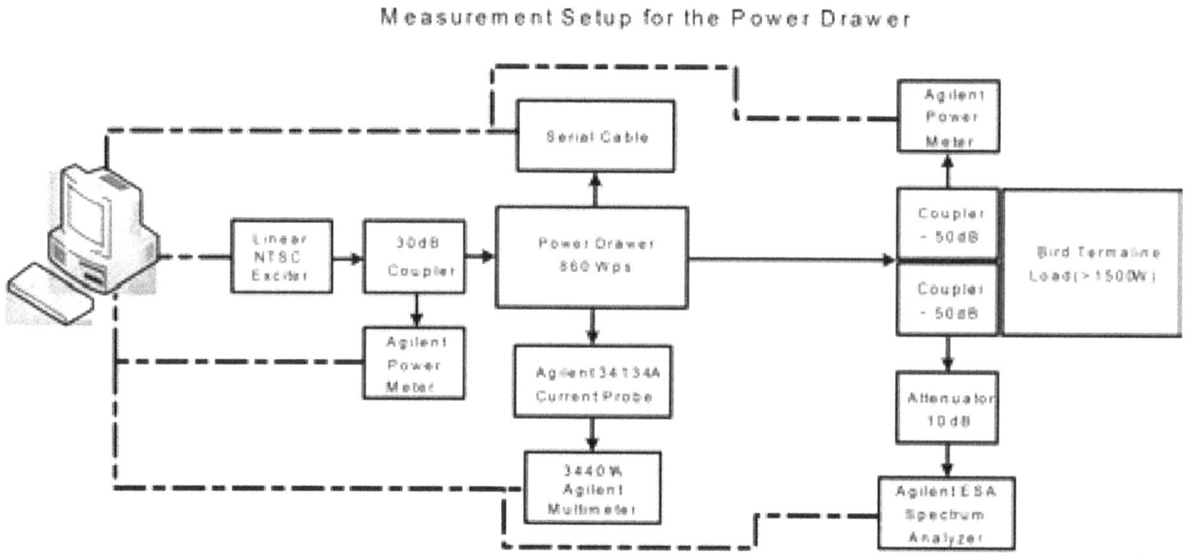

Figure 33. Design of the whole architecture for the measurement setup.

For the case that the architecture shown in **Figure 33** will be used for standard *ISDB-T* OFDM signals, some minor changes in software to measure inter-modulation distortions must be performed when the signal is applied to a power amplifier. In the references presented at the Introduction, a detailed study about *IMD* distortions caused by a power amplifier in an *ISDB-T* OFMD signal and also its peak-to-average ratio (*PAR*) characteristics is presented.

The whole architecture will serve for submit under test a wide class of power amplifiers. All reports generated in accordance with the performance analysis of each amplifier makes possible to implement a fine tuning of the device. Furthermore, it is possible to combine *push–pull* amplifiers, for the case that hybrids structures must be used, and we should expect a difference in the current of the transistors, especially in the middle of the UHF band, when the coupled and the direct port present the largest dispersion from the ideal 3 *dB* value.

Acquires a great importance the reports corresponding to the fluctuation of the temperatures of each transistor in the entire UHF band. Initially, we impose set all values of the Power (W)

modulation of 0% APL black signal, for some specified frequencies. Typically, the use of the APL 0% black signal is due to the stress conditions generated in the amplifier, by considering that this signal represents the highest output power generated by the device for an NTSC transmission. This test allows the observation of the IMD effect generated at the maximum output power in the entire band imposed for the operation of the amplifier [48–52].

Finally, corresponding to the generation of inter-modulation reports that can be used in the analysis of PAR of digital signals, we should expect a sensible broadening in the domain of applications of the measurement architecture. An alternative option for this architecture is to use it for the validation of noise analysis as well as to validate some studies for uplinks for cognate radio structures made for a wide class of OFDM systems. In conclusion, a load capable of absorbing the amount of power generated by the amplifier under test is used as termination, preceded by two couplers. The complete architecture implemented on a rack for all measurements can be visualized in **Figure 34**.

Figure 34. The whole architecture which contains all equipment used to perform all requirements measurements conveniently mounted on a rack.

The whole architecture presented in this Section can be used widely for the characterization of high-power amplifiers, and the structure was designed to be used in the production line of modern RF companies. Important parameters, such as gain, output power, inter-modulation distortion of different signals, efficiency, current, and temperature, can be measured in accordance with the intrinsic characteristics of a wide class of amplifiers. The RF electronic structure can generate very important reports that can be used to avail the performance of the amplifier under test, comparing it to other amplifiers already measured.

6. Conclusions

The theoretical research with experimental validation presented in this chapter was aimed primarily to introduce new technologies that can be used in modeling of nonlinear phenomena that attend intrinsically in the projects involving the power amplifiers used widely in the new industrial production lines of high-resolution digital systems of TV.

The competitiveness that is continuously imposed on the international scene demand that professionals involved with projects of new technologies keep always a updated vision of what can be more in terms of new emergency equipment should be projects for communication links using techniques in which the *Digital Signal Processing* (DSP), with high rates of transmission by making that information, regardless of its analog or digital form can get the link *Tx-Rx* efficiently, with optimized time, protected against intrinsic noise in transmission, with a modern computing platform that allows not only to implement the codes are written in structures *Field-Programmable Gate Array* (FPGA), as well as all the simulations that can validate the theoretical approaches formulated, in addition to a safe protection of traffic information.

This demand of designers a vigilance on construction of versatile systems, extreme precision, so that the demands imposed on business sectors may be always being supplied in real time, which keeps the heated market, strengthening thus the companies that are on the line, in the discovery of new fronts in their diverse research lines of production. Constant challenges are always objectified targets to the engineers who are involved with projects of new technologies. The space in the international market is something that outlines as one of the fundamentals of the companies who engage seriously in new lines of production equipment used in the transmission and reception of signals in the wide area of Radio Frequency (RF).

We hope this set of theoretical and experimental approaches presented in this Chapter can be our modest contribution for the development of new technologies as well as can serve as a guide for many engineers with strong interests in this field of research.

Acknowledgements

We would like to thank the Institutions: Foundation of Studies and Projects, FINEP, RJ, for all financial support received for this research; Linear Equipamentos Eletrônicos S/A, one of the best Companies in Brazil that produces RF and Microwave equipments with high performance, where were developed all projects related with a wide class of HDTV transmitters architectures; São Paulo Research Foundation, FAPESP, SP, processes with ID: 2006/01655-7 and 2013/23431-7; Minas Gerais Research Foundation, FAPEMIG, MG, by all financial supports received by the authors; Brazilian Educational Ministry, CAPES, MCT, for the PhD scholarship studies developed by an author in Austria.

Author details

Daniel Discini Silveira[1], Marcos Paulo de Souza Silva[2], Marcel Veloso Campos[3] and Maurício Silveira[4*]

*Address all correspondence to: maur.silveira@gmail.com

1 Federal University, Ufjf, Juiz de Fora, Minas Gerais, Brazil

2 São Paulo's Electricity Company, CESP, São Paulo, Brazil

3 State University, Unimontes, Montes Claros, Minas Gerais, Brazil

4 Federal University, UFSCar, São Carlos, São Paulo, Brazil

References

[1] Vuolevi, J. and Rahkonen, T., Distortion in RF Power Amplifiers, Norwood, MA: Artech House, 2003.

[2] Maas, S.A., Nonlinear Microwave Circuits, Piscataway, NY: IEEE Press, 1988.

[3] Cripps, S.C., Advanced Techniques in RF Power Amplifier Design, Norwood, MA: Artech House, 2002.

[4] Kenington, P.B., High-Linearity RF Amplifier Design, Norwood, MA: Artech House, 2000.

[5] Pothecary, N., Feedforward Linear Power Amplifiers, Norwood, MA: Artech House, 1999.

[6] Lima, J.S. and Silveira, M., Medidas de desempenho em Sistemas 8-VSB e COFDM. In: Proceedings of the International Week of Telecommunication, 2002, National Institute of Telecommunication, Santa Rita do Sapucaí, MG, Brazil.

[7] Lima, J.S., Linearization of TV transmitters with IF pre-distortion, Telecommunication Journal, National Institute of Telecommunication, 1998, Vol. 1, Brazil.

[8] Mello, A.A., Rodrigues, H.D., Lima, J.S., Silveira, M., Pereira, W.N., Ribeiro, J.A.J., Linearization of the transmitter using digital pre-distortion. In: Proceedings of the IEEE APS-URSI (APS-URSI'02); 2002, San Antonio, TE, USA.

[9] Mello, A.A., Rodrigues, H.D., Lima, J.S., Silva, M.P.S., Silveira, M., Pereira, W.N., O uso da técnica de pré-distorção digital na linearização de amplificadores de potência em RF, In: Proceedings of the International Week of Telecommunications; 2002, National Institute of Telecommunication, Santa Rita do Sapucaí, MG, Brazil.

[10] Mello, A.A., Rodrigues, H.D., Lima, J.S., Silva, M.P.S., Silveira, M., Adaptive linearization using pre-distortion experimental results, Telecommunication Journal, National Institute of Telecommunication, 2004, Vol. 3, Brazil.

[11] Mello, A.A., Rodrigues, H.D., Lima, J.S., Silva, M.P.S., Silveira, M., An efficient Numerical Approach for the Linearization of Power Amplifiers, IEEE Journal Latin America, Vol. 2, 2, 2004.

[12] Silva, M.P.S., Silveira, M., et al., An efficient analysis of the performance of nonlinear devices using as a tool the software ADS. In: Proceedings of the IEEE of Word Congress on Engineering and Technology Education (WCETE'04), 2004, Guarujá, Brazil.

[13] Mello, A.A., Rodrigues, H.D., Lima, J.S., Silva, M.P.S., Silveira, M., A new numerical approach in the linear analysis of RF amplifiers. In: Proceedings of the IEEE 33rd European Microwave Conference (EuMC'03); 2003, Munich, Germany.

[14] Mello, A.A., Rodrigues, H.D., Silva, M.P.S., Silveira, M., Adaptive digital pre-distortion

to reduce the power amplifier non-linearity. In: Proceedings of the IEEE APS-URSI (APS-URSI'03), 2003; Columbus, OH, USA.

[15] Clarke, K., Hess D., Communication Circuits: Analysis and Design, Krieger Publishing Co., 2nd ed., Malabar, FL, 1994.

[16] Kim, J. and Konstantinou, K., Digital predistortion of wideband signals based on power amplifier model with memory, Electronics Letters, 2001, Vol. 37, 23, pp. 1417–1418.

[17] Silveira, D.D., Gilabert, P.L., Lavrador P.M., Pedro J.C., Gadringer M., Montoro G., Bertran E., Magerl G., Improvements and analysis of nonlinear parallel behavioral models, Journal of RF and Microwave Computer-Aided Engineering, 2004. doi:10.1002/mmce.20385.

[18] Kenington, P.B., High Linearity RF Amplifier Design, New York, NY: Artech House, 2000.

[19] Fischer, W., Digital Video and Audio Broadcasting Technology. A Practical Engineering Guide, New York, NY: Springer, 2008.

[20] Pedro, J.C., Maas, S., A comparative overview of microwave and wireless power-amplifier behavioral modeling approaches, IEEE Transactions on Microwave Theory and Techniques, 2005, Vol. 53, 4, pp. 1150–1163.

[21] Cripps, S.C., Advanced Techniques in RF Power Amplifier Design, New York, NY: Artech House, 2002.

[22] Kenington, P.B., Analysis of instability in feed-forward loop, Electronic Letters, Vol. 33, 1997.

[23] Marmarelis, V.Z., Nonlinear Dynamic Modelling of Physiological Systems, Hoboken, NJ: J. Wiley & Sons, 2004.

[24] Cavers, J.K., Amplifier linearization using a digital pre-distorter with fast adaptation and low memory requirements, IEEE Transactions on Vehicular Technology, 1990, Vol. 39, 4, pp. 374–382.

[25] Lima, J.S., Silveira, M., The Weaver theoretical approach to generate some important TV digital signals for the transmission systems, In: Proceedings of the IEEE International Microwave and Optoelectronic Conference (IMOC'05), 2005, Belém, PA, Brazil.

[26] Rodrigues, H.D., Silveira, M., Ribeiro, J.A.J., Silva, D.G., Complex filtering for generation of SSB and VSB signals, In: Proceedings of the IEEE Asia Pacific Microwave Conference, (APMC'05), 2005, Hong Kong, China.

[27] Sedra, A.A., Smith, K.C., Microeletrônica, 5th ed., São Paulo: Makron Books, 2004.

[28] Cripps, S.C., RF Amplifiers for Wireless Communications, New York, NY: Artech House, 1999.

[29] Vuolevi, J., Analysis, measurement and cancellation of the bandwidth and amplitude dependence of intermodulation distortion in RF power amplifiers, Lecture Notes, Department of Electrical Engineering, University of Oslo, Blindern, Oslo, 2001.

[30] Wang, H., Bao, J. Wu, Z., Comparison of the Behavioral Modeling for RF Power Amplifier With Memory Effects, IEEE Microwave and Wireless Components Letters, 2002, Vol. 19, 3, pp. 179–181.

[31] Cantrell, W.H., Davis, W.A., Amplitude modulator utilizing a high-Q class-E DC-DC converter, In: Proceedings of the IEEE MTT-S Microwave Symposium Dig. (MTT-S'03), June 2003, Vol. 3, pp. 1721–1724.

[32] Elnady, A., An efficient current regulator for multilevel voltage source converter based on a simple analog control circuit, Journal of Circuits, Systems and Computers, 2013, Vol. 22, 4, pp. 1–21, doi:10.1142/S021826613500230.

[33] Hung, T.P., Metzger, A.G., Zampardi, P.J., Iwamoto, M., Asbeck, P.M., Design of high-efficiency current-mode class D amplifiers for wireless handsets, IEEE Transactions on Microwave Theory and Techniques, 2005, Vol. 53, pp. 144–151.

[34] Campos, M.V., Silveira, M., An efficient analytical approach for the non-linear analysis of RF amplifiers implemented with the FET structures, In: Proceedings of the IEEE Asia Pacific Microwave Conference, (APMC'05), Dicember 2005, doi:10.1109/APMC.2005.1606219.

[35] Campos, M.V., Silveira, M., An original approach of the effect source resistance for FET devices, In: Proceedings of the IEEE Asia Pacific Microwave Conference, (APMC'08), Dicember 2008, doi:10.1109/APMC.2008.4958523.

[36] Sevic, J.F., Statistical characterization of RF power amplifier efficiency for CDMA wireless communication systems, In: Proceedings of the IEEE Wireless Communications Conference, 1997, p. 110–113, doi:10.1109/WCC.1997.622258.

[37] GPIB tutorial, Available from: http://cp.literature.agilent.com/litweb/pdf/04396-90063.pdf.

[38] Duffy, D.G., Advanced Engineering Mathematics with Matlab, 3rd ed., New York, NY: CRC Press, 2010.

[39] Tektronix, Television measurements manual for NTSC system, Available from: http://www.tek.com/document/primer/television-measurements-ntsc-systems.

[40] Silveira, D.D.; Lima, J.S.; Silveira, M.; Dissect PA distortion from OFDM signals, Microwaves & RF Magazine, 2010; Vol. 01, pp. 88–94.

[41] Bermudez A. et al., Mathematical Models and Numerical Simulation in Electromagnetism, 1st ed., New York, NY: Springer Publisher, 2014.

[42] Varahram, P., Ali, B.M., Peak-to-average power ratio reduction and digital pre-distortion effects in power amplifiers in OFDM systems; International Journal of

Communication Systems, 2012, Vol. 25, pp. 543–552, doi:10.1002/dac.1282.

[43] Wen, J.H., Lee, S.H., Cheng, C.H., SLM-based PAPR reduction method using partial data circulation and side information insertion in OFDM systems; International Journal of Communication Systems, 2009, Vol. 22, pp. 87–100, doi:10.1002/dac.v22:1.

[44] Tu, H.C., Chiu, C.C. A novel search method to reduce PAPR of an OFDM signal using partial transmit sequences; International Journal of Communication Systems, 2007, Vol. 20, pp. 147–157.

[45] Jeng, S.S., Chen, J.M., Chang, P.H., A low-complexity companding technique using Padé approximation for PAPR reduction of an OFDM system, International Journal of Communication Systems, 2011, Vol. 24, pp. 1467–1479, doi:10.1002/dac.1231.

[46] Rozic, N., Radic, J., Theoretical analysis of iterative signal reconstruction for impulsive noise mitigation in OFDM systems, International Journal of Communication Systems, 2010, Vol. 23, pp. 1467–1479, doi:10.1002/dac.1058.

[47] Lu, Q.; Peng, T., Wang, W., Wang, W., Hu, C., Utility-based resource allocation in uplink of OFDMA-based cognitive radio networks, International Journal of Communication Systems, 2010, Vol. 23, pp. 1467–1479, DOI: 10.1002/dac.1070.

[48] Yong, J.J., Hyung, W.K., Hong, T.K.; Gi, H.R., Ji,Y.C., Kim, K., Soo, E.S., Byungdu, O., A highly efficient CDMA power amplifier based on parallel amplification architecture; IEEE Microwave and Wireless Components Letters, 2004, Vol. 14, pp. 401–403, doi: 10.1109/LMWC.2004.832050.

[49] Karthaus, U., Sukumaran, D., Schmidt, L., Ahles, S., Wagner, H., A 45 dBm balanced power amplifier module based on four fully integrated Doherty PA MMICs. A scalable solution for cellular infrastructure active antenna systems, In: Proceedings of the IEEE 42nd European Microwave Conference (EuMC'12), 2012, Vol. 1, pp. 1027–1030.

[50] Osman, A.F., Noh, N.M., Wideband LNA design for SDR radio using balanced amplifier topology, In: Proceedings of the IEEE 4th Asia Symposium on Quality Electronic Design (ASQED'02), 2012, Vol. 1, pp. 86–90, doi: 10.1109/ACQED.2012.6320481.

[51] Ji, Y.P.; Burger, J., Titizian, J., UHF band long-pulse radar power amplifiers using push-pull and balanced configurations; In: Proceedings of the IEEE International Microwave Symposium, (MTT-S'07), 2007, Vol. 1, pp. 15–18, DOI: 10.1109/MWSYM.2007.380206.

[52] Jack, K., Video Demystified: A Handbook for the Digital Engineer, 5th Ed., New York, NY: Elsevier, 2011.

6

Objectifying the Subjective: Fundaments and Applications of Soft Metrology

Laura Rossi

Abstract

The aim of the interdisciplinary research was to facilitate the understanding of a specific topic passing by different disciplinary perspectives. Soft metrology is the perfect example of a scientific field that needs that sort of approach. Seeking to provide a reproducible basis for qualifying and quantifying what are essentially 'soft' measurements (subject to human perception and interpretation) is a particularly challenging scientific endeavour. This chapter presents a theoretical overview of main concepts around soft metrology and, in the second instance, proposes a mathematical model for the measurement of a soft measurand through a dedicated index (IPER—influence on performance index).

Keywords: soft metrology, human perception, human–machine interaction, human performance

1. Introduction

'Metrology must be ideal, but practical. Rigorous, but accommodating. Demanding, but forgiving. Quantitative, but qualitative. Forward-looking, but faithful to the past'.

(John Michael Linacre, University of Sydney, 2005).

This work represents both a compendium of ideas around soft metrology intended as the set of techniques and models for measuring quantities related to perception. The knowledge of the physical world through senses is a key topic of the history of philosophy. Since Plato and Boethius, passing through all the history of Greek and Indian ancient philosophy, getting to the modern era with Descartes and Hume until reaching contemporary philosophy of mind, human perception has always been a central topic of thought as a key epistemological problem. Hereby, some specific theoretical interpretation of the relation between humans and the physical world will be presented, depending on how much they are useful to the foundations of a specific line of reasoning pertinent to the metrological scope.

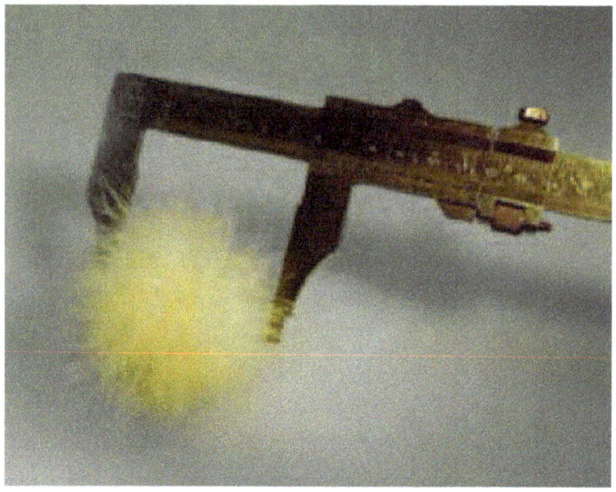

Figure 1. A metaphorical picture for soft metrology. The traditional tools/methods of metrology are not suitable for measuring quantities related to human perception.

The soft metrology expression is relatively recent and has appeared in literature the first time in 2003 in a report of the National Physical Laboratory [1] where it has been defined as 'the set of techniques and models that allow objective quantification of the properties determined by perception in the domain of all five senses'. It can be intended as the investigation about the correlation between the human subjective response and objective measurement of physical properties of the empirical world.

The main objective of this work is to contribute to the definition of the theoretic basis, the language and investigation methodology related to soft metrology. Some methodological proposals are presented as well as some experimental findings. Subjective and physiological subjects' responses to specific visual and auditory stimuli are considered, in conditions of low or high cognitive load and with an environmental setting influencing the perceptual process (presence of glare or noise). Here lies also a good question into what extent one mode of perception is affected on influenced by another, as in perceiving the flux of information coming from the empirical world cross-modal interactions normally occur (**Figure 1**).

As a matter of fact, the recent interest in metrology of perception by the industry for design and development of efficient human–machine interfaces makes the study of these issues promising interesting developments. Human–machine interface (HMI) is in fact out of the niche of basic scientific research and hi-tech applications (e.g. military, aviation and aerospace), thus making key service for consumer goods. This is attested by the recent identification of commercial brands with the HMI, for example in the automotive industry (in 2004, Citroen ASL for the notice of passing the line of road through seat vibration and APGS Lexus' 'Advanced Parking Guidance System', in 2006 Fiat 'Blue & Me', in 2009 the Toyota Head Up

Display). Soft metrology, together with cognitive science and ergonomics, can define the criteria necessary for effective planning and efficient industrial design as well.

As the aim of the interdisciplinary research is to facilitate the understanding of a specific topic passing by different disciplinary perspectives, soft metrology is the perfect example of a scientific field that needs that sort of approach.

Seeking to provide a reproducible basis for qualifying and quantifying what are essentially 'soft' measurements (subject to human perception and interpretation) is a particularly challenging scientific endeavour. The tentative to do so has been carried out by the work of MINET [2,3] as described below, in a pure and fruitful interdisciplinary perspective. An expression of this key aspect of research in this area is to involve, for example investigations of human mental and brain functions (studied primarily in psychology and neuroscience), research into how these underpin human attention, perception and cognition (psychophysics and behavioural studies) and development of measurement instrumentation and perceptual models (metrology, mathematics, modelling, computing, psychology, physics and psychophysics).

Human perception and interpretation encompasses phenomena that are of different complexity, ranging from sensory perceptions (e.g. colour, taste, odour, loudness), to environmental perceptions (e.g. soundscape, air quality, landscape), to self-perceptions (e.g. chronic pain, wellbeing, mood), to perceptual attributions (e.g. aesthetics, satisfaction, expectancies), as well as all kinds of complex interpretations and evaluations (e.g. utility, risk, maladjustment) that are based on learning and experience.

'Measuring the Impossible' is the impressive title of a recently concluded European call for new and innovative research projects, concerning the measurement of quantities and qualities related to human perception and interpretation.

As a matter of facts, many phenomena of significant interest to contemporary science are intrinsically multidimensional and multidisciplinary, with strong crossover among physical, biological and social sciences. Economic aspects, products and services appeal to consumers according to parameters of quality, comfort and beauty which are mediated by perception and culture.

A few examples of funded projects can help to get the feel of recently closed or still on-going research activities in this area.

The MONAT project [4], Measurement of Naturalness, coordinated by the National Physical Laboratory (NPL, United Kingdom), was aimed to measure the naturalness of materials, such as fabrics, on the basis on of their visual appearance (influenced by factors such as colour, texture, gloss…) and tactile feeling (thermal conductivity, hardness, friction coefficient…). Investigations into this multisensory perception and interpretation require the measurements of related perceptual, physical and psychophysical qualities and their combinations.

CLOSED [5], closing the loop of sound evaluation and design, coordinated by the Institut de Recherche Acoustique/Musique (IRCAM, France), sought to improve products by introducing, at the early design stage, information concerning the perceived sound quality of products in use.

Inspired by the MINET activities, a book on this topic has been published in 2011 and this has been the first comprehensive publication to scrutinise this measurement topic from a multi- and interdisciplinary perspective [6].

2. Fundaments of Soft Metrology

2.1. Definition and theoretical context

As far we know, the first appearance of the expression 'soft metrology' and its related definition appeared in 2003 in the NPL Report CMSC 20/03 [1]. Here, soft metrology is defined as follows:

> *The set of measurement techniques and models which enable the objective quantification of properties which are determined by human perception.? (The human response may be in any of the five senses: sight, smell, sound, taste and touch). Thus, soft metrology includes aspects of appearance (colour and gloss), noise quality, texture of food (such as creaminess) and, more broadly, topics such as biometrics and usability of systems.*

Here, the author points out that soft metrology, in its extended sense, is not yet an established branch of metrology, and today, it does not find a unique place within the structure of the National Measurement System of each country.

That does not mean that measurement scales do not exist or that research in the domain of soft metrology is not conducted, but rather that there is a lack of metrology research in considering properties of human perception and cognition as measurand of physical properties of the environment.

As a matter of facts, soft metrology requires the measurement of proper physical parameters and the development of models to correlate them to perceptual quantities. Traceable soft metrology is achievable both through the traceable measurement of the physical parameters and the development of accurate correlation models.

Figure 2. A set of boxes that are perceived to increase in SIZE: the corresponding physical measurement is of LENGTH (for example, in metres).

As an example, we would like to mention the following one, described in the NPL Report as quite explicatory of the soft metrological approach when a relation to physical stimuli and perception is investigated.

The example considers the perception and measurement of length. A series of boxes are perceived to increase in size (**Figure 2**). A human observer could be asked to indicate a number correlated to their impression of the size. Equally, a device could be constructed (a ruler) which could be used to 'measure' the size of each square, that is its area.

In this case, it is likely that the human responses, the soft measure, will simply correlate with the measurements made using the ruler, the physical measure. Hence, soft metrology can be considered the formulation of a correlation between the human response and the physical measure.

As a second example, consider the series of boxes shown in **Figure 3**.

Figure 3. A set of boxes that differ in a systematic manner.

To the human observer, something is changing as the boxes are examined from left to right—but what is it and how can it be described? Some may call it lightness, some brightness, some density.

It has been shown [7] that human observers, after a little training, can give a consistent response to the changes they see. Equally, some physical measures, for instance, the reflectance of the paper surface of every square, the density or even their luminance, can be made of each box using a suitable metre, and a scale constructed that relates the two measures.

Figure 4. The squares marked A and B are the same shade of grey.

Also, in real life, the human response is not absolute as a physical measure usually is, but it is subjectively correlated to a real or imaginary reference. This fact creates a set of interesting optical illusions, as the famous Checker illusion [8], a visual illusion published in 1995 by Edward H. Adelson, Professor of Vision Science at MIT (**Figure 4**).

Differently from the human eye, a physical system does not find luminance differences between square A and B in **Figure 4** and cannot understand the presence of the chessboards without complex elaboration of the image. The visual system is not very good at being a physical light metre, but that is not its purpose. The important task is to break the image information down into meaningful components, thereby perceiving the nature of the objects in view.

We now go rapidly through the idea of what it does imply for considering the human senses as a measuring instrument and what measuring the act of perception means. Following what said just before, soft metrology can be considered as the investigation of the correlation between human, subjective, responses and physical, objective, measures. What is generated is a measurement scale, a number series, which allows the subjective response to be predicted from the objective measure.

There is a parallelism between the process of perceiving and the one of measuring. This comes from the idea that human senses can be considered a sort of transducer. A transducer is, in fact, a tool that turns a physical quantity—as a sound pressure wave—in another physical quantity—as a voltage variation. It is assumed that the transducer has a well definite function between input and output quantities (i.e. linear) and can be calibrated, but in the case of the human being, of course, this attribute cannot be proposed.

As a matter of facts, the human being is already equipped himself with transducers (sight, hearing, smell, taste and touch) and should be considered as a filter or rather, using the language of signal, such as a black box through which a stream of input events, perceived by the senses, flows through leading to opinions, actions and output reactions.

Figure 5. The input–output model by behaviourists. The internal functioning of on organism mind (O) generates a response (R) to a stimulus (S).

This idea is borrowed from the thought of a certain strand of the early twentieth-century psychology called behaviourism. When coming to the study of human and, more broadly, its measurement, behaviourists concentrated exclusively on the study of behaviour neglecting the study of the inner workings of the mind as considered impossible to be measured in a scientifically objective way.

In contrast, behaviours, or actions produced by organisms, as external events, were regarded as objects clearly quantifiable and measurable. **Figure 5** shows the model stimulus-response

(SR) that represents the behaviourist position. In the model, a stimulus (S) in the environment affects an organism (O) and the stimulus then causes the body to produce a response (R).

In this model, the human mind is not shown. The perceiving subject is treated just as a black box, possessing unfathomable properties and internal relations impossible to be explained. In this perspective, he becomes simply an entity that is used to convert the stimuli into responses. A mathematical model between S and R can be developed using experimental results, but the model should be true only in this specific condition and eventually could not be linked to the real process involved in O.

However, the father of behaviourism, John B. Watson (1878–1958), [9] admitted himself that the decision not to investigate the sphere of mental activity was mainly caused by the limited instruments available in his time. Behaviourism was, in fact, the dominant paradigm in psychological research for nearly 50 years until the advent of cognitive science in 1960.

Therefore, while behaviourism meant the mind as a passive body operating according to simple rules of conditioning, cognitive psychology saw in it a strong, active ability, such as the selection of information from the environment and, about prior knowledge, the capability of act on the results of such selection. Fundamentally, there are three main reasons that led to the rapid growth of this new perspective:

- the failure of behaviourism in explaining language acquisition;
- the invention of new measuring devices to examine the mental activity;
- the advent of the computer and hence the widespread use of the metaphor of the mind computer.

The emergence of new investigation devices, including positron emission tomography (PET), computed tomography (CT) and magnetic resonance imaging (MRI), but also the increased use of computers, was the innovations that have most contributed significantly to the decline of behaviourism.

Psychologists realised that the mind, like a computer, could be treated as a device that represents and transforms information. This initial metaphor then flew into the actual conception of the human mind as an information processor and of the act reasoning assimilated to computing, in which the behaviourist stimulus-response model turned into input–process–output one:

- input: incoming information in mind, corresponding to the 'stimulus' of behaviourism;
- process: information converting that change and are reworked by mental processes evolving continuously with experience;
- output: output information in the form of behaviour, language, facial expressions, posture, etc., corresponding to the 'answer' or 'reaction' of behaviourism.

According to Ulric Neisser (1928–2012), an American cognitive psychologist, the study of man:

> *refers to any process in which sensory input is transformed, reduced, elaborated, stored, recovered and used* [10]

and all these modes convey exactly the meaning of the numerous potential of the human mind in the management of information from the sensory world.

Summarising, if behaviourism had the fundamental claim to limit the psychological investigation of objectively observable reactions, its strength was in the methodological need that it asserted: it is not possible to scientifically talk about what escapes any possibility of objective observation and control.

If at that time it was not possible to have this kind of control over mental processes, which is no longer the case thanks to the modern techniques of physio-neurological investigation, with the addition of the contemporary cognitive science assumption that human rational processes can be considered as an algorithm.

After this short glimpse on some relevant ideas of theories of human perception, we now go through more specific issues related to soft metrology as a part of metrological science and therefore bound to its specific approach.

2.2. Research issues and challenges

The main research issues that are currently central to the soft metrology area [11] can be grouped into three main categories:

- foundation and theory;
- instrumentation and methods;
- implementation areas and applications.

Several foundational and theoretical issues are of primary interest in the soft metrological field. These include as follows:

- language and terminology;
- concepts of measuring system (or instrument);
- the issue of measurability and uncertainty.

Language has been a major issue in twentieth-century scientific and philosophical debate. In the metrology community, an extensive revision of linguistic terms has been undertaken, starting with the publication in 1984 of the International vocabulary of basic and general terms in metrology, now at its 3rd edition [12].

As suggested [13], in any revision of terms, a revision of concepts and theories must also be considered and this would indeed be beneficial for the entire world of measurement. From a theoretical point of view, soft metrology can be assumed funding its basis on the Representational Theory of Measurement [14] where measurement is defined as

> *a process of empirical, objective assignment of symbols to attribute of objects and events of the real world, in such a way as to represent them or to describe them and also, in a shortest form, as the correlation of numbers with entities that are not numbers* [15].

In this theory, values are assigned based on correspondences or similarities between the structure of number systems and the structure of qualitative systems. A property is defined as 'quantitative' if such similarities can be established.

The concept of measurement is often misunderstood as merely the assignment of value, but it is also possible to assign a value in a way that is not a measurement. For example, a value of a person's height can be assigned, but unless it can be established that there is a correlation between measurements of height and empirical relations, it is not a measurement. To link this theory with the soft metrological research field [16], it is interesting to cite the following passage by an article by Luce and Suppes [17].

> *In the 75 or so years beginning in 1870, some psychologists (often physicists or physicians turned psychologists) attempted to import measurement ideas from physics, but gradually it became clear that doing this successfully was a good deal trickier than was initially thought.*
>
> *Indeed, by the 1940s a number of physicists and philosophers of physics concluded that psychologists really did not and could not have an adequate basis for measurement. They concluded, correctly, that the classical measurement models were for the most part unsuited to psychological phenomena. But they also concluded, incorrectly, that no scientifically sound psychological measurement is possible at all.*
>
> *In part, the theory of representational measurement was the response of some psychologists and other social scientists who were fairly well trained in the necessary physics and mathematics to understand how to modify in substantial ways the classical models of physical measurement to be better suited to psychological issues.*

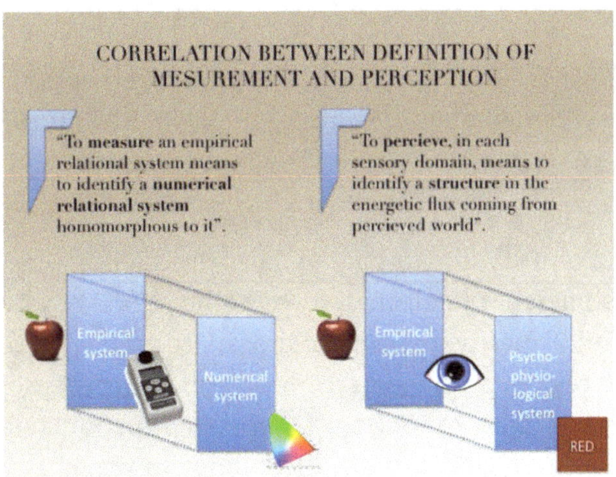

Figure 6. The two definitions of *measure* and *perceive* and streamlined representation of the passage from real word to numbers and percept through a measuring instrument (e.g. a colorimeter) or a sense (e.g. sight).

As a matter of facts, an actual correlation between the definition of measure—as coming from representational theory—and perceive [18] does exist (**Figure 6**). Both measuring and perceiving are an act of correlation between two systems.

The measuring process consists of the creation of an image (in logic: homomorphism) of the real world into a numerical system or model that maintains its proportions and inner relations, through the use of an instrument or model.

The act of perceiving can be assumed to function in the same way: from empirical stimuli, the human mind creates an interpretational model that represents proportions and relations in the real world, through the use of senses.

Concerning measurement uncertainty, a basic issue in soft metrology is to come to terms with how to handle inter-individual differences in measurement. These may be considered true differences, as postulated and measured in psychometrics, or they may be regarded as common and specific errors in measurement, as treated in psychophysics.

Instrumentation-oriented research involves the measurement of external physical events (stimuli) that more or less simultaneously give rise to perceptual and physiological (or behavioural) responses. It would include the perception and interpretation of processes and the development of sensors that, to a certain extent, would mimic human perception and interpretation.

Consider, for example, the measurement of sound. Highly accurate measurement microphones and binaural recording devices, those allow acoustic stimuli to be measured as they appear at the input of the auditory system, have been developed [5]. Sophisticated binaural reproduction or surround sound systems, with processors and algorithms, provide the required signal processing.

As concerns vision, not only luminous intensity and colour wavelengths can be measured, but also parameters of the interaction between light and matter, and surfaces properties, such as texture-topography, involving sophisticated signal processing [4]. Similar considerations also apply to the other senses.

As previously mentioned, measurements related to human perception and interpretation have a wide range of actual and potential applications [19]. Here, the areas of perceived quality (of products and services), environment, ergonomics, safety, security and clinics are briefly mentioned.

In the first part of the last century, the impact of mass production was so high that qualitative aspects of goods were somewhat neglected. Today, the shortage of energy sources and the concern for pollution may increase the demand for durable, high-quality goods. Thus, perceived quality, which results from the perception and interpretation of sensory input, may play a key role in the industrial competition. Examples of products include food, materials, simple and complex devices. A good cup of coffee, for example, is appreciated on the basis of a combination of taste, smell, sight, touch and, for the Italian recipe, of the colour of the surface cream [20].

As said, much of human behaviour is controlled by responses to the five senses. To the consumer, the whole appearance (that can be the smell, the sound, the taste...) of specific products, whether natural or man-made, is used to assess quality, both consciously and subconsciously, and hence mediate product choice.

Providing a means for reproducible measurement of parameters (such as pleasure and pain) has important implications in evaluating all kinds of products and services.

In the commercial world, a host of factors—from naturalness, aesthetics and comfort to security, service and price influences choice and gives peculiarities to a brand. Is the smell of a new car appealing for the customer? Does a fabric have a luxurious feel?

By being able to predict user perceptions and reactions through a robust modelling, can allow to save time and costs by meeting the expectations of the customers, without the need for extensive and expensive testing.

The ISO 9000 framework, however, requires the parameters able to characterise products to be measured in a precise way, together with associated tolerances, to establish a formal quality-control system. Normally, this is more easily done using instruments because of their inherent controllability, stability and repeatability. The challenge of soft metrology is to gain a comparable level of strengths.

3. The three-output model

The fundamental question to which soft metrology attempts to provide answers may be described as follows: given as input a sensory stimulus metrologically quantifiable (a special sound, a flash of light, the roughness of an object, etc.), what is the outcome of the interaction between this and humans? That is, what is the measurable output, if it exists, of this interaction?

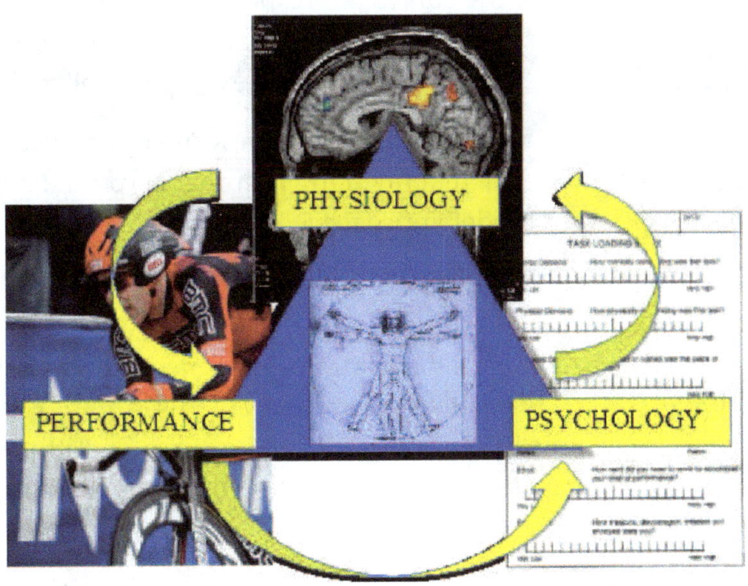

Figure 7. A schematic representation of the three-output model. Measuring the interaction between man and the objects or events of the real world means to take into account three kinds of outcomings from physiology, from evaluation of performance and psychology.

To answer this question, the output of the perceptual interaction between subject and world objects or events shall be defined at first. According to the tools and analytical models available, three main outputs (**Figure 7**) have been identified [21]:

- physiological output,
- performance,
- psychological output.

The first output coming from human–world interaction is called 'physiological', and it can be quantified through various clinical techniques that monitor some physiological parameters considered as important indicators of changes in the level of perception, emotion or cognition (e.g. the measurement of heart rate, blood pressure, pupil dilation, sweating, brain electrical activity).

The second output is referred to human performance (we will refer to it also as 'operational') and is quantified by measuring the changes in the attitudes of a subject doing a particular task. These can be for example attention, fatigue or accuracy and can be assessed via various parameters, such as reaction times, the number of correct answers, test duration, etc.

The third output is determined and quantified by evaluation scales, questionnaires and interviews (we will refer to it as the 'psychological' output), and this is one of the most popular methods for subjective data acquisition. The response of subjects to a specific question about their opinion on a particular aspect of a stimulus is defined 'psychological' as it is mediated by the psychology (culture, mood, prejudices…) of the person to whom the question is asked.

The method of investigation proper to soft metrology must consist in the comparative analysis of these three main outputs. Through an organic synthesis of these data, is it possible to get reliable objective information about the subjective response to a particular sensory stimulus [22]. An example of analysis of this sort will be given through the description of the influence on performance (IPER) index, correlated by a numerical example explicating its applications.

Before moving further, it is important to introduce a review of key terms of the three fields of investigation. When necessary they are redefined in order to incorporate in the soft metrological terminology, the three-output model approach.

3.1. Redefinition hypothesis and terminology

As previously mentioned for what regards soft metrology is necessary to establish a common language and a definition of key terms to simplify the organisation of concepts of this new branch of metrology. Hereby, a list of important terms is given with the related definitions congruent with the model given by the last edition of the International Vocabulary of Metrology [23].

A first and remarkable tentative is the review by Goodman about the measurement of physical parameters in sensory science [24]. Goodman summarises the definitions and descriptions of some relevant physical parameters useful for soft metrological applications. As we interact with our environment, and with objects within that environment, through our five senses, the physical measurements that are most relevant for sensory science are those relating to the parameters that are sensed by our sensory transducers. As reported in Goodman's article, these are listed in **Table 1**.

The object of the subjective measurement is the interaction between a subject and the physical word, and the schematic representation of **Figure 8** shows the interaction between subject and physical word displaying the key terms and expressions that is important to define.

Sensory modality	*Physical parameter and SI units*	*Sensory response*
Vision	***Luminance* (candela per metre squared, cd·m^{-2}):** light reflected from or emitted by a surface	Brightness
	***Gloss* (dimensionless):** light reflected in specific directions relative to the incident direction	Shininess
	***Chromaticity for non-self-luminous surfaces* (dimensionless):** spectral reflectance of surface combined with spectral irradiance of illuminating light source	Colour
	***Chromaticity for self-luminous surfaces* (dimensionless):** spectral radiance of surface	Colour
	***Dimensional characteristics* (metre, m):** length, volume, etc.	Size and shape
	***Chroma, saturation, hue and other colour appearance measures* (dimensionless):** derived from spectral reflectance and spectral radiance/irradiance measurements, defined in terms of various colour measurement systems	Colour appearance in the context of the visual environment
	***Goniometric and spatial surface characteristics* (dimensionless):** spectral reflectance as a function of position and angle combined with spectral irradiance of illuminating light source as a function of position and angle	Visual texture and pattern
	***Light scattering characteristics* (dimensionless):** spectral transmittance as a function of position, angle and thickness	Transparency, clarity, haze, translucency
Touch	***Surface topography* (metre, m):** height of surface as a function of position	Roughness/smoothness
	***Friction* (Newton, N):** force experienced when moving a fingertip over the surface	Stickiness, slipperiness
	***Hardness*(dimensionless):** resistance to indentation (measured on various defined ratio scales)	Hardness
	***Tensile strength, elasticity* (pascal, Pa, or newton per metre squared, N·m^{-2}):** resistance to deformation	Stretchiness, bendability, drape, compressibility
	***Thermal effusivity* (joule per metre squared per kelvin**	Coldness, wetness

Sensory modality	*Physical parameter and SI units*	*Sensory response*
	per square-root second, $J \cdot m^{-2} \cdot K^{-1} \cdot s^{-1/2}$ or watt square-root second per metre squared per kelvin, $W \cdot S^{1/2} \cdot m^{-2} \cdot K^{-1}$): ability of a material to exchange thermal energy with its surroundings	
Sound	***Acoustic pressure* (pascal, Pa, or newton per metre squared, $N \cdot m^{-2}$):** sound wave amplitude	Loudness
	***Acoustic intensity* (watt per metre squared, $W \cdot m^{-2}$):** sound power per unit area	Loudness
	***Acoustic frequency* (hertz, Hz):** sound wave frequency	Pitch, sharpness, tone quality, timbre
	***Acoustic impedance* (decibel, dB—note this is accepted for use with SI, but is not an SI unit):** attenuation of sound waves through a medium	Muffled
Taste and smell	***Chemical composition* (mole per metre cubed, $mol \cdot m^{-3}$)**	Flowery, fruity, salty, sweet, bitter, sour....

Table 1. Key physical parameters for sensory transduction as reviewed by T. Goodman (2012), pp. 59–60.

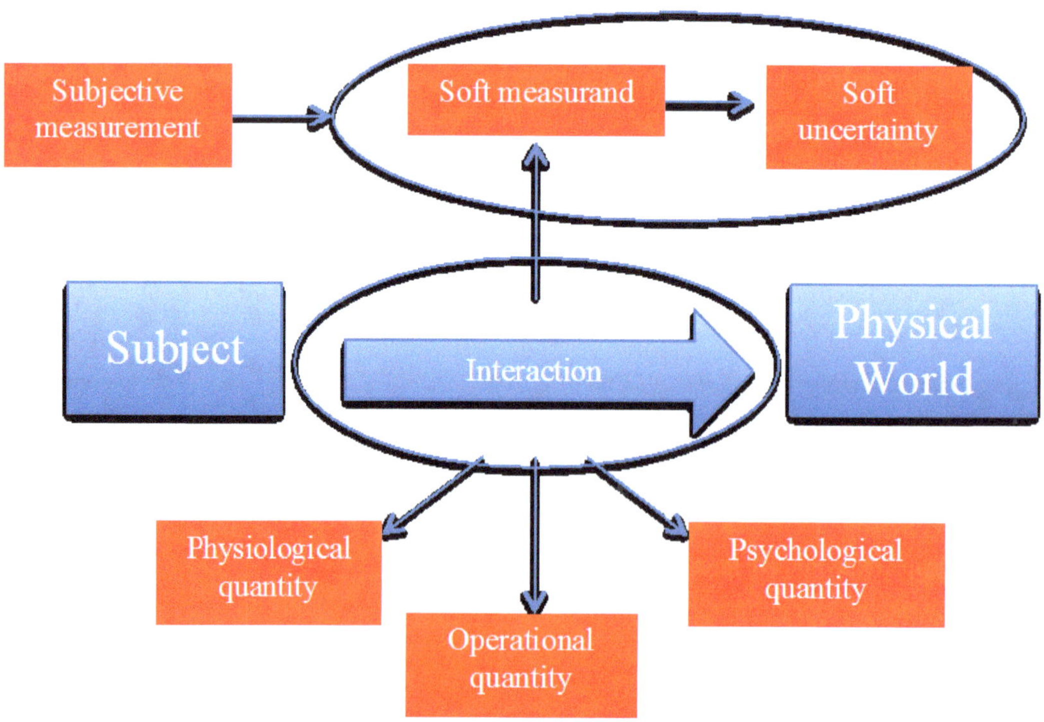

Figure 8. Schematic representation of relationships between key terms in the description and measurement of the interaction between the subject (S) and the physical world (PW).

Subject: human being directly involved in the subjective measurement process (an interesting parallelism is the definition given in VIM: 2008—3.8 of sensor: element of a measuring system that is directly affected by a phenomenon, body, or substance carrying a quantity to be measured).

Soft metrology: set of models and techniques that allow subjective measurements intended as the process of experimentally obtaining one or more quantity values that can be attributed to human physiological, operational and psychological quantity—as defined by soft metrology—directly involving one human being.

Subjective measurement: process of experimentally obtaining one or more quantity values that can be attributed to human physiological, operational and psychological quantity—as defined by soft metrology—directly involving one human being.

Physiological quantity in soft metrology: property of a phenomenon belonging to human body that has a magnitude expressible as a number and a reference.

Operational quantity in soft metrology: property of a human-centred task that has a magnitude expressible as a number with proper unit of measurement, in terms of distance (difference) from an optimal or ideal execution of the same task.

Psychological quantity in soft metrology: property of the attribution of a judgment by a human to an object, event or phenomenon—concerning his/her opinions, emotions and sensations—that has a magnitude expressible as a number and a reference in a given scale.

Soft measurand: a characteristic of a specific object to be measured, in a specific situation. This object outcomes from the interaction between a human and the physical world—event not necessarily repeatable and directly quantifiable—and related to perception, cognition and volition.

Soft uncertainty: non-negative parameter characterising the dispersion of the quantity values being attributed to a soft measurand.

Intrasubjective repeatability: comparability level of subjective measure under a set of repeatable measurement conditions, related to the same subject.

4. Influence on performance (IPER) index

4.1. IPER index definition

We give now a formulation proposal within the soft metrological field of research, as we developed a set of experiments aimed to investigate how noise can affect the performance of humans involved in a task. However, the primary object of these kinds of studies is more general and lies in the possibility of testing a theoretical method of analysis peculiar to soft metrology. The idea is to acquire a robust tool that permits a reliable interpretation of subjective responses to sensorial stimuli and get a quantification of non-metrologically definable objects, such as the 'influence of external stimuli on human performance'. To get this quantification, three kinds of outputs—coming from subjective measurement—have been identified as significant. They are characterised as described hereinafter (**Figure 9**).

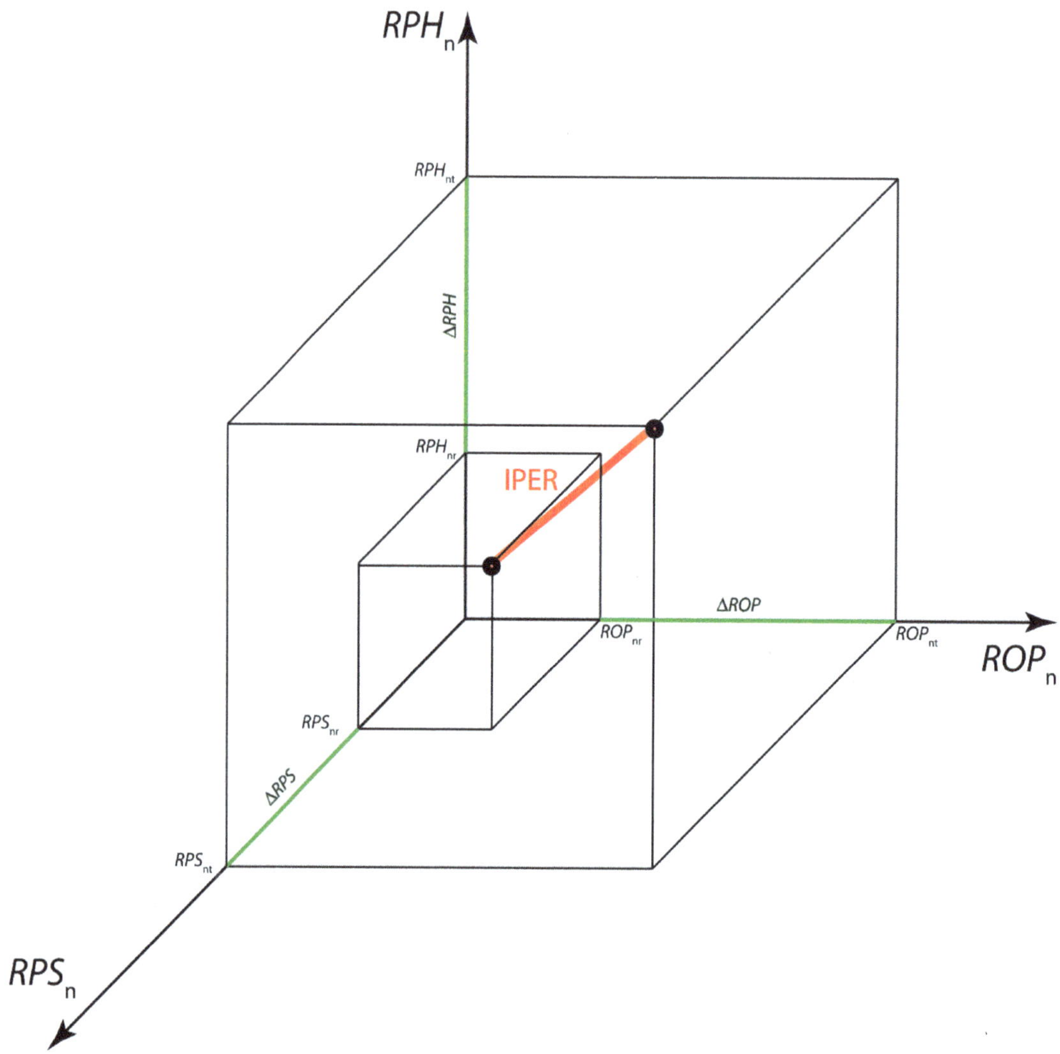

Figure 9. Geometrical description of IPER.

The idea of synthesising in a single number these three outputs in a significantly and properly weighted way comes from the need of metrologically describe a complex and variegated object of study or, in general, what is called soft measurand. Due to its specific synthesising nature, such single number allows prompt and easy comparisons between different complex and multifaceted conditions, events or setups where several subjective parameters are involved. This single number is the here-proposed IPER index.

The main characteristic of IPER index is that all its components are dynamically weighted in order to picture—as plausibly as possible—the actual behaviour of the interaction between a subject and a stimulus together from a physiological, operational and psychological point of view. The 'influence on performance' is the soft measurand considered, and this influence can be due to different phenomena or events from time to time.

The IPER index combines the three outputs R described before:

RPH = Physiological output;

ROP = Operational output;

RPS= Psychological output.

The physiological output (from now RPH) is the first output coming from human-–world interaction, and it can be quantified through various clinical techniques that monitor some physiological parameters considered as important indicators of changes in the level of perception, emotion or cognition (e.g. the measurement of heart rate, blood pressure, pupil dilation, sweating, brain electrical activity). It depends on:

1. the subject;
2. other influencing external parameters:
    1. the test stimulus;
    2. other stimuli imposed as constants;
    3. stimuli present during the test that are erroneously considered as non-influencing and that are not controlled;
    4. experimental methodology (e.g. the order in which the subject carries out some tasks part of the experiment).
3. measurement uncertainty:
    1. referred to the subject;
    2. referred to stimuli;
    3. referred to measurement instruments used for the measure of the physiological parameter.

The operational output (from now ROP), is referred to human performance and is quantified by measuring the changes in the attitudes of a subject doing a particular task. These can be for example attention, fatigue or accuracy and can be assessed via various parameters, such as reaction times, number of correct answers, test duration, etc. It depends on:

1. the subject;
2. evaluation method: the parameter chosen for evaluation the performance (e.g. response time, errors percentage…);
3. difficulty of the task assigned to the subject:

   objective (e.g.: memorising a list of five digit or nine digit) and subjective (e.g.: memorising a list of number in a language that for the subject is the native or not);

4. measurement uncertainty associated to the evaluation method chosen:
    1. instrumental;
    2. subjective (learning and fatigue).

The psychological output (from now RPS) is the subject's response to a specific question about his opinion on a particular aspect of a stimulus, determined and quantified by evaluation scales, questionnaires and interviews. This specific output is evaluated through the subject's reply to a question about his opinion on the perceived quality of the performance in doing the task given. RPS depends on:

1. the structure of the questionnaire;
2. the subject's interpretation of each question (a good questionnaire should avoid as much as possible the influence of personal interpretation of questions and terms on subject's answers. Anyway, this is a kind of bias that, even if not considered, must be mentioned as reputedly existent);
3. the subject's answers reliability (e.g. same answer to same question posed in a different time of the test);
4. the subject's capability of evaluating proportions.

As IPER aim is to combine together these three R, in order to make them comparable, a normalisation procedure is needed.

The normalisation criteria chosen consist in determine the measurability range (R_{MIN} and R_{MAX}) for each R and then reconduct each R to a range between 0 and 1. Measurability range can be the one given by literature (e.g. the pupil can dilate from a minimum of 2.2 mm to a maximum of 7.9 mm), or derive from test conditions (e.g. regarding the percentage of errors we have 0% corresponding to any error and 100% corresponding to any correct answer). On the other hand, different kinds of ranges, and therefore normalisation criteria, can be chosen. For example, it is possible to choose as measurability range the minimum and maximum values reaches by the specific set of subjects involved in the experiment (e.g. regarding the percentage of errors we impose a minimum of 20% deriving from the performance of the subject that in the test group did the best, and the same with the maximum of 85% deriving from the worst performance). Another possibility could be to set measurability ranges on the parameters of each subject, but of course in this way IPER will not be directly comparable between different subjects.

In order to go from each R to the corresponding R_n (where R_n stands for 'normalised output'), the following linear procedure is adopted, but more complex normalisation rules could be adopted too:

$$RPH_{n} = \frac{RPH_{MIN} - RPH}{RPH_{MIN} - RPH_{MAX}}$$

$$ROP_{n} = \frac{ROP_{MIN} - ROP}{ROP_{MIN} - ROP_{MAX}}$$

$$RPS_{n} = \frac{RPS_{MIN} - RPS}{RPS_{MIN} - RPS_{MAX}}$$

Each one of these output (R_n) corresponds to a parameter monitored in a reference condition (r) and in one affected by the presence of an influencing phenomena (t). So, after the normalisation procedure, we obtain the following six a-dimensioned parameters:

$$RPH_{nr} \qquad RPH_{nt}$$

$$ROP_{nr} \qquad ROP_{nt}$$

$$RPS_{nr} \qquad RPS_{nt}$$

Where:

the subscript 'nr' means normalised response in *'reference'* condition, that is the condition where the task done by the subject is supposed not influenced by something external.

the subscript 'nt' means normalised response in *'test'* condition, or the condition where the task done by the subject should be affected by the presence of something external (i.e. disturbing, distracting, annoying…).

Then, in order to compare the two conditions, the difference between the two kinds of outputs (reference and test) is considered:

$$\Delta RPH = RPH_{nt} - RPH_{nr}$$

$$\Delta ROP = ROP_{nt} - ROP_{nr}$$

$$\Delta RPS = RPS_{nt} - RPS_{nr}$$

where

> ΔRPH = Finite increment between physiological response in test condition and the physiological response in reference condition.
>
> ΔROP = Finite increment between operational response in test condition and the operational response in reference condition.
>
> ΔRPS = Finite increment between psychological response in test condition and the psychological response in reference condition.

After the normalisation and the calculation of the difference between reference condition and test condition, IPER is:

$$\text{IPER} = \sqrt{\Delta RPH^2 + \Delta ROP^2 + \Delta RPS^2}$$

IPER goes from a minimum value of 0 to a maximum value of v3. In order to obtain a more legible and directly understandable value, IPER is then transposed to a scale between 0 and 100 using the following formula and becoming $IPER_{0,100}$.

$$IPER_{0,100} = 100 \cdot \frac{IPER}{\sqrt{3}}$$

Mathematically, the IPER index is the Euclidean distance between two points in a 3D Cartesian space where the axis are the normalised physiological output (RPH_n), operational output (ROP_n) and psychological output (RPS_n).

The first point represents the test results in reference conditions (r), the second point in test conditions (t). If the subject responses show no differences between the two testing condition, the $\text{IPER}_{0,100}$ index is 0 (i.e. no measurable influence on performance). If the differences are the maximum permissible for each response (depending from range chosen for each parameter), the $\text{IPER}_{0,100}$ index will be 100.

Using a scalar index makes easier the evaluations of differences, but information on the contribution of the three outputs is lost. This aspect is described after where a vectorial version of IPER is given.

At the same time, the IPER index is very effective for comparing the influence of different external stimuli. A similar approach is adopted in colorimetry to quantify colour differences using the CIE Laboratory colour system. In the CIE Laboratory system, the difference between two colours is given as the distance between their representation on a 3D space where the axes are correlated with lightness, chroma and hue. Where *lightness* is the brightness of an area judged relative to the brightness of a similarly illuminated area that appears to be white or highly transmitting; *chroma* is the colourfulness, of an area judged as a proportion of the brightness of a similarly illuminated area that appears white or highly transmitting; *hue* is the attribute of a visual sensation according to which an area appears to be similar to one of the perceived colours, red, yellow, green, and blue, or to a combination of two of them [25].

The last issue regarding the IPER index has to do with the fact that the final value looses the information regarding the contribution of each single component to the final result. For example, two subjects in a specific experiment reach an $IPER_{0,100}$ value of 48 but the first deriving especially by the fact that he made lots of mistakes in the given task and the second especially because of a great physiological reaction. In order to highlight this differences and at the same time understand the dynamic of the experiment, the computation of the percentage of importance of each R on the final $IPER_{0,100}$ value is required.

So, given the previous example, the first subject can have an IPER = 48, composed for the 10% by physiology, 60% by performance and 30% by psychology, while for the second subject the same value is composed by 70% physiology, 5% performance and 25% psychology.

These values are computed as follows:

$$IPER_{\mathrm{ph}} = 100 \cdot \frac{\Delta RPH^2}{IPER^2}$$

$$IPER_{\mathrm{op}} = 100 \cdot \frac{\Delta ROP^2}{IPER^2}$$

$$IPER_{\mathrm{ps}} = 100 \cdot \frac{\Delta RPS^2}{IPER^2}$$

Where:

$\boldsymbol{IPER_{ph}}$ = percentage of importance on $IPER_{0,100}$ value of the physiological parameter;

$\boldsymbol{IPER_{op}}$ = percentage of importance on $IPER_{0,100}$ value of the operational parameter;

$\boldsymbol{IPER_{ps}}$ = percentage of importance on $IPER_{0,100}$ value of the psychological parameter;

4.2. Numerical example and uncertainty evaluation

Suppose the following experiment: the subject A is involved in a mnemonic task (e.g. memorise and repeat the correct order of a set of numbers), while an environmental disturbing factor is present (e.g. acoustic noise at two different levels). In reference condition, the acoustic noise is not present. The soft measurand is the influence of noise on mnemonic performance. The three outputs are:

Parameter	Measurement unit	Range		Measured value	Uncertainty	Range	Normalized value	Results	
		Min	Max					Value	Uncertainty
RPH_r	mm	2,2	7,9	5,0	0,10	-5,70	0,49		
RPH_t				6,1	0,11		0,68		
ROP_r	%	0	100	80,0	0,00	-100,00	0,80		
ROP_t				52,0	0,00		0,52		
RPS_r	1	0	10	8,0	0,29	-10,00	0,80		
RPS_t				5,0	0,29		0,50		
ΔRPH	mm							0,19	
ΔROP	%							-0,28	
ΔRPS	1							-0,30	
$IPER$	1							0,45	0,06
$IPER_{ph}$	%							18,1	9,1
$IPER_{op}$	%							38,1	9,8
$IPER_{ps}$	%							43,8	14,1
$IPER_{0,100}$	1	0	1,73					**26,2**	3,4

Table 2. Spread sheet of $IPER_{0,100}$ for subject A, in test condition 1.

- *RPH* corresponds to the measurement of pupil diameter;
- *ROP* corresponds to the percentage of correct answer in the mnemonic task;
- *RPS* corresponds to the answer to the question 'How much well you did your task?' from 0 to 10 (where 0 means 'very badly' and 10 means 'very good').

The measuring range for these three *R* is:

- for ***RPH***: 2.2 ÷ 7.9 mm corresponding to physiological dilation range of a healthy pupil [26];
- for ***ROP***: 0 ÷ 100% corresponding to the percentage range of correct answer; and
- for ***RPS***: 0 ÷ 10 corresponding to the answer scale range to the question 'How much well you did your task?'.

These conditions are considered:

- **Reference condition:** Mnemonic task in silent condition;
- **Test condition1:** Mnemonic task in the presence of background noise of type1;
- **Test condition2:** Mnemonic task in the presence of background noise of type2.

As shown in **Table 2**, for the subject A in test condition1 $IPER_{0,100}$ is 26.2 ± 3.4 (a detailed description of how uncertainty is calculated is given in the following paragraph § 5.3).

Parameter	Measurement unit	Range		Measured value	Uncertainty	Range	Normalized value	Results	
		Min	Max					Value	Uncertainty
RPH_r	mm	2,2	7,9	5,0	0,10	-5,70	0,49		
RPH_t				6,2	0,11		0,70		
ROP_r	%	0	100	80,0	0,00	-100,00	0,80		
ROP_t				40,0	0,00		0,40		
RPS_r	1	0	10	8,0	0,29	-10,00	0,80		
RPS_t				3,0	0,29		0,30		
ΔRPH	mm							0,21	
ΔROP	%							-0,40	
ΔRPS	1							-0,50	
$IPER$	1							0,67	0,06
$IPER_{ph}$	%							9,8	4,7
$IPER_{op}$	%							35,2	6,6
$IPER_{ps}$	%							55,0	8,5
$IPER_{0,100}$	1	0	1,73					**38,9**	3,6

Table 3. Spread sheet of $IPER_{0,100}$ for subject A, in test condition 2.

In **Table 3**, the value for $IPER_{0,100}$ in test condition 2 is calculated. The subject A shows a similar physiological reaction to test condition1 but a different behaviour for what regards operational and psychological parameters giving a greater final value of 38.9 ± 3.6.

From this, we deduce that the noise of type 2 produces a greater influence on subjects A performance than noise of type 1, both referring to the silent condition.

At this point, many other considerations and analysis can be done. For example, supposing to have a panel of 50 subjects tested we can get a total $IPER_{0,100}$ mean value for the noise of type1 and type 2 as $IPER_{0,100}$ is directly comparable between subjects and get for example that the second one has an influence on the performance of 35% greater than the first one.

It is also possible to create some homogenous groups of subjects, for example, considering one parameter at time and then to test the possibility of a correlation with variables associated to the subjects (e.g. age, iris colour, sex, nationality…).

Each output involved in the IPER index evaluations comes from measurements or other procedures that can be assimilated to measurements. For example, the answer to a questionnaire is a measure of a subjective sensation or opinion on a psychological scale and an operational output is a measure of the capability to perform a given task.

The three constituents of the *IPER* index are known with a given uncertainty, evaluated following the methodologies describes above.

It is important to evaluate the uncertainty of *IPER* in order to clearly understand weak points in the experiment setup (e.g. if the uncertainty of a component is too high respect to the others) or to give the correct interpretation to numerical values.

As clearly explained in the International Vocabulary of Metrology (VIM), every number in the uncertainty interval has the same probability to represent the true value that if considered unique is, in practice, unknowable. Every comparison of results or deductions considering values inside the uncertainty interval has poor physical meaning.

According to the European Accreditation (EA) guide [27], all the expanded uncertainty specified before considers a rectangular distribution and should be reduced to a normal distribution before starting the uncertainty evaluation.

The coefficient of sensibility c_i where i is the generic parameter in $IPER$ is:

$$c_{RPH_r} = \frac{RPH_r - RPH_t}{\left(RPH_{max} - RPH_{min}\right)^2 \sqrt{\frac{\left(RPH_r - RPH_t\right)^2}{\left(RPH_{max} - RPH_{min}\right)^2} + \frac{\left(ROP_r - ROP_t\right)^2}{\left(ROP_{max} - ROP_{min}\right)^2} + \frac{\left(RPS_r - RPS_t\right)^2}{\left(RPS_{max} - RPS_{min}\right)^2}}}$$

$$c_{RPH_t} = -c_{RPH_r}$$

$$c_{ROP_r} = \frac{ROP_r - ROP_t}{\left(ROP_{max} - ROP_{min}\right)^2 \sqrt{\frac{\left(RPH_r - RPH_t\right)^2}{\left(RPH_{max} - RPH_{min}\right)^2} + \frac{\left(ROP_r - ROP_t\right)^2}{\left(ROP_{max} - ROP_{min}\right)^2} + \frac{\left(RPS_r - RPS_t\right)^2}{\left(RPS_{max} - RPS_{min}\right)^2}}}$$

$$c_{ROP_t} = -c_{ROP_r}$$

$$c_{RPS_r} = \frac{RPS_r - RPS_t}{\left(RPS_{max} - RPS_{min}\right)^2 \sqrt{\frac{\left(RPH_r - RPH_t\right)^2}{\left(RPH_{max} - RPH_{min}\right)^2} + \frac{\left(ROP_r - ROP_t\right)^2}{\left(ROP_{max} - ROP_{min}\right)^2} + \frac{\left(RPS_r - RPS_t\right)^2}{\left(RPS_{max} - RPS_{min}\right)^2}}}$$

$$c_{RPS_t} = -c_{RPS_r}$$

Similar formulas are used for the each component of the IPER index.

5. Conclusions

The field of 'classical' metrology (the one related exclusively to physical quantities) is still formally lacking all the concepts around soft metrology. As a matter of facts, this is still considered a newborn branch of research that need a robust methodological and theoretical architecture to be recognised.

However, due to what said before and described in this chapter, the interest on this topic is growing significantly. This is demonstrated by the recent interest by the fact, for example, many international congresses on metrology are adding specific sessions dedicated to human perception, appearance and in general about the quantification of subjective aspects of human–world interaction. Moreover, the Committee in charge for the new revision of GUM is recently willing to include definitions and models concerning soft metrology.

Acknowledgements

The author developed the ideas presented in this chapter during the 3 years research for the Ph.D. in Metrology. This time has been incredibly challenging and represents the best period of all the time spent at University as a student and as a researcher. The author wants to thanks the entire professors and researchers team with whom the greatest work and ideas exchange have been done.

Author details

Laura Rossi

Address all correspondence to: laura.rossi@altran.com

Altran S.p.A., Bologna, Italy

References

[1] M.R. Pointer, Report to the National Measurement System Directorate. New directions – Soft Metrology, NPL Report CMSC 20/03, 2003.

[2] European Commission, Measuring the Impossible, EUR 22424, European Communities, Bruxelles, 2007.

[3] E. Galanter, K. Hossain, B. Berglund, P. Bourgine, M.C. Morrone, Measuring the Impossible, Report of the MINET High-Level Expert Group, EU NEST, Bruxelles, 2010.

[4] T. Goodman et al., The measurement of naturalness, in: Proceedings of the 12th IMEKO TC1 and TC7 Symposium On Man, Science and Measurement, Annecy, France, September, 3–5, 2008, pp. 31–36.

[5] P. Susini, et al., Closing the loop of sound evaluation and design, in: Proceedings of the 2nd ISCA/DEGA Tutorial & Research Workshop on Perceptual Quality of Systems, Berlin, Germany, September, 4–6, 2006.

[6] B. Berglund, G.B. Rossi, J.T. Townsend, L.R. Pendrill, (Eds), Measurement with Persons – Theory, Methods and Implementation Areas, Taylor & Francis, New York, 2011.

[7] P. Iacomussi, G. Rossi, L. Rossi, A comparison between different light sources induced glare on perceived contrast, *Light & Engineering*, 2012, 20(1).

[8] E.H. Adelson, Checkershadow illusion,2005, Retrieved 2007-04-21. http://web.mit.edu/persci/people/adelson/checkershadow_illusion.html.

[9] J.B. Watson, Psychology as the behaviourist views, *Psychological Review*, 1913, 20, 158-177.

[10] U. Neisser, *Cognitive Psychology*, Prentice-Hall, Englewood Cliffs, NJ, USA, 1967.

[11] G.B. Rossi, Measurement of quantities related to human perception, in: F. Pavese et al. (Eds), Advanced Mathematical and Computational Tools in Metrology and Testing (AMCTM VIII), World Scientific, Singapore, 2009.

[12] ISO IEC Guide 99, *International Vocabulary of Metrology – Basic and General Concepts and Associated Terms*, 3rd edition, ISO, Geneva, 2007.

[13] G.B. Rossi, Cross-disciplinary concepts and terms in measurement, *Measurement*,2009, 42, 1288–1296.

[14] L. Finkelstein, Theory and philosophy of measurement, in: P.H. Sydenham (Ed.) Handbook of Measurement Science, vol. 1: Theoretical Fundamentals, Wiley, Chicester, 1982.

[15] E. Nagel, Measurement, *Erkenntnis*, December 1931,2(1), 313–335, published by Springer, the Netherlands.

[16] G. Iverson, R.D. Luce, The representational measurement approach to psychophysical and judgmental problems, in: M.H. Birnbaum (Ed.), Measurement, Judgment, and Decision Making, Academic Press, New York, 1998.

[17] R.D. Luce, P. Suppes, (2001), Representational measurement theory, in: J. Wixted, H. Pashler (Eds), Stevens' Handbook of Experimental Psychology, 3rd edition, vol 4: Methodology in Experimental Psychology, Wiley, New York, 2001, pp.1-41.

[18] S. Mannarini, *"Psicometria"*, Il Mulino, Manuali, 2003.

[19] F.Pavese et al. (Eds), *Advanced Mathematical and Computational Tools in Metrology and Testing (AMCTM VIII)*, World Scientific, Singapore, 2009.

[20] C. Spence (2012), Auditory contributions to flavour perception and feeding behaviour, *Physiology and Behaviour*, November 5, 2012, 107(4) 505–515.

[21] L. Rossi, A. Schiavi, Oltre la psicoacustica. Gli effetti del rumore sull'uomo nella prospettiva della soft metrology, Rivista Italiana di Acustica, 2012, 36(1), Gennaio-Marzo.

[22] L. Rossi, A. Schiavi, A. Astolfi, Measurement of subjective reaction to noise through performance evaluation, in: Proc. 10th Int. Congr. Noise as a Public Health Problem (ICBEN 2011), July, 24–28, London, 2011.

[23] JCGM 200:2008, International vocabulary of metrology – basic and general concepts and associated terms (VIM3). This document is available on-line on BIPM web site: http://bipm.org.

[24] T. Goodman, Measurement of physical parameters in sensory science, in: B. Berglund, G.B. Rossi, J.T. Townsend, L. Pendrill, Measurement with Persons, Taylor & Francis, New York, 2012.

[25] ISO 11664-4:2008(E)/CIE S 014-4/E:2007 Colorimetry - Part 4: CIE 1976 L*a*b* Colour Space.

[26] C.J.K. Ellis, The pupillary light reflex in normal subjects, *British Journal of Ophthalmology*, 1981,65,754–759.

[27] EA-4/02 • Expression of the Uncertainty of Measurement in Calibration.

7

Uncertainty of Measurement in Medical Laboratories

Paulo Pereira

Abstract

The "Guide to the Expression of Uncertainty in Measurement" (GUM) is not systematically used in medical laboratories, for what the laboratorian should understand the Uncertainty Approach and its importance to recognize the level of realism of results. This chapter presents, discusses, and recommends the models fulfilling GUM principles. An example is given to a single test for an easier understanding of the determination of measurement uncertainty. All the practice uses a freeware. Results with larger measurement uncertainty intervals have a significant probability of being unrealistic, arising a high risk of the uncorrected clinical decision. A flow chart to the selection of models for the determination of measurement uncertainty in a medical laboratory is recommended.

Keywords: bias, bottom-up, GUM, measurement uncertainty, medical metrology, precision, top-down, trueness

1. Introduction to the "Guide to the Expression of Uncertainty in Measurement"

"The lack of consensus in the international scientific community regarding the expression of measurement uncertainty occurred in 1977. Two years later, the Bureau International des Poids et Mesures (BIPM) and 21 laboratories agreed upon it was important to develop an international procedure for expressing measurement uncertainty and for combining individual uncertainty components into a single total uncertainty measurement in chemistry and physics. However, there was no consensus regarding the calculus for expression of measurement uncertainty. In 1980, the BIPM Working Group on the Statement of Uncertainties developed the Recommendation INC-1 "Expression of Experimental Uncertainties" [1], approved in the Comité International des Poids et Mesures (CIPM) in 1981 [2] and reapproved 5 years after [3]. The CIPM suggested to the International Organization for Standardization (ISO) the development of a measurement uncertainty master document based on the Working Group recommendation. The ISO Technical Advisory Group on Metrology (TAG 4) assumed the responsibility to write the guideline. Seven organizations participated in the work of TAG 4: BIPM, International Electrotechnical Commission (IEC), ISO, International Organization of Legal Metrology (OIML), International Union of Pure and Applied Chemistry (IUPAC), International Union of Pure and Applied Physics (IUPAP), and International Federation of Clinical Chemistry and Laboratory Medicine (IFCC), representing the medical laboratory. TAG 4 defined the Working Group 3 (ISO/TAG 4/WG 3) with a committee of experts from BIPM, OIML, IEC, and ISO. The guideline "Guide to the Expression of Uncertainty in Measurement" (GUM) was first published in 1993 and 2 years later was corrected and reprinted. Measurement uncertainty is defined as the "nonnegative parameter characterizing the dispersion of the quantity values being attributed to a measurand, based on the information used" (entry 2.26 of [4]). It characterizes "the quality of a result of a measurement" expressed in uncertainty (quantitative indication). It was not intended to be applied to another quantity than numerical, for what it cannot be a standard to the determination of measurement uncertainty in all estimations. GUM is an open access document as it was republished with minor correction by BIPM in 2008 [5]. The terminology to the Uncertainty Approach is part of the International Vocabulary of Metrology (VIM), which is also freely available from the Bureau [4].

Measurement uncertainty provides information on the level of confidence on the measurement result. One example where the uncertainty is needed is when comparing a result to a clinical decision value. Measurement uncertainty is the quantifiable expression of the doubt related with the outcome. The expanded uncertainty U provides an interval within which the value of the measurand is assumed to be determined by a defined level of confidence. It is the result of multiplying the standard combined uncertainty u_c by a coverage factor k. The choice of the factor k is established on the level of confidence desired. For an approximate level of confidence of 95%, k is usually set to 2, and with a confidence higher than 99%, k is typically set to 3 when degrees of freedom for the combined uncertainty are more than 20.

Commonly, the measurement uncertainty result is expressed as value±expanded uncertainty. For example, in a screening immunoassay, a ratio of 1.0±0.3 (expanded uncertainty k=2) corresponds to the interval 0.7 to 1.3 considering the clinical decision ("cutoff") ratio is equal to 1. It is interpreted as that the measured value±expanded uncertainty covers the "cutoff" value, and the result is in a measurement uncertainty where the result positive or negative cannot be declared (i.e., it is considered indeterminate).

Pereira et al. [6] claimed that a valid uncertainty evaluation should consider:

- "A clear definition of the measurand (i.e., the quantity to be measured),
- A comprehensive specification of the measurement procedure and the measurement objects, and

- A comprehensive analysis of the effects impacting the measurement results. From a laboratory view, the main effects are intermediate precision (within-laboratory precision) and any residual measurement bias."

According to the European Federation of National Associations of Measurement, Testing and Analytical Laboratories (Eurolab) Technical Report 1/2007 [7], there are four main approaches to estimating measurement uncertainty fulfilling Uncertainty Approach principles (see 2.1):

- Modeling,
- Single laboratory validation [including quality control (QC)],
- Interlaboratory comparisons, and
- External Quality Assessment (EQA) [Proficiency Testing (PT)].

Dybkaer [8], past IFCC President and a recognized supporter of GUM approach, argued in a paper that "this International Standard will outline the principles of estimating measurement uncertainty according to GUM (1993, 1995), ways of simplification in routine measurement, and possibilities of reporting. This International Standard applies to:

- Routine medical laboratories wishing or required to provide values with measurement uncertainty;
- Medical laboratories seeking accreditation according to ISO/IEC 17025 (testing) or ISO 15189;
- Organizations providing accreditation; and
- Manufacturers of *in vitro* diagnostic devices wishing to provide guidance on measurement uncertainty in their information documents" [8].

Eurachem/CITAC published in 2000 the "Quantifying Uncertainty in Analytical Measurement" (QUAM; revised for the third time in 2012) intended to be applied uniquely to measurement uncertainty in chemistry [9]. This document takes into account the practical experience of estimation of measurement uncertainty in chemistry laboratories. It emphasizes that procedures introduced by chemistry laboratories to estimate their measurement uncertainty should be integrated with existing quality assurance measurements, as these measures frequently provide much of the information required to evaluate the measurement uncertainty. It provides unequivocally for the use of validation and related data in the estimation of measurement uncertainty in full compliance with GUM principles. Different from GUM, it purpose is for not only modeling approaches ("bottom-up") but also empirical approaches ("top-down"). The "top-down" approaches satisfy the need of chemistry laboratories looking to an alternative to modular models, given that these models are often inapplicable in laboratories' methods such as medical laboratories' methods. Its approach is consistent with the ISO/IEC 17025 and even with ISO 15189 requirements. Because medical laboratories methods are mainly chemical, QUAM is an adequate reference to support the estimation of measurement uncertainty in this field, and it is referred in this Chapter to describe the basic concepts of models to the estimation of measurement uncertainty (see 2.2).

The Nordtest TR 537 "Handbook for Calculation of Measurement Uncertainty in Environmental Laboratories" [10], along with Eurolab TR 1/2007 "Measurement Uncertainty Revisited: Alternative Approaches to Uncertainty Evaluation" [7], proposes "top-down" approaches easily practicable in laboratories, including medical laboratories. Such as Eurachem/CITAC documents these are open access publications. The Finnish Environment Institute (SYKE) released a freeware, MUKit, featuring TR 537 mathematical models [11].

The National Pathology Accreditation Advisory Council (NPAAC) published in 2007 a public document entitled "Requirements for the Estimation of Measurement Uncertainty" to support the Australian accreditation of medical laboratories [12]. This guideline recommends a set of "top-down" models to use in this field.

The Clinical and Laboratory Standards Institute (CLSI)/IFCC C51-A "Expression of Measurement Uncertainty in Laboratory Medicine" was published in January 2012 (the code changed to EP29-A without text revision) [13] and it is the only global guideline for the determination of measurement uncertainty. The CLSI EP29 Working Group is chaired by Anders Kalnner, another recognized GUM expert in medical laboratories. Its application to medical laboratories is not consensual [14, 15]. According to this guideline, it "describes the principles of estimating measurement uncertainty and guides clinical laboratories and *in vitro* diagnostic device manufacturers on the specific issues to be considered for implementation of the concept in medical laboratory. This document illustrates the assessment of measurement uncertainty with both 'bottom-up' and 'top-down' approaches. The 'bottom-up' approach suggests that all possible sources of uncertainty are identified and quantified in an uncertainty budget. A combined uncertainty is calculated using statistical propagation rules. The 'top-down' approach directly estimates the measurement uncertainty results produced by a measuring system. The methods to estimate the precision and bias are presented theoretically and in worked examples."

The ISO/PDTS 25680 "Medical Laboratories—Calculation and Expression of Measurement Uncertainty" [16] was written initially as an International Standard and later as a Technical Specification and was canceled on June 2011. The contents of the last draft were analogous to the CLSI C51-P.

Currently, the determination of measurement uncertainty is required in ISO/IEC 17025 and ISO 15189 [17]. It is mandatory in Australia, Latvia, and France after November 1, 2016.

The laboratorian should understand that several uncertainty components are immeasurable in Uncertainty Approach, for example, diagnostic uncertainty in tests with binary results (positive/negative). For further details about other sources of uncertainty, please refer to [18], and for diagnostic uncertainty, please refer to [19].

This Chapter presents a discussion of the estimation models of measurement uncertainty and its determination in a single medical laboratory test. Preferably, the reader should have basic statistical skills to understand the concepts and mathematical models. There are cited several references that should be revised for a deeper understanding of the Uncertainty Approach and also to review the examples of estimation in other medical laboratory tests.

2. Principles of GUM

2.1. Principles of uncertainty approach

GUM is recognized as the master document on measurement uncertainty throughout the field of metrology, for what it is also called "uncertainty bible." The measurement uncertainty evaluation is recognized to be applied to all types of quantitative assay results in physics and chemistry. The theoretical modeling follows the Uncertainty Approach published in 1980. The principles are as follows:

a. "The uncertainty in the result of a measurement generally consists of several components, which may be grouped into two categories according to the way in which their numerical value is estimated:

Those which are evaluated by statistical methods and

Those which are evaluated by other means.

There is not always a simple correspondence between the classification into categories A or B and the previously used classifications into "random" and "systematic" uncertainties. The term "systematic uncertainty" can be misleading and should be avoided.

Any detailed report of the uncertainty should consist of a complete list of the components, specifying for each the method used to obtain its numerical value.

b. The components in category A are characterized by the estimated variances (or the estimated standard deviation s_i) and the number of degrees of freedom v_i. Where appropriate, the covariances should be given.

c. The components in category B should be characterized by quantities (, which may be considered as approximations to the corresponding variances, the existence of which is assumed. The quantities may be treated like variances and the quantities u_j like standard deviations. Where appropriate, the covariance should be treated in similar ways.

d. The combined uncertainty should be characterized by the numerical value obtained by applying the usual method for the combination of variances. The combined uncertainty and its components should be expressed in the form of "standard deviations."

e. If, for particular applications, it is necessary to multiply the combined uncertainty by a factor to obtain an overall uncertainty, the multiplying factor used must always be stated [1].

These principles provide a methodology where:

- Uncertainty evaluation should combine the major sources of uncertainty;
- Uncertainties of random and systematic error components should be treated identically;
- Major contributions coming from Type A and Type B errors should be identified and considered as standard uncertainty; and

- Uncertainty components are expressed as combined standard uncertainty or by the result of combined standard uncertainty multiplied by a defined coverage factor, designated expanded uncertainty [7].

The approach is based on a model designed to consider the interrelation of all sources of uncertainty that significantly affect the measurand. All known systematic errors must be corrected if significant, and they are not included in the calculations. All the principal sources are combined according to the laws of propagation of uncertainty. The calculated result is the combined uncertainty±value associated with a quantitative measured value. The uncertainty budget expresses all measurement sources as well as the combined calculus. A Pareto chart can be used to display visually and compare different sources and their weight on combined uncertainty. This step requires the medical laboratory to define a mathematical model, which could require the competencies of a mathematician or a statistician with special training in measurement uncertainty. The combined uncertainty is multiplied by a Student's t value and its result, referred as "expanded uncertainty," is expressed by a confidence interval (e.g., k=2 to a confidence interval equal to 95%) [20, 21]. For further details about principles of Uncertainty Approach, please refer to [5].

2.2. Fundamental concepts of models to the estimation of measurement uncertainty

2.2.1. *Stochastic mathematical equation of the measurement method*

The equation should assure the stochasticity of the reaction. Nonstochastic equations cause an incorrect estimate of measurement uncertainty. This is a complex task when designing a modeling approaches, requiring principally in chemical expertise in medical laboratories' tests. It is uniquely applied to modeling approaches.

2.2.2. *Distribution type in the estimation of standard deviations*

For each uncertain variable, the probable results with a probability distribution should be defined. The conditions contiguous to the variable determine the category of certain distribution. The categories consist of the following:

a. Discreet

- Bernoulli (i.e., success/failure in a single experiment);
- Binomial (i.e., the number of success/failure in some experiments);
- Multi-nominal (i.e., the frequency registered a certain result in some experiments); and
- Poisson (i.e., number of events of a certain occurrence from a period zero to a period t).

b. Continuous variables: exponential and normal.

Two types of estimations could be made: Type A or Type B. These different estimates are merged in the combined standard uncertainty. Before the combination, all uncertainty sources must be expressed as standard uncertainties (i.e., as relative standard deviations). The standard deviation shall be measured according to data distribution, which could not be a normal distribution due to the number of different uncertainty sources that could include not only chemical but also physics. The following *rules* for distribution could be applied to most of the uncertainty sources in medical laboratory:

Rule 1 for distribution: Normal distribution with known *n*

"Where the uncertainty component was evaluated experimentally from the dispersion of repeated measurements, it can readily be expressed as a standard deviation. For the contribution to uncertainty in single measurements, the standard uncertainty is simply the observed standard deviation; for results subjected to averaging, the standard deviation of the mean is used."

The standard deviation of the mean of *n* values taken from a population is given by *n*:

$$u(x_i) = s_{\bar{x}} = s / \sqrt{n} \tag{1}$$

Rule 2 for distribution: Normal distribution with known confidence interval

"Where an uncertainty estimate is derived from previous results and data, it may already be expressed as a standard deviation. However, where a confidence interval is given with a level of confidence *p*% (in the form ±*a* at *p*%), then divide the value *a* by the appropriate percentage point of the normal distribution for the level of confidence given to calculate the standard deviation."

$$u(x_i) = a / (t - value) \tag{2}$$

Rule 3 for distribution: Rectangular distribution

"If limits of ±*a* are given without a confidence level and there is reason to expect that extreme values are likely, it is normally appropriate to assume a rectangular distribution, with a standard deviation of"

$$u(x_i) = a / \sqrt{3} \tag{3}$$

Rule 4 for distribution: Triangular distribution

"If limits of ±*a* are given without a confidence level, but there is reason to expect that extreme values are unlikely, it is normally appropriate to assume a triangular distribution, with a standard deviation of"

$$u(x_i) = a / \sqrt{6} \tag{4}$$

Rule 5 for distribution: Estimation made by judgment

"Where an estimate is to be made on the basis of judgment, it may be possible to estimate the component directly as a standard deviation. If this is not possible, then an estimate should be made of the maximum deviation that could reasonably occur in practice (excluding simple mistakes). If a smaller value is considered substantially more likely, this estimate should be treated as descriptive of a triangular distribution. If there are no grounds for believing that a small error is more likely than a large error, the estimate should be treated as characterizing a rectangular distribution" (entry 8.1 of [9]).

2.2.3. *Combined standard uncertainty according to the law of the propagation of uncertainty*

VIM defines combined standard uncertainty as the "standard measurement uncertainty that is obtained using the individual standard measurement uncertainties associated with the input quantities in a measurement model. Note: In case of correlations of input quantities in a measurement model, covariances must also be taken into account when calculating the combined standard measurement uncertainty (…)" (entry 2.31 of [4]).

Combined uncertainty is the result of calculus according to the law of the propagation of uncertainty, which combines the sources of uncertainty in a single value of uncertainty. The law of the propagation of uncertainty is derived based on the Taylor series expansion of a functional relationship regularly used in differential calculus. A measurement process can be modeled mathematically using the function:

$$y = f(x_1, x_2, \dots, x_n) \tag{5}$$

When sources are not correlated (independent variables), the calculus is done according to the variance rules; when sources are correlated (dependent variables, share of variance), the calculus is done according to covariance [(entry Chapter 8 of [20], 22].

2.2.4. *Independent variables*

"The general relationship between the combined standard uncertainty $u_c(y)$ of a value y and the uncertainty of the independent parameters $x_1, x_2, \dots, x_n$ on which it depends on is

$$u_c(y) = (y(x_1, x_2, \dots)) = \sqrt{\sum_{i=1,n} c_i^2 u(x_i)^2} = \sqrt{\sum_{i=1,n} u(y, x_i)^2} \tag{6}$$

where $y(x_1, x_2, \dots)$ is a function of several parameters, $x_1, x_2, \dots$, c_i is a sensitivity coefficient evaluated as $c_i = \partial y / \partial x_i$, the partial differential of y on x_i, and $u(y, x_i)$ denotes the uncertainty in y arising from the uncertainty in x_i. Each variable's contribution $u(y, x_i)$ is just the square of the associated uncertainty expressed as a standard deviation multiplied by the square of the relevant sensitivity coefficient. These sensitivity coefficients describe how the value of y varies with changes in the parameters x_1, x_2, etc." (entry 8.2.2 of [9]).

2.2.5. *Dependent variables*

"Where variables are not independent, the relationship is more complex:

$$u\left(y\left(x_{i,j\ldots}\right)\right)=\sqrt{\sum_{i=1,n} c_i^2 u\left(x_i\right)^2+\sum_{\substack{i,k=1,n \\ i \neq k}} c_1 c_k u\left(x_i, x_k\right)} \tag{7}$$

where $u(x_i, x_k)$ is the covariance between x_i and x_k, and c_i and c_k are the sensitivity coefficients (…). The covariance is related to the correlation coefficient r_{ik} by

$$u(x_i, x_k)=u\left(x_i\right) u\left(x_k\right) r_{ik} \tag{8}$$

where $-1 \leq r_{\mathrm{ik}} \leq 1$" (entry 8.2.3 of [9]).

QUAM recommends the Kragten spreadsheet to the determination of measurement uncertainty [23].

2.2.6. *Simpler expressions for independent variables*

It is possible to avoid the use of complex calculus and to use simpler expressions taken from the expression for the general relationship of variables. Two rules are proposed [root-sum-of-squares (RSS)]:

Rule 1 for combination of uncertainty: Sum or difference

"For models involving only a sum or difference of quantities [e.g., $y=(p+q+r+\ldots)$], the combined standard uncertainty $u_c(y)$ is given by

$$u_{\mathrm{c}}\left(y\left(p, q \ldots\right)\right)=\sqrt{u\left(p\right)^2+u\left(q\right)^2+\ldots} \tag{9}$$

Rule 2 for combination of uncertainty: Product or quotient

"For models involving only a product or quotient [e.g., $y=(p \cdot q \cdot r \cdot \ldots)$] or

$$y=p /\left(q \cdot r \cdot \ldots\right) \tag{10}$$

The combined standard uncertainty $u_c(y)$ is given by

$$u_c(y) = y\sqrt{\left(u(p)/p\right)^2 + \left(u(q)/q\right)^2 + \ldots} \tag{11}$$

where $u(p)/p$, etc., are the uncertainties in the parameters, expressed as relative standard deviations. Note: Subtraction is treated in the same manner as addition, and division in the same way as multiplication" (entry 8.2.6 of [9]).

Theoretically, because RSS does not take into account the partial derivatives, it results in an inaccurate uncertainty result. However, the inaccuracy of determination is usually considered nonsignificant to the estimated result.

2.2.7. *Expanded uncertainty*

VIM defines expanded uncertainty as the "product of a combined standard measurement uncertainty and a factor larger than the number 1. Notes: (1) The factor depends on the type of probability distribution of the output quantity in a measurement model and on the selected coverage probability. (2) The term "factor" in this definition refers to a coverage factor. (3) Expanded measurement uncertainty is termed "overall uncertainty" in paragraph 5 of Recommendation INC-1 (1980) (see the GUM) and simply "uncertainty" in IEC documents" (entry 2.35 of [4]).

The final step in the evaluation of measurement uncertainty is the calculus of expanded uncertainty. Its purpose is the designation of an interval that may be expected to include a large fraction of the distribution of values, which could reasonably be attributed to the measurand. Its calculus is according to the formula:

$$U = k \cdot u \tag{12}$$

where k is a coverage factor according to the type of probability distribution and u is the combined uncertainty. The choice of k value should be done according to factors such as the level of confidence required, any knowledge of the underlying distributions, or any knowledge of the number of values used to estimate random effects.

The value for k in a medical laboratory usually is taken from a one- or two-tailed normal distribution for Student's t. When the effective degrees of freedom v_{eff} are higher than about 6, usually k is equal to 2, which correspond to 95% confidence; when v_{eff} are less than about 6, they shall be defined. The European Co-Operation for Accreditation (EA) recommends in EA4-02 "Expression of the Uncertainty of Measurement in Calibration," a formula for v_{eff} calculus [24]:

"Estimate the effective degrees of freedom v_{eff} of the standard uncertainty $u(y)$ associated with the output estimate y from the Welch-Satterthwaite formula:

$$v_{eff} = \left(u^4(y)\right) / \left(\sum_{i=1}^{N} u_i^4(y) / v_i\right) \tag{13}$$

where $u_i(y)$ (i=1, 2, …, N), defined in the equation, are the contributions to the standard uncertainty associated with the output estimate y resulting from the standard uncertainty associated with the input estimate x_i, which are assumed to be mutually statistically independent, and v_i is the effective degrees of freedom of the standard uncertainty contribution $u_i(y)$."

When a standard uncertainty is measured requiring Type A evaluation, the degrees of freedom v_i are measured according to a simpler formula:

$$v_i = n - 1 \tag{14}$$

The table with k values set from a t distribution evaluated for a coverage probability of 95% to be used for Type A evaluation is featured in (entry Annex G of [5]). If v_{eff} is not an integer, which is usually the case, truncate v_{eff} to the next lower integer. When a standard uncertainty is measured requiring Type B evaluation, the calculus is more complex. The common practice is to carry out such assessments in a mode that guarantees that any underestimation is avoided. Comsidering this practice is followed, the degrees of freedom of the standard uncertainty $u(x_i)$ acquired from a Type B evaluation may be taken to be . The estimation of k could be easily done in Microsoft® Excel® using the function=TINV(probability;deg_freedom) (note: for a 95% confidence, the probability is equal to the difference between 1 and 0.95). For further details about v_{eff} and levels of confidence in measurement uncertainty, please refer to (entry Annex G of [5]).

2.2.8. Reporting the measurement uncertainty

a. The report

The measurement uncertainty report should include:

- "A description of the methods used to calculate the measurement result and its uncertainty from the experimental observations and input data;
- The values and sources of all corrections and constants used in both the calculation and the uncertainty analysis; and
- A list of all the components of uncertainty with full documentation on how each was evaluated."

Note: These elements could be also referred to documented sources.

The data and analysis should be presented in such a way that its important steps can be readily followed and the calculation of the result repeated if necessary (…)."

(…) "When reporting the results of routine analysis, it may be sufficient to state only the value of the expanded uncertainty and the value of *k*" (entry 9.2 of [9]).

The combined standard uncertainty or expanded uncertainty is reported. The majority of assay and calibration certificates describe expanded uncertainty. QUAM proposes the following format for reporting:

Combined standard uncertainty: "(Result): x (units) [with a] standard uncertainty of u_c (units) [where standard uncertainty is as defined in the ISO/IEC "Guide to the Expression of Uncertainty" and corresponds to 1 standard deviation]." Note: The use of the symbol ± is not recommended when using standard uncertainty as the symbol is commonly associated with intervals corresponding to high levels of confidence. Terms in parentheses [] may be omitted or abbreviated as appropriate" (entry 9.3 of [9]).

Expanded uncertainty: "Unless otherwise required, the result x should be stated together with the expanded uncertainty U calculated using a coverage factor k=2 (…). The following form is recommended: "(Result): (x} U) (units) [where] the reported uncertainty is [an expanded uncertainty as defined in the International Vocabulary of Basic and General terms in Metrology, 2nd ed., ISO 1993 (note: this Chapter suggests the use the present VIM edition referred in [4]), calculated using a coverage factor of 2 [which gives a level of confidence of approximately 95%]." Terms in parentheses [] may be omitted or abbreviated as appropriate. The coverage factor should, of course, be adjusted to show the value actually used" (entry 9.4 of [9]).

The customers and consumers of this information must have the skills to understand the purpose of measurement uncertainty and to take appropriate actions based on this evaluation. Such skills are often lacking for the customers and consumers of laboratory test (physicians, patients, and agencies).

b. Reporting in medical laboratory scope

The information required to report the result of measurement depends on its intended use. In the medical laboratory, the final consumer is the patient, blood donor, or other, who is not responsible for the diagnosis and follow-up. The primary customer is the physician or someone else with the responsibility for the technical action (screening, diagnosis, follow-up, or other). The personnel who takes a decision on the result must understand the purpose of measurement uncertainty and its value for the judgment. If this does not happen, reports of measurement uncertainty could generate doubts that could compromise the clinical decision. These skills are rare in physicians or other healthcare professionals. It is not requested by the physician who probably does not understand its concept. Therefore, the majority of hospital laboratories do not report measurement uncertainty because it does not add value to clinical decisions and may only cause indecision. The report must be fit for its purpose [(entry 9.4 of [9]), 25]. **Figure 1** summarizes the stages of measurement uncertainty methodology.

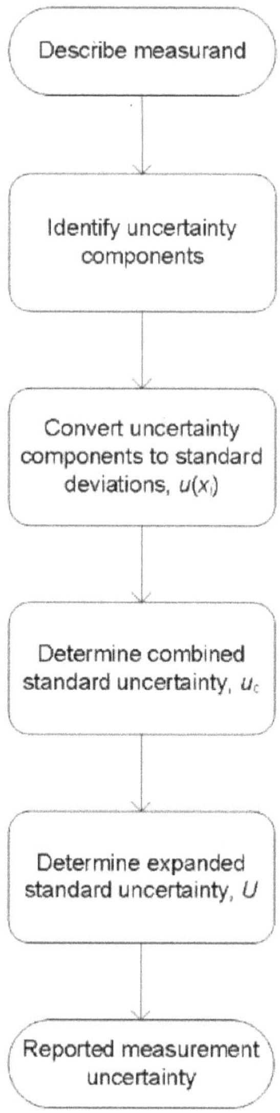

Figure 1. Methodology to determine and report measurement uncertainty.

2.2.9. Metrological traceability

Metrological traceability is defined as the "property of a measurement result whereby the result can be related to a reference through a documented unbroken chain of calibrations, each contributing to the measurement uncertainty" (entry 2.41 of [4]). Therefore, measurement uncertainty result should be metrologically traceable, assuring the comparability of outcomes in a metrological traceability chain, which is defined as the "sequence of measurement standards and calibrations that are used to relate a measurement result to a reference" (entry 2.42 of [4]). **Figure 2** shows an example of a metrological chain for a medical laboratory test. The measurement uncertainties and bias are determined according to the metrological traceability chain. The accuracy, level of accreditation laboratory, instability, and cost per material increase significantly from medical laboratories to the top of the hierarchy. On the contrary, the measurement uncertainty, bias, and availability of materials decrease from bottom to top.

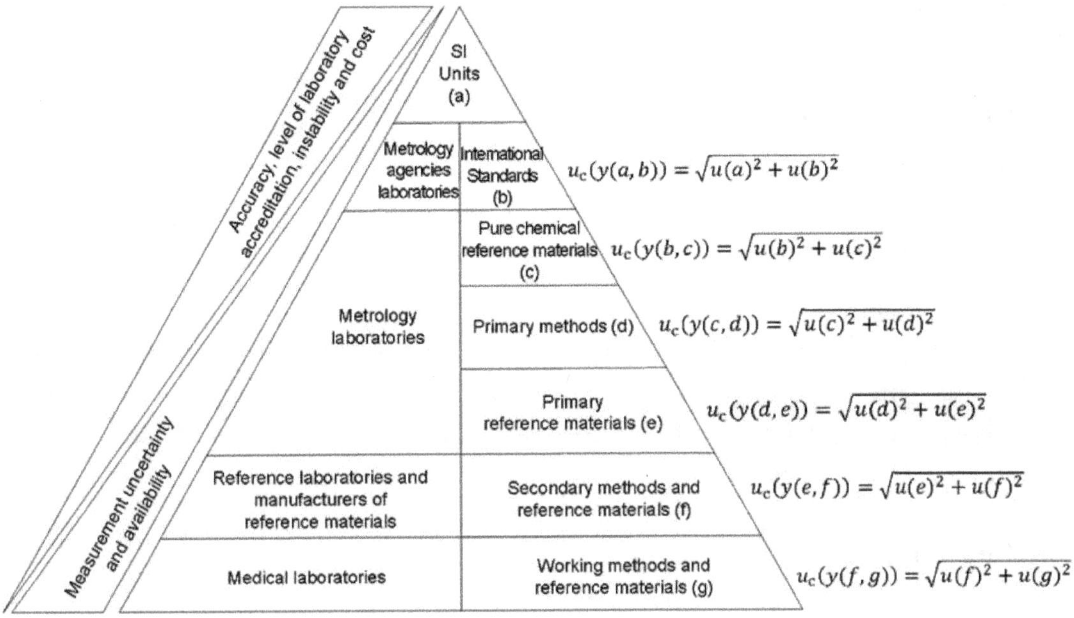

Figure 2. A metrological traceability chain of measurement results in a medical laboratory test.

Although properly implemented in general metrology, it is not widely employed in most of medical laboratory tests due to the unavailability of reference materials and reference methods. Also, the "medical traceability" is hard to achieve due to the "physicochemical complexity" of human samples caused principally by the within-individual and interindividual biological variation [26]. For further details about metrological traceability, please refer to [27].

2.2.10. *Effect of bias in measurement uncertainty*

Significant systematic effects should be corrected (entry Annex D of [5]). Where bias is statistically significant and is uncorrected, it should be reported with the measurement uncertainty (entry 7.16 of [9]).

Dybkaer [8] claimed that minimizing a diagnostic misclassification required that "trueness obtained through metrological traceability based on a calibration hierarchy." On the article, he argued that the reduction of bias should happen according to seven approaches:

- "The type of quantity that is to be measured must be defined sufficiently well. This is particularly demanding when analyte isomorphs or speciation are involved.
- The principle and method of measurement must be carefully selected for analytical specificity.
- A practicable measurement procedure including sampling must be exhaustively described.
- A calibration hierarchy must be defined to allow metrological traceability, preferably to a unit of the International System of Units (SI). Traceability involves plugging into a reference measurement system of reference procedures and commutable calibration materials.
- An internal QC (IQC) system must be devised to reveal increases in bias.

- Any correction procedures must be defined and validated.
- Where possible, there should be participation in EQA (PT) using material with reference measurement values" [28].

Magnusson and Ellison [29] published a paper on the treatment of uncorrected measurement bias when determining measurement uncertainty, suggesting a correction process and determination of uncertainty for corrected results.

In some medical laboratory fields, such as virology, some biases cannot be corrected as seronegative "window period," verification bias, spectrum bias, and length bias. Therefore, estimates of measurement uncertainty may not adequately describe the variability that is observed (entry 1.2.5 of [30]). For further details about dealing with bias in measurement uncertainty, please refer to [29].

2.3. Modeling approach

2.3.1. Fundamental concepts

This model is usually expressed in an equation accounting for the interrelation of the input quantities affecting the measurand pooled essentially with empirical data. Commonly, it is used a combination of different approaches to determine measurement uncertainty. The model incorporates a correction to minimize the effect of identified systematic error components. The law of propagation of uncertainty or propagation of distributions by Monte Carlo simulation method allows the computing of the combined standard uncertainty u_c of the result. The combined standard uncertainty is determined, and factor k is selected according to the chosen confidence level to determine expanded uncertainty. The partial derivative method requires a more complex model equation including partial derivatives of standard uncertainties. For further details about the modeling approach for the evaluation of uncertainty, please refer to (entry Chapter 8 of [5]).

2.3.2. Identification of sources of uncertainty

a. Cause-and-effect diagram

The cause-and-effect diagram is valuable when the mathematical model is being formulated. It describes different groups or sources that may cause statistically significant contributions to measurement uncertainty (important sources of uncertainty), for example, sampling, storage conditions, instrument effects, reagent purity, assumed reaction stoichiometry, measurement conditions, sample effects, computational effects, blank correction, operator effects, and random effects (entry 6.7 of [9], entry Stage IV of [31]).

The groups of causes diverge in each assay, and the subcauses could be different even for the same commercial assays in different laboratories (i.e., some subcauses could cause significant uncertainty in one medical laboratory, and they should be expressed in mathematical module but, in a different medical laboratory, may not cause significant uncertainty and should be not expressed). This diagram should be carefully formulated for each measurement uncertainty evaluation, and it could be a complex task. The diagram should not include bias measurement because it is not allowed in the calculation of measurement uncertainty, but when bias is statistically significant the contribution of bias to uncertainty (bias uncertainty) should be included. It should be clear that the mathematical model must be carefully formulated to include important causes but also avoid duplications of sources, which would increase estimates of uncertainty. The skills to formulate a correct mathematical model are not common in medical laboratory scientists or researchers. **Figure 3** shows an example of a cause-and-effect diagram that identifies known significant sources of uncertainty for a certain test statistically. This diagram describes the groups of sources (or standard uncertainty intended for measurement of combined standard uncertainty) as well as the causes of those sources. The result has a measurement uncertainty associated. The groups of causes (and subcauses) that contribute to this measurement uncertainty were defined after a depth research that included equipment process (workflow), service manual, reagents literature, assay development guidelines, manufacturer perspective of significant uncertainty causes, and scientific journals (entry Stage IV of [31]).

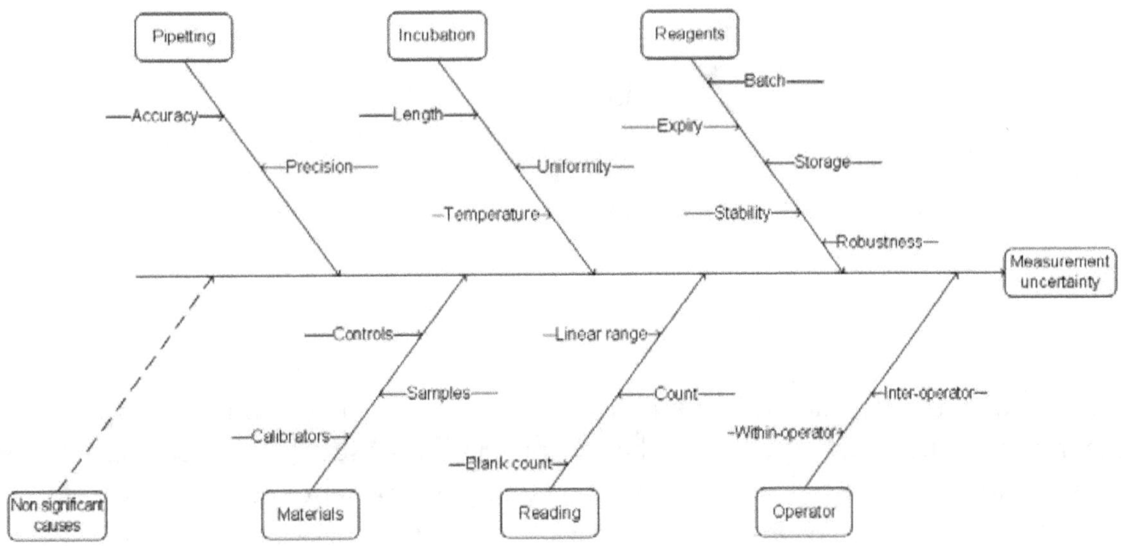

Figure 3. A cause-and-effect diagram for the determination of measurement uncertainty in a hypothetical medical laboratory test.

b. Pareto diagram

The purpose of a Pareto diagram (bar diagram) is to highlight the critical sources of uncertainty, recognized as those with a significant contribution to measurement uncertainty. Usually, it is linked to the cause-and-effect diagram, allowing an easy detection of the primary sources of uncertainty. Typically, it is considered that approximately 80% of the effects come from 20% of the causes (Pareto principle also known as the "80-20 rule"), illustrating that a small part of the causes has the significant contribution to the effect. **Figure 4** shows an example of a Pareto diagram showing the "operator," "material," and "reagents" as the components with a significant contribution to the measurement uncertainty of a certain test, contributing to 92%

of the uncertainty. When it is using a modular approach, the selection of the uncertainty components that will be combined according to a model equation should be selected according to this principle. In this condition, the unmeasured components are statistically considered nonsignificant. Thompson and Ellison [32] claimed that the mostly determined measurement uncertainty was statistically significantly lower when compared to the standard deviation in reproducibility conditions, indicating a lower estimation of the uncertainty. The uncertainty not recognized was invoked to as "dark uncertainty" [32]. The mathematical models of empirical models already consider this principle, for what the Pareto diagram is useless for the laboratorian using these models.

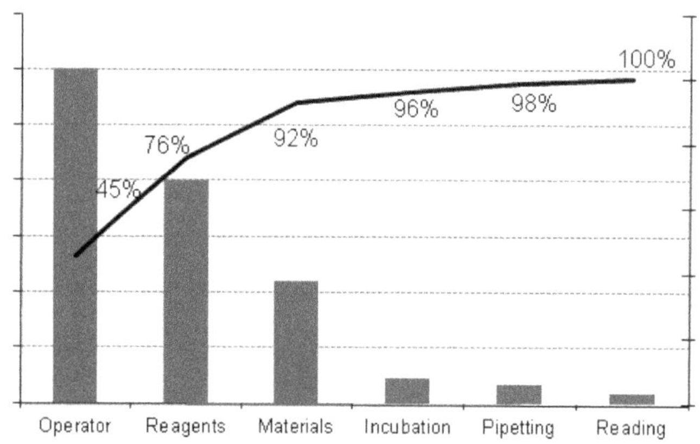

Figure 4. A Pareto diagram for the determination of measurement uncertainty in a hypothetical medical laboratory test.

2.3.3. Partial derivative method

The partial derivative method was used during the years close to GUM publication, for what is recognized as the "GUM method." The laboratorian must identify the critical sources of uncertainty and combine in a stochastic model equation. This required depth skills in chemistry, mathematics, and statistics, which are rarely available in medical laboratories, represent a high cost. Usually, the output of the modeling approach is an "uncertainty budget" summarizing the determination of the combined standard uncertainty and the uncertainty components. Pareto diagrams are also used to compare input values with the combined standard uncertainty. Usually, model equations are not developed in medical laboratories but in a reagent manufacturer's laboratory during the test research and development (R&D). Even when medical laboratories are using "in-house" tests with published model equations, generally it is not practical for the determination of measurement uncertainty because, at least, it will require staff with advanced statistical skills of Uncertainty Approach, which is also not common in medical laboratories. It is easy to recognize the role of this approach in the manufacturer, because it is possible to identify and control the major sources of uncertainty, allowing the manufacturer to reduce certain uncertainty sources. This allows the new test to have a smaller measurement uncertainty (i.e., a high probability of the results cannot be significantly different from *in vivo* results). However, this role is not intended to a medical laboratory, where the major uncertainty sources are associated principally with good laboratory practices (e.g., training of staff and storage conditions). The manufacturer considers this a useful method for assessing the influence of reference value uncertainties to the pooled uncertainty related to the final result of measurement.

Some other well-known disadvantages should be considered in the decision to use this method and in the interpretation of measurement uncertainty result. Preanalytical sources are not usually considered, including biological sources, which causes a misestimation of measurement uncertainty. In medical laboratory tests, measurement uncertainty is composed uniquely of analytical components. This is a critical lack of the accuracy of the determination. The application of inadequate model equation could lead to misestimation or overestimation; for example, unsuspected covariance could give rise to an overestimation of measurement uncertainty. Also, the statistical distribution might not be the same distribution than the one associated with method, which makes the evaluation inaccurate. The complex calculation of partial derivatives is time-consuming and expensive, requiring skilled statisticians. For further details about the partial derivative method, please refer to [5].

2.3.4. Propagation of distributions by Monte Carlo simulation method

The propagation of distribution method is an alternative to partial derivate method, and it is expected to be recommend in the next revision of GUM as the primary option when a modeling approach is considered. Currently, GUM proposes the propagation of uncertainties through the mathematical model measurement. The propagation of distribution substitutes the propagation of uncertainties method, correcting some obvious lacks of the propagation of uncertainties methods such as the linearity of the model and the normal distribution of the random variable representing the possible values of the measurand. This limitation is associated with the application of the partial derivative method in some tests. The propagation of distribution is determined commonly using Monte Carlo simulation method (Monte Carlo). BIPM published in 2008 a Supplement 1 to the GUM "Propagation of Distributions Using a Monte Carlo Method" [33], considering a correct application of Monte Carlo in the determination and evaluation of uncertainties. This method does not request a complex calculation of partial derivatives different from the partial derivative method. Monte Carlo simulation involves a random sampling of a probability distribution.

Contrary to the partial derivative method, Monte Carlo can be used in linear and nonlinear equations. It is used to simulate measurement results based on the input quantities of the probability density function (PDF). PDF describes the relative likelihood for a continuous random variable to take on a given value. Monte Carlo produces the propagation of PDF of the input through the mathematical model measurement, providing results in a PDF describing the values of the measurand consistent with the available data. Nonlinear model equations, asymmetric distributions, and other problems to the partial derivative method are not significant. Another difference to the propagation of uncertainties is the use of Welch-Satterthwaite formula in the estimation of expanded uncertainty, which is unnecessary. The limits of a symmetrical range of coverage are estimated by the values of the 2.5th and 97.5th percentiles for p=95%. An extreme asymmetry could indicate the need to increase the M. Monte

Carlo is not adequate when the output distribution is not symmetrical. The confidence of the results is based on the demonstration of the model equation and the number of simulated measurements.

The effect of the number of simulated measurements M in the sampling error for estimates is major. Supplement 1 to the GUM request at least 10^5 repetitions/trials of the model equation to achieve a statistically acceptable result. At present, it is easy to obtain results with higher M, such as 10^6, due to the available software and hardware. A simple test could be performed in Microsoft® Excel® 2007 or later release using =NORMINV(RAND();mean;standard_dev) function and histograms. The reader will observe that M=10 graph does not show a bell-shaped curve (see Figure 5, histogram A), simulations with M≥10 show a bell-shaped curve (see **Figure 5**, histograms A–F), and M=10^5 (see **Figure 5**, histogram E) and M=10^6 (see Figure 5, histogram F) bell-shaped curves are closer from what is represented in the infinite number of samples. This task was not possible for most of the users a decade ago. Model equations with unreliable quantities can produce a sampling error that cannot be reduced increasing M. Monte Carlo estimates are considered reasonably accurate when repeated simulations deliver values of $u_c(y)$ that do not diverge from each other in the second significant number. Some other known limitations of Monte Carlo are that the calculated uncertainties vary from one run to the next because of the intentionally random nature of the simulation and that is hard to identify the most significant contributions to the combined uncertainty without repeating the simulation.

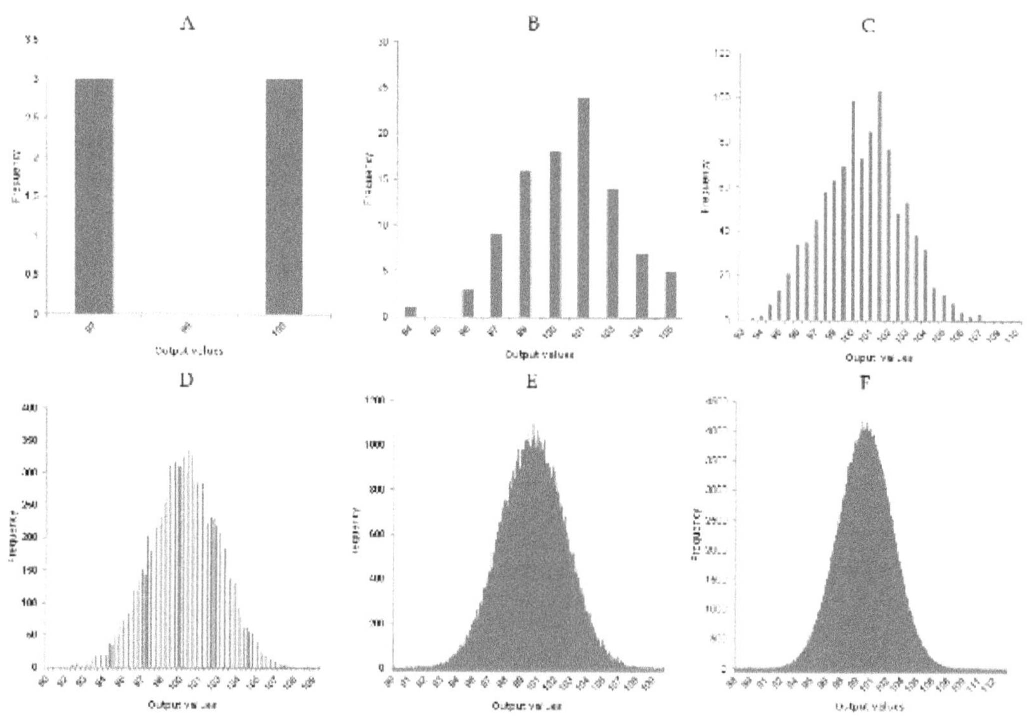

Figure 5. Histograms from six simulations with data normally distributed: (A) 10, (B) 10^2, (C) 10^3, (D) 10^4, (E) 10^5, and (F) 10^6 trials.

Because, in the case of most medical laboratory tests, any model equation to describe the measurement is not known, the modeling approach cannot be applied in this case using partial derivative or Monte Carlo for what their role should be principally at the reagent manufacturer level. For further details about propagation of distributions by Monte Carlo simulation method, please refer to [33].

2.4. Empirical approaches

2.4.1. *Single laboratory validation (including QC)*

Single laboratory validation and QC, the major components of variance, can regularly be calculated by an "in-house" validation study pooled with internal QC using repeated determinations of reliable control samples. The evaluation of bias uncertainty is commonly done using a certified reference material (CRM) or comparing using a reference method. The laboratorian easily associates this approach to a method evaluation model. For further details about single laboratory validation for the evaluation of measurement uncertainty, please refer to Nordtest TR 537 [10] and ISO 11352 [34]. These technical guidelines use general approaches that can also be applied to results of medical laboratories' tests.

The within-laboratory reproducibility uncertainty s_{Rw} is calculated by pooling the repeatability standard deviation s_{r} arising from replicate measurements of human samples and the intermediate standard deviation s_{I} from between runs as in Equation (15) (entry 4 of [10]).

$$s_{\mathrm{Rw}} = \sqrt{s_{\mathrm{r}}^2 + s_{\mathrm{I}}^2} \tag{15}$$

Equation (16) (entry 5 of [10]) is used to compute the bias uncertainty u_b if a CRM is available. Bias b is the result of the mean deviation of measurement results of replicates from the corresponding reference value; s_b is the bias standard deviation, is the reference value standard uncertainty, and m is the number of replicate determinations:

$$u_b = \sqrt{b^2 + \left(s_b / \sqrt{m}\right)^2 + u_{C_{\mathrm{ref}}}^2} \tag{16}$$

To obtain the combined standard uncertainty, the uncertainty due to precision and that due to bias are combined as in Equation (17) (entry 1.2.2 of [7]):

$$u_{\mathrm{c}}(y) = \sqrt{s_{\mathrm{Rw}}^2 + u_b^2} \tag{17}$$

This approach considers the within-laboratory reproducibility standard deviation according to two different methods:

(a) Method validation protocol intended for validating the precision of numerical quantity tests. It is recommended for most of the quantitative tests—the approach described in CLSI EP15-A3 protocol to evaluate the precision (and bias) [34]. A series of five analytical runs with three replicates per run is suitable. For further details about precision evaluation, please refer to [(entry Chapter 2 of [35]), (entry 1.2.2 of [7])].

(b) Using data from between-run variation. The data arise from long-term determination, usually using IQC data. For some tests, such as IQC sample's batches, the expiration date is only 1 month; for others, the expiration could be 1 year. Data coming from batches using a larger period are preferable, when available. The statistical power of the determination is critically influenced by this period. IQC should also be chosen when the mobile mean and variance are stable (test in control). For further details about precision evaluation, please refer to (entry 1.2.2 of [7]).

2.4.2. Interlaboratory comparisons

In the interlaboratory validation approach, the principal sources of variability can often be assessed by interlaboratory studies performed and evaluated according to ISO 5725 [36]. This approach to estimating uncertainty is fully described in ISO 21748 [37].

The approach requires the determination of the between-laboratories reproducibility standard deviation s_R from the results of an interlaboratory trial according to ISO 5725. In a standardized method, these precision data are usually given in an Appendix. For further details about interlaboratory comparisons approach, please refer to (entry 1.2.3 of [7]).

2.4.3. EQA (PT)

EQA programs are proposed to verify periodically the performance of a laboratory test based on data of a laboratory group using proficiency tests [36]. Medical laboratory could also use the comparison data to determine the measurement uncertainty. EuroLab Technical Report 1/2007 [7] presents an approach to laboratories to evaluate measurement uncertainty. Accordingly, the standard uncertainty measured from the results of the group's participants could to a common test be treated as an early evaluation of the combined standard uncertainty. The reproducibility standard deviation could be determined using the results of the group's laboratories with a combined standard uncertainty. For further details about EQA (PT) approach, please refer to (entry 1.2.4 of [7]).

2.5. Practical application

2.5.1. Test and measurand

To exemplify the determination of measurement uncertainty, an *in vitro* chemiluminescent immunoassay was chosen to determine the concentration of antibodies to the hepatitis C virus (HCV), Abbott Prism® HCV (Abbott Diagnostics, Abbott Park, IL, USA) [38], in the Transmissible Agents Laboratory, Portuguese Institute of Blood and Transplantation (IPST; blood establishment/blood bank). The measurand is the immunoglobulin concentration in the serum or plasma samples, which binds to solid-phase particles attached to recombinant antigens of the Core, NS3, NS4, and NS5 regions of the HCV genome.

The results of samples were derived from the photons emitted from the chemiluminescent reaction. They are expressed in a number of photons over a certain period. The number of photons is proportional to the number of antibody-antigen complexes released. This count is corrected by subtracting the number of photons in the dark. The equation to determine the test's results is defined by the manufacturer, validated using true negative and true positive human samples, and permitted by the national agencies. The "cutoff" n_c is estimated per analytical run using positive and negative calibrators considering , where n_c is the number of photons of "cutoff" value, is the number of photons of the negative calibrator expressed as the average result of the two lowest replicates out of three, is number of photons of the positive calibrator expressed as the average result of the two highest replicates out of three, and 0.55 is a factor critical for the false-negative rate. The results are verified in an ordinal scale using the ratio of sample corrected number of photons divided by "cutoff" corrected number of photons. Considering ratio value, the "cutoff" is a constant equal to 1. Ratios equal or higher than the "cutoff" are classified as positive, and ratios lower than the "cutoff" are classified as negative. A multiparametric positive sample is used at the end of each analytical run to control the run. This sample is supplied uniquely by the manufacturer and the acceptance criterion for anti-HCV is very large with ratio value from 1.02 to 6.00.

According to reagent manufacturer, "no qualitative performance differences were observed" for the immunoassay in controlled studies using anti-HCV nonreactive and reactive specimens when testing the following potentially interfering substances at the specified levels: bilirubin (≤20 mg/dL), hemoglobin (≤500 mg/dL), red blood cells (≤0.4% v/v), triglycerides (≤3.000 mg/dL) or protein (≤12 g/dL) [38].

Other possible sources of false results are not stated in reagent insert and could be identified by review of the scientific literature, particularly Young's extensive summaries of the effects of preanalytic and analytic variables:

a. Preanalytical components

- Hepatitis C IgG antibodies (Serum no effect analytical): Stability of specimen: "No effect observed when serum stored at ambient temperature for 14 days, refrigerated for 31 days, or frozen for 2 months."
- HCV antibodies (Serum no effect analytical): Stability of specimen: "No effect observed when serum stored at ambient temperature for 14 days, refrigerated for 31 days, or frozen for 2 months."
- HCV core protein (Serum positive physiological): Test association: "In patients with chronic HCV infection, no significant positive correlation of r=0.081 with serum tumor necrosis factor receptor p55; in patients with chronic HCV infection, no significant positive correlation of r=0.141 with serum tumor necrosis factor receptor p75" [39].
- Disease, drug, and herbs and natural medicine effects

b. There is no related association to disease effects [40].

HCV antibodies (Serum positive):

- Hepatocellular carcinoma: "In 100 patients with hepatocellular carcinoma, 92 were positive for HCV antibodies."
- Chronic hepatitis: "In 21 patients with chronic hepatitis, 18 were HCV antibody positive."
- Cirrhosis of liver: "In 24 patients with cirrhosis of liver 22 were HCV antibody positive" [41].

There is no related association with natural medicine effects [42].

When the medical laboratory receives samples from patients having potentially interfering substances, it is desirable to make a new measurement using an assay that has not affected the potential cause or to request a collection of a sample during a period when the potential cause is not present. Correlation with past results, as well as other assays, could also be helpful. Results affected by preanalytic and analytic variables could be unrealistic, and the probability of incorrect clinical decision could be critical. In IPST, during the interview of candidates as blood donors, candidates with hepatocellular carcinoma and chronic or cirrhosis of the liver were rejected.

Because the model equation is unknown, uniquely empirical methods are used in this example. Interlaboratory comparisons were not determined for Abbott Prism® immunoassay results due to the method validation to be normally and uniquely intralaboratorial for screening immunoassays.

2.5.2. Bias uncertainty

Data for laboratory bias evaluation are lacking due to the unavailability of a CRM or a reference laboratory. The mean difference between two laboratories of the IPST was used as an approximation for bias estimation. The bias due to the different usage of reagent batches is already included in the within-laboratory reproducibility standard deviation. For further details about single laboratory validation approach, please refer to (entry 1.2.2 of [7]).

2.5.3. Results

Table 1 summarizes the measurement uncertainty determinations for the intralaboratory and EQA approaches. It is evidenced that the standard deviation decreases according to the ratio, and the highest relative standard deviation is observed for samples close to the "cutoff" value. Standard deviation results under repeatability conditions used replicate testing from 0.5 to 1.5.

2.5.4. Discussion

The first intralaboratory approach (EP15) has precision estimates similar to what is claimed in a manufacturer's precision study [38] tested in five runs over 5 days, where s_{Rw} is from 5.7%

(average ratio equal to 3.17) to 8.6% (average ratio equal to 0.17) under intralaboratory conditions (within-laboratory reproducibility). On the contrary, in the second intralaboratory approach using replicated results and between-run "cutoff" data, the estimated s_{Rw} is 16.5%. The contribution of precision of the replicate results was the major uncertainty component, as it has selected unique results with a ratio close to the "cutoff" value. The relative standard deviation is lower at higher ratios, as the EP15 results demonstrate it with an average ratio equal to 3.70.

Approach	u_{Rw}							Bias method	u_b	u_c	U
	Method	Ratio	s_r (method)	s_r	s_I (method)	s_I	s_{Rw} (Equation 3)				
Intralab	EP15 validation data[a]	2.2–4.2	NA	NA	NA	NA	8.5%	Control sample measured at two laboratories[a]	6.0%	10.4%	21%
	Validation and QC data	0.5–2.8	Replicates from ratio 0.5 to 1.5	14.4%	Between-run "cutoff"[b]	8.5%	16.5%	Control sample measured at two laboratories[a]	6.0%	17.8%	36%
EQA[c]		6.2–10.1	NA	NA	NA	NA	NA	NA	NA	13.9 %	28 %

NA, not applicable.
[a]Sample Accurun-1 Series 2400 batch no. 10017751 (Seracare Life Sciences, Inc., Milford, MA, USA); average ratio is equal to 3.70 (15 replicate measurements per run).
[b]Average ratio equal to 1 (299 results per run).
[c]For this approach, the obtained reproducibility standard deviation is set as the combined uncertainty (sample UK NEQAS no. 9316); average ratio is equal to 2.67 (results from nine laboratories).

Table 1. Estimated measurement uncertainty values for the Abbott Prism® HCV method using data from intralaboratory and EQA approaches.

The between-run data are more reliable due to much higher degrees of freedom of this estimation. This is evidence comparing the usage of EP15 precision data and the usage of the between-run data. **Table 1** shows that the between-run data are derived from approximately 300 results determined during a long period, and EP15 data were calculated uniquely using 15 results for 5 days from December 16 to 20, 2014.

The EQA results from nine laboratories tested in the Abbott Prism® gave an estimate of a relative expanded uncertainty of 28% [35] compared to the second intralaboratory approach with an estimated uncertainty of 36%. The higher measurement uncertainty was caused principally by the heterogeneity of the group's laboratories. It used a sample with an average ratio equal to 8.2, for what if it used a sample with a ratio close to the "cutoff" value, the measurement uncertainty should be higher, closer to the second intralaboratory approach percentage.

Supposedly, it is an alternative to the between-run precision using "cutoff" raw data for the use of a long-term IQC data, given that the whole analytical process is covered. Nevertheless, using a QC sample with an average ratio equal to 1 makes it a second suggestion. For example, the usage of IQC data from March 23, 2010 to June 11, 2011 determined from sample Seracare Accurun-1 Series 2400 batch no. 116406 (Seracare Life Sciences Inc. Milford (MA)) was not a primary choice considered the between-run "cutoff" data because the average ratio was statistically significantly higher. The average ratio of 293 samples was equal to 2.42, and the intermediate standard deviation was equal to 11.1%.

The example should not be understood as applicable to all tests in the medical laboratory. Although the mathematical models could be applied to any quantitative test, some adaptations must be done in different tests. For example, in clinical chemistry, one of the major concerns is the lack of biological components when determining measurement uncertainty. This problem happened already with the Westgard et al. [43] total analytical error (TAE) [43], which is the sum of the absolute value of bias with a *k* multiplied by the standard deviation. Fraser [26] expanded the TAE concept to the total biological error (TE_{ba}) combining not only analytical error components but also biological precision and accuracy. These models are based in the Error Approach (also recognized as the Traditional Approach or True Value Approach). The Australian NPAAC guideline [12] considers not only analytical sources of precision but also the individual biological variation (CV_I) when it is known using GUM empirical estimations and Fraser's proposal. The Westgard QC provides quality requirements tables with CV_I for most of the chemistry tests [44]. For further details about the empirical determinations of measurement uncertainty combining CV_I, please refer to [12].

Although the determination of measurement uncertainty is practical in a medical laboratory, as it was already demonstrated [5], there is a serious lack of consensus on the use of the Uncertainty Approach, and the medical laboratory staff rarely understands it. Although TAE does not fulfill part of the Uncertainty Approach principles (i.e., it cannot be a model for the determination of measurement uncertainty), it is accepted in medical laboratory practice and recognized by some as a role analogous to measurement uncertainty [15, 45].

Medical laboratory results affect 60% to 70% of clinical decisions, affecting 100% of validation of blood donations, cells, and tissues. Consequently, results with higher measurement uncertainty have a significant probability of being unrealistic arising from a high risk of the uncorrected clinical decision. Measurement uncertainty has been demonstrated to be a tool to verify the level of confidence in medical laboratory results. Results with large measurement uncertainty intervals have a significant chance to be distant from the true value (i.e., *in vivo* value). Therefore, measurement uncertainty could also be used as a method validation estimation. The laboratorian could consider it as analogous to measurement uncertainty but not confusing for both concepts, differing principally in the use of bias.

3. Recommended models in medical laboratories

According to what was explained in Section 2.3, the use of modeling approach is intended principally to the R&D of medical laboratory tests. Its use in medical laboratories is complex and expensive and the measurement uncertainty results could be misestimated because not all uncertainty components are determined, different what happens using empirical data. The reagent manufacturer should choose the Monte Carlo simulation instead of partial derivative method because the propagation of distribution result is considered more reliable.

The application of empirical models is recommended for medical laboratory tests, as most of its theory and practice are already known and the results are more realistic when compared to the modeling approach. However, the laboratorian should select the model according to the chance to produce more realistic results. **Figure 6** represents in a flow chart the steps to the selection of measurement uncertainty models in medical laboratories. The first model to select is the intralaboratory approach using repeatability and between-run precision using long-term data. This is usually applied to tests used in the long-term. The second choice is the intrala-

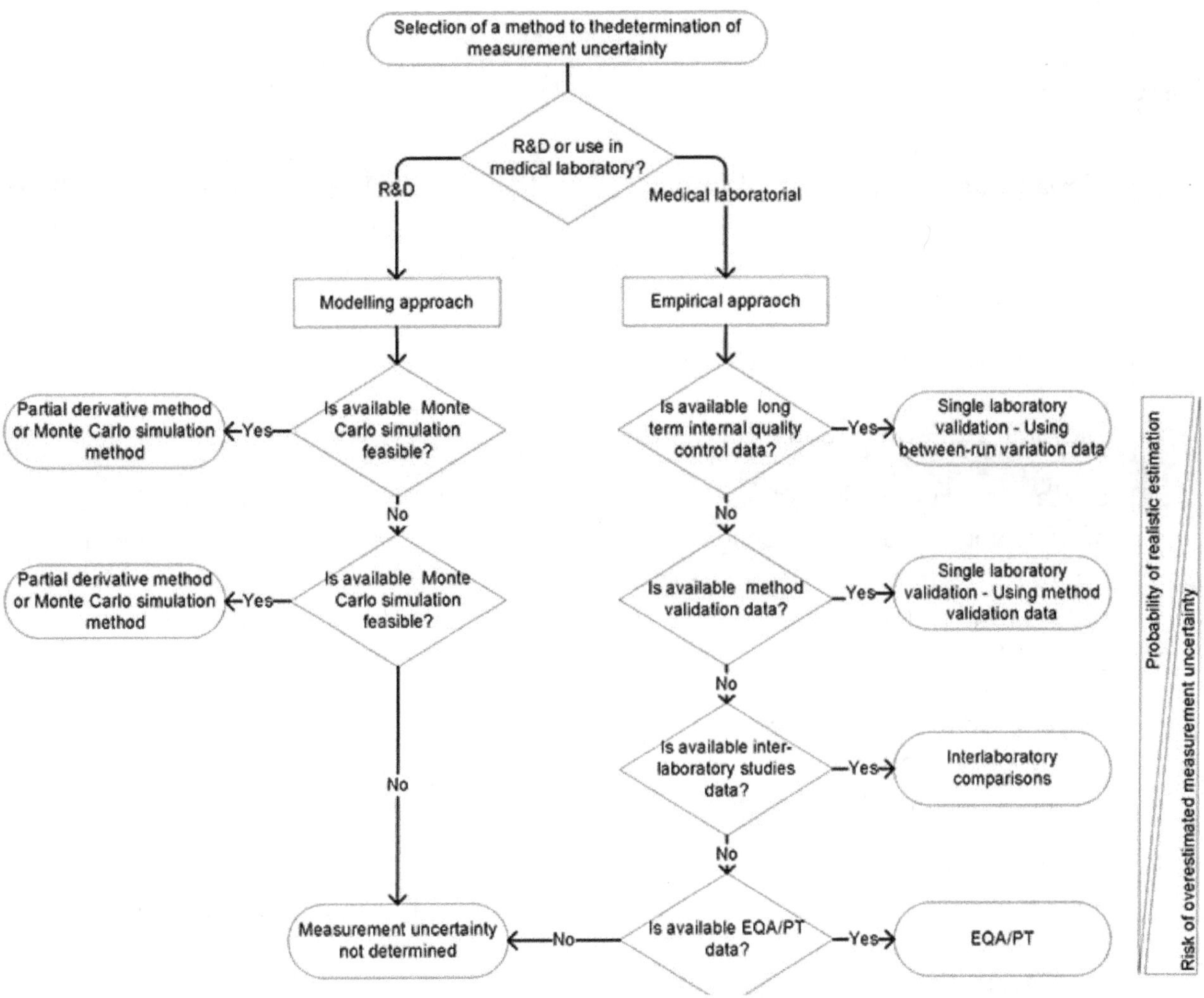

Figure 6. Flow chart to select a model for the determination of measurement uncertainty in a medical laboratory.

boratory approach using EP15 data. It is expected to be used in initial uncertainty estimation in a brand-new test. Interlaboratory comparisons are the third choice, and it is principally used when the comparator test is external. The fourth and last option is the EQA (PT). Due to the heterogeneity of the group's data, this estimation should be carefully evaluated. There is a high risk of an unrealistic estimation with the overestimation of measurement uncertainty and in some cases could be useless.

Author details

Paulo Pereira

Address all correspondence to: paulo.pereira@ipst.min-saude.pt

Quality Management Department, Portuguese Institute of BloodTransplantation, Lisbon, Portugal

References

[1] Bureau International des Poids et Mesures. Rapport BIPM-80/3, Report on the BIPM Inquiry on Error Statements. Sèvres: BIPM; 1980. Retrieved from: http://www.BIPM.org/utils/common/pdf/rapportBIPM/1980/03.pdf. Accessed: January 4, 2016.

[2] Giacomo P. News from the BIPM. Metrologia 1982; 18(1): 43–44.

[3] Giacomo P. News from the BIPM. Metrologia 1987; 24(1): 45–50.

[4] Bureau International des Poids et Mesures. JCGM 200:2012 International Vocabulary of Metrology—Basic and General Concepts and Associated Terms, 2008 Version with Minor Corrections. 3rd ed. Sèvres: BIPM; 2012. Retrieved from: http://www.BIPM.org/utils/common/documents/jcgm/JCGM_200_2012.pdf. Accessed: January 4, 2016.

[5] Bureau International des Poids et Mesures. Evaluation of Measurement Data—Guide to the Expression of Uncertainty in Measurement. JCGM 100:2008, GUM 1995 with Minor Corrections. Sèvres: BIPM; 2008. Retrieved from: http://www.BIPM.org/utils/common/documents/jcgm/JCGM_100_2008_E.pdf. Accessed: January 4, 2016.

[6] Pereira P, Magnusson B, Theodorsson E, Westgard J, Encarnação P. Measurement uncertainty as a tool for evaluating the "grey-zone" to reduce the false negatives in immunochemical screening of blood donors for infectious diseases. Accred Qual Assur 2015; 21(1): 25-32. DOI: 10.1007/s00769-015-1180-x

[7] European Federation of National Associations of Measurement, Testing and Analytical Laboratories. Technical Report No. 1/2007 Measurement Uncertainty Revisited:

Alternative Approaches to Uncertainty Evaluation. EuroLab; 2007. Retrieved from: http://www.eurolab.org/documents/1-2007.pdf. Accessed: January 4, 2016.

[8] Dybkaer R. From total allowable error via metrological traceability to uncertainty of measurement of the unbiased result. Accred Qual Assur 1999; 4(9): 401–405. DOI: 10.1007/s007690050395

[9] Ellison SLR, Williams A. Quantifying Uncertainty in Analytical Measurement. 3rd ed. Eurachem/CITAC; 2012. Retrieved from: https://www.eurachem.org/images/stories/Guides/pdf/QUAM2012_P1.pdf. Accessed: January 4, 2016.

[10] Magnusson B, Näykk T, Hovind H, Krysell M. NordTest NT TR 537 Handbook for Calculation of Measurement Uncertainty in Environmental Laboratories. 31st ed. Oslo: Nordic Innovation; 2011. Retrieved from: http://www.nordtest.info/index.php/technical-reports/item/handbook-for-calculation-of-measurement-uncertainty-in-environmental-laboratories-nt-tr-537-edition-3.html. Accessed: January 4, 2016.

[11] SYKE. MUkit—Measurement Uncertainty Kit Finnish Environmental Institute; 2014. Retrieved from: http://www.syke.fi/en-us/Services/Calibration_services_and_contract_laboratory/MUkit__Measurement_Uncertainty_Kit. Accessed: January 4, 2016.

[12] National Pathology Accreditation Advisory Council. Requirements for the Estimation of Measurement Uncertainty. Barton, ACT: NPAAC; 2007. Retrieved from: http://www.health.gov.au/internet/main/publishing.nsf/Content/B1074B732F32282DCA257BF0001FA218/$File/dhaeou.pdf. Accessed: January 4, 2016.

[13] Clinical and Laboratory Standards Institute. EP29-A Expression of Measurement Uncertainty in Laboratory Medicine. Wayne, PA: CLSI; 2012.

[14] Westgard J. Update on Measurement Uncertainty: New CLSI C51A Guidance. Retrieved from: http://www.westgard.com/CLSI-c51.htm. Accessed: January 4, 2016.

[15] Westgard J. Managing quality vs. measuring uncertainty in the medical laboratory. Clin Chem Lab Med 2012; 48(1): 31–40. DOI: 10.1515/CCLM.2010.024

[16] International Organization for Standardization. ISO/PDTS 25680 Medical Laboratories —Calculation and Expression of Measurement Uncertainty. Geneva: ISO; 2008.

[17] Pereira P, Westgard J, Encarnação P, Seghatchian J, Sousa G. Quality management in the European screening laboratories in blood establishments: a view on current approaches and trends. Transfus Apher Sci 2015; 52(2): 245–251. DOI: 10.1016/j.transci.2015.02.014

[18] Pereira P, Westgard J, Encarnação P, Seghatchian J, de Sousa G. The role of uncertainty in results of screening immunoassays in blood establishments. Transfus Apher Sci 2015; 52(2): 252–255. DOI: 10.1016/j.transci.2015.02.015

[19] Pereira P, Westgard J, Encarnação P, Seghatchian J. Evaluation of the measurement uncertainty in screening immunoassays in blood establishments: Computation of

diagnostic accuracy models. Transfus Apher Sci 2015; 52(1): 35–41. DOI: 10.1016/j.transci.2014.12.017

[20] Kimothi S. The Uncertainty of Measurements. Milwaukee, WI: ASQ; 2002.

[21] Hibbert D. Measurement uncertainty. In: Quality Assurance for the Analytical Chemistry Laboratory. Oxford: Oxford University Press; 2007, pp. 161–202.

[22] Struik D. A Source Book in Mathematics 1200–1800. Cambridge, MA: Harvard University Press; 1969.

[23] Kragten J. Calculating standard deviations and confidence intervals with a universally applicable spreadsheet technique (Tutorial review). Analyst 1994; 119 (10): 2161–2165. DOI: 10.1039/AN9941902161

[24] European Co-operation for Accreditation. EA-4/02 M: 2013 Expression of the Uncertainty of Measurement in Calibration. 2nd ed. Europe: EA; 2013. Retrieved from: http://www.european-accreditation.org/publication/ea-4-02-m-rev01-september-2013. Accessed: January 4, 2016.

[25] Petersen P, Jørgensen L, Jensen E, Kynde K, Brandslund I, et al. Er GUM skadelig ? - Eller blot overflødig? (Is GUM harmful? - Or become redundant?). Klinisk Biokemi: Norden 2003; 15(2): 27–31.

[26] Fraser C. Biological Variation: From Principles to Practice. Washington, DC: AACC Press; 2009.

[27] Vesper H, Thienpont L. Traceability in laboratory medicine. Clin Chem 2009; 55(6): 1067–1075. DOI: 10.1373/clinchem.2008.107052

[28] Dybkaer R. Setting quality specifications for the future with newer approaches to defining uncertainty in laboratory medicine. Scand J Clin Lab Invest 1999; 59(7): 579–584. DOI: 10.1080/00365519950185391

[29] Magnusson B, Ellison SL. Treatment of uncorrected measurement bias in uncertainty estimation for chemical measurements. Anal Bioanal Chem 2008; 390(1): 201–213. DOI: 10.1007/s00216-007-1693-1

[30] M. Pepe. Oxford Statistical Science Series 28. The Statistical Evaluation of Medical Tests for Classification and Prediction. Oxford: Oxford University Press; 2003.

[31] Krouwer J. Assay Development and Evaluation. Washington, DC: AACC Press; 2002.

[32] Thompson M, Ellison S. Dark uncertainty. Accred Qual Assur 2011; 16(10): 483–487. DOI: 10.1007/s00769-011-0803-0

[33] Bureau International des Poids et Mesures. JCGM 101 Evaluation of Measurement Data —Supplement 1 to the "Guide to the Expression of Uncertainty in Measurement"—Propagation of Distributions using a Monte Carlo Method. Sèvres: BIPM; 2008.

Retrieved from: http://www.BIPM.org/utils/common/documents/jcgm/JCGM_101_2008_E.pdf. Accessed: January 4, 2016.

[34] International Organization for Standardization. ISO 11352 Water Quality—Estimation of Measurement Uncertainty Based on Validation and Quality Control Data. Geneva: ISO; 2012.

[35] Clinical and Laboratory Standards Institute. EP15-A3 User Verification of Precision and Estimation of Bias. 3rd ed. Wayne, PA: CLSI; 2014.

[36] International Organization for Standardization. ISO 5725-2 Accuracy (Trueness and Precision) of Measurement Methods and Results—Part 2: Basic Method for the Determination of Repeatability and Reproducibility of a Standard Measurement Method. Geneva: ISO; 1994.

[37] International Organization for Standardization. ISO 21748 Guidance for the Use of Repeatability, Reproducibility and Trueness Estimates in Measurement Uncertainty Estimation. Geneva: ISO; 2010.

[38] Abbott Diagnostics Division. Abbott Prism HCV. Code: 84-5342/R9. Wiesbaden: Abbott Laboratories; 2008.

[39] Young D. Effects of Preanalytical Variables on Clinical Laboratory Tests. 3rd ed. Washington, DC: AACC Press; 2007.

[40] Young D. Effects of Disease on Clinical Laboratory Tests. Volumes 1 & 2. 4th ed. Washington, DC: AACC Press; 2001.

[41] Young D. Effects of Drugs on Clinical Laboratory Tests. Volumes 1 & 2. 5th ed. Washington, DC: AACC Press; 2000.

[42] Narayanan S, Young D. Effects of Herbs and Natural Products on Clinical Laboratory Tests. Washington, DC: AACC Press; 2007.

[43] Westgard J, Neil Carey R, Wold S. Criteria for judging precision and accuracy in method development and evaluation. Clin Chem 1974; 20(7): 825–833.

[44] Westgard J. Quality Requirements. Retrieved from: http://www.westgard.com/quality-requirements.htm. Accessed: January 4, 2016.

[45] Pereira P, Westgard J, Encarnação P, Seghatchian J. Analytical model for calculating indeterminate results interval of screening tests, the effect on seroconversion window period: A brief evaluation of the impact of uncertain results on the blood establishment budget. Transfus Apher Sci 2014; 51(2): 126–131. DOI: 10.1016/j.transci.2014.10.004

8

Electrical Conductivity Measurements in Agriculture: The Assessment of Soil Salinity

Fernando Visconti and José Miguel de Paz

Abstract

Soil salinity is an important issue constraining the productivity of irrigation agriculture around the world. The standard method for soil salinity assessment is based on a laboratory method that is cumbersome and gives rise to limitations for data-intensive works. The use of sensors for the assessment of the apparent electrical conductivity (EC) of soils offers a way to overcome these constraints. These sensors are based on three electromagnetic phenomena, namely, electrical resistivity, electromagnetic induction, and reflectometry. Each class of sensors presents its own advantages and drawbacks. In the following chapter, these are presented along with the most popular commercial EC sensors used in nowadays agriculture, equations for the assessment of soil salinity on basis sensor measurements, some examples of application, and present and future development trends.

Keywords: electrical conductivity, soil salinity, agriculture, irrigation, electrical resistivity, electromagnetic induction, time domain reflectometry, amplitude domain reflectometry, frequency domain reflectometry

1. Introduction

Soil salinity is the concentration in the soil pore water of major dissolved ions. These are mainly Na^+, Mg^{2+}, Ca^{2+}, Cl^-, SO_4^{2-}, HCO_3^-, and in some instances also CO_3^{2-}. In agricultural lands, K^+ and NO_3^- also become major ions and thus, significantly contribute to salinity. All these ions build up in soils as a consequence of both evaporation and plant transpiration, which extracts almost pure water from soils while leaving its salts behind, and also as a consequence of fertilization practices. As soil salinity increases, the potential of the soil pore water decreases, thus obliging plants to overcome an increasingly high energy gap for soil water uptake. Additionally, various ions, e.g., Na^+, Cl^-, may also cause specific toxicity effects on plants, impair their nutritional balance, and/or decrease the permeability of soils with further indirect deleterious effects on crops. The development of these stressful conditions poses a remarkable threat for the sustainability of agriculture, mainly under irrigation. Of the world cultivated land, which amounts to roughly 1500 Mha, about 340 Mha, i.e. 23%, is salt affected [1]. Soil salinization has been estimated to cause income losses of about 12 billion US$ per year globally [2]. In order to cope with soil salinity and the soil salinization issue, reliable and fast techniques for water and soil salinity evaluation are needed. The rigorous assessment of soil salinity requires, first, the extraction of representative samples of the soil pore water, and second, the subsequent determination of the concentrations of the aforementioned ions at a standard soil water content for their eventual summation in a single parameter known as total soluble salts (TSSs). Although rigorous, this method is expensive, labor-intensive, soil-destructive, and results-deferred, therefore posing severe constraints for data-demanding works both in space and time.

The carrying out of electrical conductivity (EC) measurements offers a way to overcome some of the limitations of the sampling and laboratory method. Specifically, the second part consisting in the analytical determinations of the major ion contents in the soil pore water can be replaced by one single measurement of EC, thus remarkably decreasing work and expenses in the laboratory, and shortening the obtaining of data. This methodological change has been globally so successfully adopted, that nowadays the standard for soil salinity assessment is the EC at 25°C of soil extracts at water saturation, in this chapter abbreviated as $EC_{e,25}$. Besides, since not only the soil pore water conducts electricity, but also the bulk soil, its apparent electrical conductivity (EC_a) can be used as a proxy for the assessment of soil salinity. However, some caution is in order when using EC_a in this regard, because EC_a depends on other soil properties, importantly water content, texture, structure, and mineralogy, in addition to salinity itself, therefore complicating the interpretation of measurements [3]. Furthermore, the soil EC_a can be measured by means of various different techniques, which are based on i) potential drop or electrical resistivity (ER), ii) electromagnetic induction (EMI), and iii) reflectometry, either time (TDR), amplitude (ADR), or frequency (FDR) domain. These techniques and the corresponding sensors feature different abilities to sample soils for EC_a measurements and therefore give rise to EC_a values that present differences.

In this chapter, the foundations of the EC techniques nowadays available for the assessment of soil salinity in agriculture will be presented, along with the specific models developed to make estimations of soil salinity. The main commercial sensors used to make measurements of EC in agricultural soils will be presented, too, commenting on their strengths and weaknesses. Examples of the practical application of sensors for soil salinity assessment in agriculture will be given. Finally, we will try to envisage the future trends in the development and applications of this technology.

2. Electrical conductivity

The cause of the electrical conductance is the existence of particles with electric charge which, from a microscopic point of view, are loosely bound to specific positions within materials and thus, are capable of conveying electric charge. The materials featured by these particles are able to conduct electricity and thus are known as conductors. Liquid water is a conductor because under natural conditions, it contains dissolved ions which are movable charged particles. Soil, the other material that concern us here the most, is a composite conductor in which water, solid particles, and air are present in variable quantities and arrangements, and in which the electric charge carriers are water dissolved and solid particle loosely adsorbed ions. These ions are, specifically, the major inorganic ions in the aqueous systems of the earth crust, i.e., cations Na^+, Mg^{2+}, Ca^{2+}, and K^+, and anions Cl^-, SO_4^{2-}, HCO_3^-, CO_3^{2-} and NO_3^-.

When waters or soils are exposed to no electric fields, ions move randomly within, and macroscopically no net electric current is observed. On the contrary, when an electric field is applied, cations move to lower potentials whereas anions move to higher ones, and therefore, the water or soil system conducts electricity. The EC (σ) features the proportionality factor between the current density (**J**, A/m^2) and the electrical field (**E**, V/m): $\mathbf{J} = \sigma\ \mathbf{E}$ and thus, it measures the increment of the electric current through each unit area of a surface perpendicular to its flowing direction per unit increase of an externally applied electric field. Therefore, the EC of a material is a physical quantity that indicates its ability to conduct electric current. It is the reciprocal of the material's resistivity (ϱ): $\sigma = 1/\varrho$, and its units in the SI are the S m^{-1}.

3. Standard EC for soil salinity assessment

The EC of soil materials measured as such is known as bulk or EC_a (σ_a) because the measured EC corresponds to an EC-equivalent homogeneous single-phase material [3]. Even more important is to note that EC_a is different from the EC of the soil pore water (EC_p or σ_p), i.e. the EC of the soil solution separated from the soil solids. The convenient way of soil salinity estimation using EC instead of chemical analysis would involve the measurement of EC_p. However, such direct measurement of EC_p is never made because, on the one hand, of a practical issue, and on the other hand, of the need for standardization.

From a practical point of view, representative samples of the soil solution at usual field soil water contents are difficult or impossible to obtain [4]. Besides, soil salinity continuously varies as so does the soil water content, thus demanding a specific soil water content for standardization. By international agreement, the standard soil water content for salinity assessment is soil water saturation. On the one hand, water saturation is the lowest soil water content from which a sample of the soil solution can be easily obtained and, on the other hand, water saturation is the highest soil water content attainable under field conditions and thus, representative of the soil salinity to which plants are exposed. Therefore, the universal standard for soil salinity appraisal is the EC of the soil saturation extract at 25°C [5], which is abbreviated as $EC_{e,25}$ in this chapter. The salt tolerance of all crops is expressed in terms of $EC_{e,25}$ [6, 7], and therefore, in agriculture, all measurements obtained with whichever other method have to be converted to $EC_{e,25}$ in order to be useful for soil salinity assessment. All $EC_{e,25}$ data are obtained through the preparation of saturated pastes by equilibration of a disturbed soil sample with deionized water, sampling of the aqueous phase by vacuum extraction, and eventual EC measurement [8]. To correctly interpret $EC_{e,25}$ values in soil studies and agriculture, various environmental factors must be taken into account and specifically for agriculture, the salt tolerance of available crops must be known. Since interpretation of $EC_{e,25}$ is not within the scope of this chapter, the interested reader is led to [5] and [9].

4. Temperature effects on EC

As indicated, in soil studies and agriculture, EC data have to be expressed at a standard temperature of 25°C. This is because the EC of all materials depends on temperature. Specifically, the EC of waters and also soils increases as temperature increases. Thus, unless the temperature of all waters and soils is adjusted to 25°C by equilibration in thermostatic baths, all measurements must be corrected to 25°C. This is done by measurement of the temperature (T) of the material under test, and subsequent application of an adequate equation to convert from EC at T (EC_T or σ_T) to EC at 25°C (EC_{25} or σ_{25}). The function that relates EC with T depends on the specific salt composition, i.e. on the concentration of every major ion in solution. Since the sum of all ion concentrations is exactly what we estimate through the EC measurement, an empirical function is needed to assess EC_{25} from EC_T. There are two functions in major use for this assessment provided T is between 3 and 50°C. The ratio model (Eq. 1) is based on the EC evolution of KCl 0.01 M solutions with temperature, while the exponential model (Eq. 2) was developed by [5] by taking EC measurements on different soil saturation extracts and various salt solutions at different temperatures [10, 11]:

$$\sigma_{25} = \frac{\sigma_T}{1 + 0.0191(T - 25)} \tag{1}$$

$$\sigma_{25} = \sigma_T\left(0.4470 + 1.4034e^{-T/26.815}\right) \tag{2}$$

The use of empirical equations has a practical consequence in that the difference between true and corrected EC_{25} depends on T, and this difference decreases as T is closer to 25°C, being null at T = 25°C. Therefore, the EC measurements must be taken as close to 25°C as possible to avoid this empirical bias; usually, a range of 25 ± 5°C is enough to have differences under 1% in most instances. This warning extends to all EC measurements in soil studies and agriculture either in waters or soils.

5. EC measurement methods

EC measurements in soil studies and agriculture can be readily made with various types of sensors based on different electromagnetic phenomena which are classified into i) ER, ii) EMI, and iii) reflectometry, either TDR, ADR, or FDR.

5.1. Electrical resistivity (ER)

When an electric field is applied to a piece of a conductor, and current develops, the system is featured by resistance $R = l/(\sigma A)$, where l and A are, respectively, the length and the cross-sectional area of the piece of material under test. On basis, the inverse proportionality between σ and R, measurement of EC can thus be done by measuring the resistance of a piece of soil or water of known dimensions. Modern ER measurements are taken by using alternating currents (ACs) of extremely low to super low frequencies, usually below 30 Hz. At low frequencies, i.e. below 1 MHz, capacitance and electrolytic effects on the measurement electrodes and besides, amplifier distortions, are avoided, while resistive effects overwhelmingly contribute to the signal like in a purely direct current (DC) measurement [12, 13].

Nowadays, the simplest device to measure EC is a digital ohmmeter, which is composed of a probe made of a pair of metal electrodes, a power supply able to provide a standard constant current (I_{std}) to the electrodes and the soil in-between, and a voltmeter able to measure the potential drop across (ΔV). Thus, the soil resistance is simply calculated applying the Ohm's law as $R = \Delta V/I_{std}$, and hence EC as

$$\sigma = kI_{std} / \Delta V \tag{3}$$

where $k = l/A$ is known as the cell constant of the probe. The cell constant depends on the probe design, it has units of reciprocal of length, and can be analytically assessed in some simple cases.

In soil studies and agriculture, all EC measurements in waters, including irrigation waters and soil extracts such as the saturation extract, are taken by means of ER using laboratory bench or handheld conductimeters. These instruments are calibrated with EC standards to determine their cell constant. In agriculture most EC values lay in the range between 0.01 and 1.6 S m^{-1} which, in order to take unbiased measurements, determines both the EC of the standards needed for calibration and the appropriate cell constant of the instrument. For agricultural applications, the EC standards cover two orders of magnitude, usually with specific EC values of 0.0147, 0.1413, and 1.288 S m^{-1} at 25°C, which correspond to aqueous solutions 0.001, 0.01, and 0.1 M in KCl, whereas the appropriate cell constants are around 1 cm^{-1}.

The two-point ER method just presented has some constraints that are revealed as non-negligible when the method is used for soils. This is because the two-point ER method not only measures the conductor's resistance but also the resistance of the probe electrodes and wiring and besides, is altered by the Galvani potential difference that develops across the contacts of the electrodes with the conductor. To overcome these limitations, the four-point method developed by [14] for geophysics sounding is used instead of the simple two-point method.

Contrary to the two-point method, in which the potential drop is measured with the same electrodes used to inject the current into the soil, in the four-point method, each function is carried out with its own pair of electrodes. By using separate electrodes, neither the electrodes resistances, nor the contact resistances between the metal electrodes and the soil show up in the measured EC, thus making interpretation of data more straightforward. The electrodes can be arranged according to various configurations, which differ in geometry and electrode spacing [15].

5.2. Electromagnetic induction (EMI)

The EC_a of soil materials can be estimated with no contact by means of EMI. In EMI instruments, there is one transmitting (Tx) and one or more receiving (Rx_1, Rx_2, etc) coils (**Figure 1**). The Tx is connected to an oscillator operating at very low frequencies (VLF), specifically in the range between 1 and 100 kHz in which the conductivity soil response has been found to be almost independent from frequency [16]. This way, the Tx generates a primary time variable magnetic field (Hp). By the EMI phenomenon, this time variable magnetic field induces a varying electric field in the soil and in response, many alternating eddy currents are generated within. The amplitude of the total alternating electric current is proportional to i) the EC of the soil, ii) the rate of change of the primary magnetic field (Hp), and iii) the orientation and proximity of the instrument to the soil. The ACs generated in the soil lead in turn, by the same EMI phenomenon, to the creation of a secondary magnetic field (Hi). The resulting total magnetic field (Hp + Hi) induces again, by the same EMI phenomenon, a current in the receiver(s) coil(s) of the instrument. The amplitude of the quadrature-phase of this total field (real component) is the one which is related to a depth-weighted soil EC (EC_a*) [17]. Importantly, note that EC_a* is different from EC_a. This signal is amplified and formed into an output voltage, which is shown as such EC_a* value to the user [18].

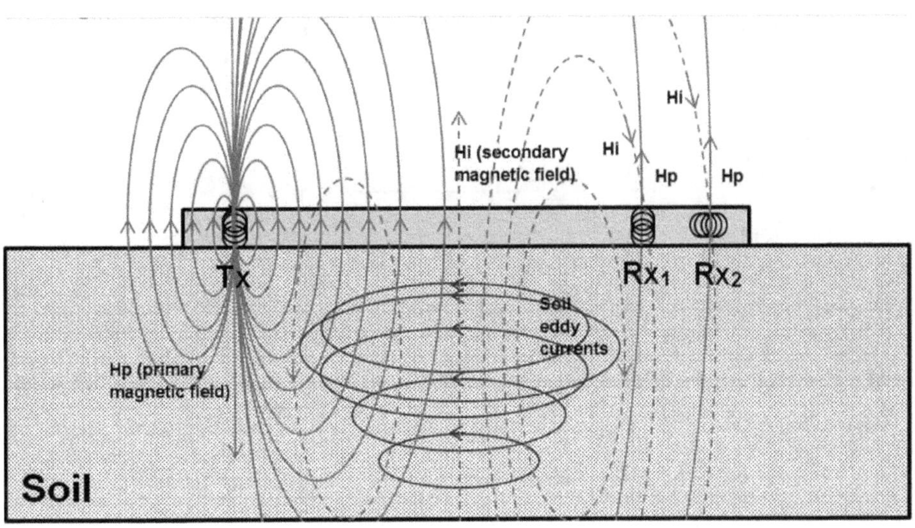

Figure 1. Schematic overview of the functioning of an EMI sensor, specifically the DUALEM-1 in which one receiving coil is coplanar (Rx_1) and the other is perpendicular (Rx_2) to the transmitting coil (Tx).

5.3. Reflectometry

Reflectometry is based on the effects soil has on primary alternating electric currents transmitted into the soil via embedded electrodes. In reflectometry, the characteristics of ACs change in response to the dielectric properties of the soil medium, and therefore other alternating signals are generated. These secondary signals are recorded, and their speed, amplitude, or frequency are analyzed, thus giving rise to three different techniques. These are two genuine reflectometry methods such as TDR and ADR, in addition to FDR, which is based on the electrical resonance of RLC circuits.

Reflectometry EC_a measurements are based on the fact that conductivity is one of the main mechanisms through which electromagnetic signals transmitted into the soil lose energy. This energy loss is represented by the imaginary part of the soil dielectric permittivity which is conveniently represented by a complex frequency-dependent variable: $\varepsilon^*(f) = \varepsilon'(f) - j\,\varepsilon''(f)$, where $j = \sqrt{-1}$ is the imaginary unit, and $\varepsilon'(f)$ and $\varepsilon''(f)$ are, respectively, the real and imaginary parts of ε^*. The real part of permittivity represents energy storage and is mainly, but not only, related to the soil water content (θ, m^3/m^3) because of the remarkably higher relative dielectric permittivity of pure water (80) regarding soil solids (3–5) and soil air (1). The imaginary part of permittivity depends on various energy loss mechanisms such as dipole relaxation ($\varepsilon_{rel}''(f)$) and importantly EC_a, through:

$$\varepsilon''(f) = \varepsilon''_{rel}(f) + \frac{\sigma_a}{2\pi f \varepsilon_0} \tag{4}$$

where $\varepsilon_0 = 8.85418\ 10^{-12}$ F/m is the vacuum permittivity. In reflectometry, EC_a is assessed from energy losses usually assuming that the other loss effects encompassed in $\varepsilon_{rel}''(f)$, are negligible regarding the conductivity loss, i.e. $\sigma_a/2\pi f\varepsilon_0$. Since frequency similarly affects both $\varepsilon_{rel}''(f)$ and $\sigma_a/2\pi f\varepsilon_0$, EC_a measurements will be barely affected by frequency changes. On the contrary, the assessment of θ is made on basis the apparent permittivity (ε_a), which depends on both energy storage and loss:

$$\varepsilon_a = \frac{\varepsilon'(f)}{2}\left[1 + \sqrt{1 + \tan^2\left(\frac{\varepsilon''(f)}{\varepsilon'(f)}\right)}\right] \tag{5}$$

As a result, since energy losses decrease as frequency increase, ε_a approaches the real permittivity (ε') as frequency increases and therefore, as frequency increases θ estimations will be more accurate.

5.3.1. Time domain reflectometry (TDR)

TDR is a broadband high-frequency technique originally applied to soil studies as a means for the in situ fast estimation of θ [19]. Reflectometry in the time domain is based on the reflection primary precisely timed electrical pulses undergo when sent along a transmission line (TL) ending in various electrodes inserted into the soil. The three essential parts of a TDR instrument are i) a TL which is formed, in turn, by a coaxial cable ending in a probe formed by two-four metal rods, ii) a fast-rise signal generator operating at high frequencies between 0.02 and 3

GHz, and iii) a fast oscilloscope (**Figure 2** left). The oscilloscope is fast enough to sample the voltage level of the TL at intervals down to around 100 ps and, hence, obtain the TDR waveform [21] (**Figure 2** right). This way, the fast-rise electromagnetic pulse composed of a wide range of frequencies transmitted to the TL is partially reflected back and forth at the end and at the beginning of the TL giving rise to an electromagnetic oscillation whose voltage amplitude is sampled. The main characteristics of the primary and reflected pulses that are useful in this regard are i) traveling back and forth velocity and ii) attenuation. The first characteristic is related to ε_a, and thus mainly to θ, whereas the second one is mainly related to just the imaginary part of permittivity and thus to EC_a [22, 23]. The voltage at late time (**Figure 2** right) is usually used to derive the EC_a by means of the Giese and Tiemann equation [24] as indicated in [25]. However, in practice, several different voltage values in the TDR attenuation curve can be reliably used and therefore, various different equations have been proposed to calculate EC_a from TDR waveforms [26, 27].

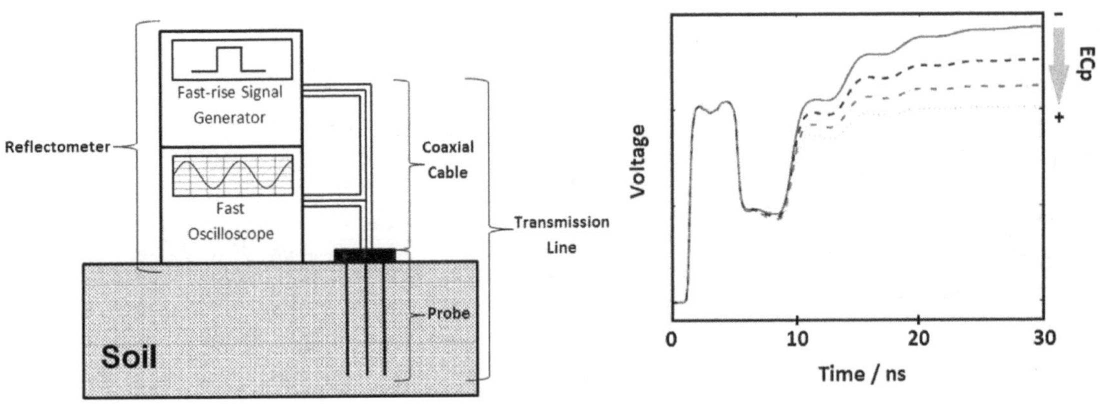

Figure 2. Schematic diagram of a TDR instrument (left) and graph showing the attenuation effect of soil salinity on TDR waveforms after [20] (right).

5.3.2. *Amplitude domain reflectometry (ADR)*

Reflectometry in the amplitude domain is another genuine reflectometry technique like TDR. Similar elements to those previously described for TDR equipment are used for ADR. In ADR, however, the measurement is based on the amplitude features of the standing electromagnetic oscillation in different parts of the TL [28]. Besides, since in ADR the signal generator operates in a frequency range between 10 and 100 MHz, i.e. significantly lower than in TDR, instrumentation prices decrease.

5.3.3. *Frequency domain reflectometry (FDR)*

Contrary to TDR and ADR, FDR is not based on the analysis of reflected electromagnetic pulses but on the resonance features of RLC circuits in which a capacitor is formed by two electrodes and the in-between and surrounding soil. The RLC circuit in FDR instruments is formed by a signal generator, plus resistor, inductive and capacitive elements, including the lossy capacitor involving the soil (**Figure 3**). This lossy capacitor is characterized by a resistance (R_1) and a capacitance (C_1) which depends, in addition to the soil, on its specific design and ε_a according to $C_1 = g\varepsilon_a$, where g is the capacitor design factor in length units. The soil sensing capacitor is connected in parallel with the parasitic capacitance due to the circuit board and connections (C_2), and in series with the circuit board which is characterized by a well-known capacitance (C_3) [29, 30]. Therefore, g and C_2 are instrument-specific, and even sensor-specific, and their values must be obtained by calibration using liquids of known dielectric properties.

In FDR, the signal generator operates at frequencies below TDR, which are in the range from 10 to 200 MHz. At these relatively low frequencies, ε_a measurements are more dependent on other soil properties in addition to θ and therefore, FDR is less reliable than TDR for θ estimations [31]. In FDR instruments, the frequency of the AC is usually varied within a narrow range until resonance is achieved in the circuit, i.e., until amplitude is maximum. The resonant frequency depends mostly on ε_a, while the amplitude depends mostly on EC_a. Therefore, in FDR, the EC_a is assessed from the amplitude at resonance [12].

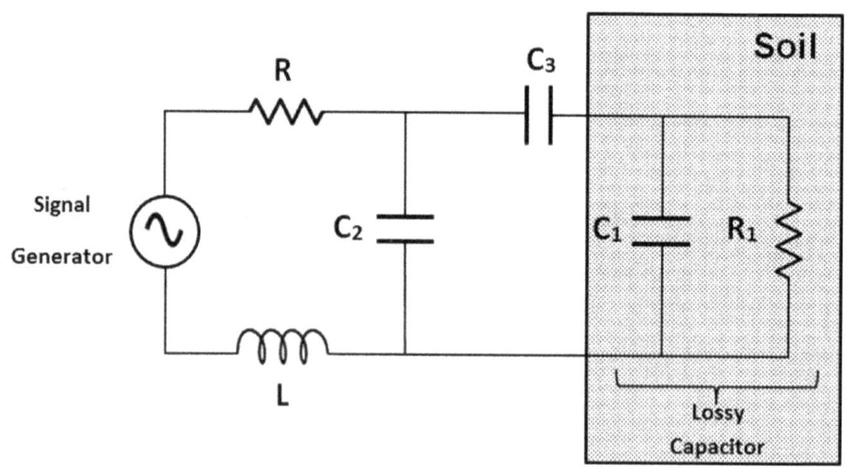

Figure 3. Circuit diagram for a single-probe FDR sensor.

6. Assessment of soil salinity from EC measurements

As indicated in Section 3, $EC_{e,25}$ is the standard for soil salinity assessment; however, the method of the saturation extract is labor-intensive and soil-destructive. To overcome either one or both of these drawbacks, several alternative methodologies have been proposed. These methodologies can be grouped into two classes: i) those based on more readily prepared soil water extracts and ii) those based on EC_a measurements. The first class of methods constitute alternative sampling and laboratory methods and will not be dealt with in this chapter. The interested reader is referred to the literature on the subject, e.g. [32] and references therein.

In the second class of methods, the interpretation of EC_a measurements is made by means of models relating EC_a with $EC_{e,\,25}$. Unfortunately, universally valid equations do not exist for this transformation. Such relationship must be assessed almost in every instance by using two different methods. The first one consists of two steps: i) the assessment of EC_p from EC_a and then ii) the assessment of EC_e from EC_p. The second one is based on the assessment of EC_e directly from EC_a. Each of these methods presents its own advantages and drawbacks.

6.1. Two-step estimation of EC_e from EC_a

The first part of this method, i.e. the calculation of EC_p from EC_a, is based on what is known about how soils behave as conductors. In the second part, EC_e is related to EC_p either through modeling of soil solution dilution and concentration processes, or by means of empirical equations.

6.1.1. First step: estimation of EC_p from EC_a

Since soils are composite materials made of solids, water, and air, electric charge is conveyed through them by means of three different paths acting in parallel [33]. These are i) a continuous liquid pathway in which dissolved ions are the charge carriers, ii) an alternate solid-liquid pathway in which both exchangeable and dissolved ions are the charge carriers, and iii) a pathway formed by solid particles in direct and continuous contact with one another in which exchangeable ions are the charge carriers. Since just the continuous liquid (i) and continuous solid pathways (iii) can be straightforward parameterized as separate units, the following equation with just two summands representing, respectively, the continuous liquid plus continuous solid pathways, can be used to model EC_a as a function of soil properties [34]:

$$\sigma_a = \theta(a\theta + b)\sigma_p + (a\theta + b)B\rho_b CEC \tag{6}$$

In Eq. 6, the alternate liquid-solid pathway is somewhat included into both summands, $(a\theta + b)$ is a factor known as tortuosity or transmission coefficient where a and b are fitting parameters which depend on soil texture and structure, B is the equivalent conductance of the counterions on the soil exchange complex (S $m^2/mmol_C$), ρ_b is the soil bulk density (kg/m^3), and finally, CEC is the soil cation exchange capacity ($mmol_C/kg$), which depends on soil texture, clay mineralogy, and organic matter content. In Eq. 6, there are three soil-specific parameters (a, b, and B) that must be assessed by calibration by taking several EC_a measurements and analyzing the soil for θ, EC_p, ρ_b, and CEC. Once the parameters (a, b, and B) of Eq. 6 have been calibrated, σ_p can be isolated to estimate the EC of the soil solution by means of Eq. 7.

$$\sigma_p = \frac{\sigma_a - (a\theta + b)B\rho_b CEC}{\theta(a\theta + b)} \tag{7}$$

In addition to the model parameters, an adequate estimation of EC_p from EC_a by means of Eq. 7 requires knowing the values of ρ_b, CEC, and θ in the same soil volume. The former two (ρ_b, CEC) can be assumed to be barely spatial and overall time variable and thus, as steady soil properties for many applications. On the contrary, the soil water content (θ) is usually very variable both in space and time and can virtually never be considered as steady for survey and monitoring applications. Therefore, to correctly estimate EC_p from EC_a measurements, θ must be determined along with EC_a in the soil volume under test. This determination could rigorously be done by sampling of the measured soil volume and subsequent laboratory analysis. However, this direct method would override most of the advantages gained by working with EC_a sensors. To overcome this limitation, the soil water content could either be estimated by means of the various nowadays available indirect sensing techniques such as i) neutron thermalization, ii) gamma ray attenuation, iii) those related to soil thermal properties, and iv) electromagnetic methods [35]. Nevertheless, the first three methods provide θ estimations not for the same soil volume surveyed by the EC_a sensor but for an adjacent soil volume, thus giving rise to errors. With electromagnetic methods and specifically with reflectometry, the requirement of θ estimations in exactly the same soil volume under test has been addressed because most of these instruments can measure EC_a in addition to estimate θ. For use with such instruments, various simple empirical equations have been proposed to assess EC_p. Eq. 8 uses the empirical linear relationship that has been revealed to exist between σ_a and ε_a as the water content changes while EC_p is kept constant, to assess EC_p on basis just EC_a and ε_a sensor measurements [36]:

$$\sigma_p = \frac{\sigma_a - \sigma_s}{m\left(\varepsilon_a - \varepsilon_{ss}\right)} \tag{8}$$

where the parameters σ_s, ε_{ss} can be interpreted, respectively, as the particle surface EC, and the soil solids dielectric permittivity. They both along with m have to be assessed by calibration and depend on soil properties such as texture, mineralogy, and organic matter. Note that in Eq. 8, and other models, EC_p estimations are expected to improve provided ε_a is replaced by ε' [37], and that temperature corrections are still pending. The Hilhorst [38] model (Eq. 9) constitutes a simplified version of Eq. 8, where just one parameter, i.e. the soil dielectric permittivity at zero EC_a ($\varepsilon_{\sigma a} = 0$), is needed, and $\varepsilon_w(T)$ is the dielectric permittivity of water, whose value depends just on temperature.

$$\sigma_p = \frac{\sigma_a \varepsilon_w(T)}{\varepsilon_a - \varepsilon_{\sigma a=0}} \tag{9}$$

Provided the relatively low variability of ϱ_b and CEC regarding θ, a simpler version of the model represented by Eq. 7 has been developed for assessing EC_p along with the use of instruments capable of both EC_a measurements and θ estimations [39]:

$$\sigma_p = \frac{\sigma_a - \sigma_s}{(a\theta + b)\theta} \tag{10}$$

where σ_s is a lumped coefficient, i.e., equal to the second summand ($(a\theta + b) B \varrho_b$ CEC) in Eq. 6, representing the particle surface EC, and θ is usually empirically assessed with a third-order polynomial of the form: $\theta = a_3\varepsilon_a^3 + a_2\varepsilon_a^2 + a_1\varepsilon_a + a_0$ [19]. An alternative to this polynomial is the simplified dielectric mixing (SDM) model, which is more theoretically based [40]. Thus, the following equation (Eq. 11), which combines Eq. 10 with the SDM model has been proposed to estimate EC_p at 25°C ($\sigma_{p,25}$) from sensor measurements of EC_a, ε_a and T [41]:

$$\sigma_{p,25} = \frac{\sigma_a h(T) - \sigma_s}{\left[a\frac{-b_0 + \sqrt{\varepsilon_a}}{b_1} + b\right]\left[\frac{-b_0 + \sqrt{\varepsilon_a}}{b_1}\right]} \tag{11}$$

where $h(T) = \sigma_{25}/\sigma_T$ is a function of temperature as given by the ratio (Eq. 1) or the exponential model (Eq. 2).

6.1.2. *Second step: estimation of EC_e from EC_p*

Once EC_p has been estimated by means of the previous equations, it can be related to EC_e by means of either process-based or functional models, or even a mixture of both. The process-based models simulate the dilution of the soil solution from the field water content at which measurements have been taken (θ_f) to the soil saturation water content (θ_s). This model can be as simple as a dilution ratio giving rise to the following estimation: $EC_{e,25} = (\theta_f \varrho_{bs} EC_{p,25})/(\theta_s \varrho_{bf})$, where ϱ_{bs} and ϱ_{bf} are the bulk densities of, respectively, the saturated paste and field soil [42, 43], or either a complex model such as SALSOLCHEMEC [44], that requires, in addition to θ and ϱ_b data, the likely major ion contents and CEC of the soil. These complex models, though more accurate, can be regarded as less appealing because they require more data and further elaboration of results. The functional models, on the contrary, are based on the statistical relationship between $EC_{e,25}$ and $EC_{p,25}$, which must be obtained by calibration beforehand. A mixture of both approaches can be applied considering the dilution ratio to calculate a proxy of $EC_{e,25}$, i.e. $EC_{e,25}'$, whose statistical relationship with the true $EC_{e,25}$ must also be obtained by calibration beforehand. Nevertheless, an advantage of this latter approach over the purely statistical one is that estimation errors will diminish.

6.2. One-step estimation of EC_e from EC_a

EC_e can be directly estimated from EC_a measurements using purely empirical calibrations. Most of these calibrations have been developed to interpret the EC_a measurements taken with ER and EMI techniques. Calibrations are usually obtained by multiple linear regression (MLR), principal components regression (PCR), partial least squares regression (PLSR), or either kriging (KR) or co-kriging regressions (CKR). Regardless of the specific regression technique, in these models, a profile EC_e average can be assessed as a mathematical function of several EC_a sensor measurements, e.g. vertical and horizontal configuration EMI measurements (EC_{a-v}*, EC_{a-h}*), in addition to an n number of soil properties ($P_1, P_2, \ldots, P_n$) including other sensor measurements, coordinates, micro-topography features, etc:

$$EC_e = h\left(\sigma_{a-v}^*, \sigma_{a-h}^*, \ldots, P_1, P_2, \ldots, P_n\right) \quad (12)$$

Profile EC_e averages are not the only data we can obtain with the use of these techniques. With many ER and certainly most EMI instruments we have got the ability to delineate the one-dimensional (1D) soil EC_a, and therefore, we can elaborate on calibrations for EC_e variations along the soil vertical coordinate.

7. Commercial sensors for measuring EC

7.1. Electrical resistivity sensors

ER gives rise to the simplest of techniques for the assessment of EC_a in agriculture with essentially two classes of ER systems: static and mobile instruments. Two-point ER measurements can be done with commercial handheld digital multimeters (DMM, DVOM). Provided the length (L), radius (r), and spacing (d) between the pair of test leads driven into the soil are known EC_a can be measured by less than 50 €. However, two-point ER measurements are usually made with sensors specifically developed for soil applications. The capacitance-conductance (CC) combined 5TE and GS3 sensors by Decagon (Decagon Devices, Inc., Pullman, Washington, USA) provide EC_a measurements using ER. Besides, both sensors also measure ε_a through capacitance, which is the simplest sensor technique for θ estimation, and T through a thermistor in roughly the same soil volume, and are worth between 200 and 300 €. Subsequent elaboration of EC_a, ε_a, and T data with the equations in Section 6 along with equations in Section 4 allows for the assessment of $EC_{p,25}$.

In the 1970s, a four-point ER probe based on a Wenner's array was developed by Rhoades and van Schilfgaarde [45]. On basis of this design, various commercial devices were developed next in order to make discrete and continuous EC_a measurements in agriculture [33]. The Martek (Martek Instruments, Raleigh, North Carolina, USA) soil salinity sensors adequately served for both profile soundings and continuous burial measurements. Models such as the SCT-10 included a temperature sensor, and thus provided both raw and temperature corrected readings for improved data interpretation. The commercialization of this sensor was discontinued by Martek Instruments; however, a very similar device including temperature measurement can still be bought from Eijkelkamp (Giesbeek, The Netherlands) for roughly 4500 €.

Since the works of Wenner, the static four-point ER sensors early evolutioned to mobile EC_a instruments [46]. Nowadays, there are instruments sold by Veris (Veris Technologies, Inc., Salina, Kansas, USA), in which sensors take the shape of coulters mounted in a trailer that is towed by a vehicle through the field under test. These instruments are worth between 12,000 and 24,000 €, and integrate global positioning system (GPS) and data-logging utilities. Besides, the newest Veris models (V2000XA, V3100, V3150) present two pairs of potential measurement electrodes instead of just one, which jointly enable a simple soil profiling of EC_a with two depths. The coulter electrodes are also sold individually and therefore users can build their own craft ER sensor systems [47].

The most critical issues in order to make reliable EC_a measurements with ER sensors are i) knowing accurately the probe cell constant value and ii) assuring good contact between all the electrodes and the soil. The cell constant can be analytically assessed in most instances as shown in ref. [48]. This is the only option when performing four-point measurements with mobile systems. However, for most applications with commercial two-point and four-point static sensors, the probe cell constant is empirically assessed by means of calibration using EC standards (Section 5.1).

7.2. Electromagnetic induction sensors

The EMI instruments most commonly used in nowadays agriculture include the DUALEM-1, DUALEM-2, and DUALEM-21 (Dualem, Inc., Milton, Ontario, Canada); the EM38, EM38-DD, and EM38-MK2 sensors (Geonics Ltd., Mississauga, Ontario, Canada), and the Profiler EMP-400 (Geophysical Survey Systems, Inc., Salem, New Hampshire, US). The simplest EMI instruments are the EM38 and the DUALEM-1, in which there are just one transmitting and one or two receiving coils 1 m apart, and are worth between 11,000 and 14,000 €. As a consequence, these instruments are sensitive to the soil EC_a down to a depth between 0.5 and 1.5 m, which is where most plant's roots develop and thus, the most interesting for soil studies and agriculture.

The soil depth response to EMI instruments depends on the separation and orientation of transmitting and receiving coils, and on their distance over ground in the following ways: i) as the separation between both coils increases, the soil depth contributing to the sensor signal increases; ii) when one coil, either the transmitter or the receiver, is turned from vertical to horizontal, the soil depth contributing to the sensor signal decreases; and iii) as the height over ground increases, the soil depth contributing to the sensor signal decreases.

The EM38 presents two parallel coils and measurements can be performed with both coils either vertical (V-V) or horizontal (H-H) to the soil. When the sensor is laid onto the soil, 70% of the cumulative sensor signal is provided by the upper 1.55 m in the V-V orientation, and by the upper 0.75 m in the H-H orientation. The DUALEM-1 presents three coils, one transmitter and two receivers (**Figure 1**). While the transmitter coil is vertically oriented, one receiver is 1 m apart and parallel to the transmitter (vertical orientation), and another is 1.1 m apart and perpendicular to the transmitter (horizontal orientation). When the sensor is laid onto the soil, 70% of the cumulative sensor signal is provided by the upper 1.5 m in the V-V orientation, and by the upper 0.5 m in the V-H orientation. As described, two EC_a^* measurements can be made with both the EM38 and DUALEM-1 sensors. Both measurements can then be elaborated to have qualitative information about how EC_a changes with soil depth. For example, if the V-V EC_a^* (EC_{a-v}^*) is higher than the H-H or V-H EC_a^* (EC_{a-h}^*), it is because EC_a increases with soil depth. On the contrary, if EC_{a-v}^* is lower than EC_{a-h}^*, it is because EC_a decreases with soil depth. Even though the EM38 and DUALEM-1 sensors work at slightly different frequencies, which are, respectively, 14.7 and 9 kHz, both provide the same data for EC_{a-v}^* and EC_{a-h}^* and can then be used interchangeably [49].

The EM38-DD is an evolution of the EM38, in which two EM38 sensors are bolted together and electronically coupled. Besides, one is vertically and the other horizontally oriented, so as to have simultaneous EC_{a-v}* and EC_{a-h}* measurements. The EM38-MK2 is a further evolution of the EM38 in which there are two receiving coils instead of just one. The receivers are parallel to the transmitter and, respectively, 0.5 and 1.0 m apart from it and, as a consequence, to have 70% of cumulative sensor signal, between 0.75 and 1.55 m are required in the V-V-V orientation, and between 0.4 and 0.75 m are required in the H-H-H orientation [50].

In the DUALEM-2, the separation between the transmitter and receivers has been doubled to be 2 m and, as a consequence, the depth for 70% cumulative signal is 3 m in the V-V orientation and 1 m in the V-H orientation. The DUALEM-21 is a combination of the previous DUALEM-1 and DUALEM-2 in which one pair of receivers are 1 m away from the transmitter, and another pair of receivers are 2 m away. This configuration allows DUALEM-21 users to have four depths of simultaneous EC_a sounding.

An instrument similar to the EM38 and the DUALEM-1 is the EMP-400, in which there are just two parallel coils, one transmitter and one receiver separated by 1.22 m to take measurements in the V-V or H-H orientations. The EMP-400 is a multi-frequency instrument able to take measurements at whichever three frequencies between 1 and 16 kHz; however, to have EC_a* measurements similar to those provided by similar EMI instruments, users have to work at 15 kHz. The EMP-400 is worth roughly 16,000 €.

All the aforementioned EMI instruments can be used along with data-logging and positioning systems (GPS) to have georeferenced and continuous EC_a* measurements. In fact, DUALEM instruments and Profiler EMP-400 present internal GPS receivers [50]. Dualem also sells the sensors of their instruments individually under denominations 1S, 2S, etc., to allow practitioners to build their own customized equipment.

7.3. Reflectometry sensors

ER instruments provide EC_a measurements, while EMI instruments provide EC_a estimations through EC_a*. However, since EC_a strongly depends on θ in addition to soil salinity, and secondarily on other soil properties, reliable estimations of EC_p require, at least, a reliable estimation of θ in the soil volume under test. With maturing of reflectometry techniques, reliable estimations of both θ and EC_a can be made in exactly the same soil volume, giving rise to effective EC_p estimations [22, 51]. Another advantage of reflectometry is that contact between soil and electrodes is important but not so critical than with ER.

7.3.1. Time domain reflectometry sensors

Nowadays, there are three types of TDR measurement systems. The first type consists of four parts: i) a compact reflectometer that includes a signal generator, a fast oscilloscope, and a microcontroller; ii) one datalogger; iii) one or more multiplexors; and iv) several TDR probes that can be monitored at once. A remarkable example of an affordable compact reflectometer with wide multiplexing capabilities is the TDR100 (Campbell Scientific Inc., Logan, Utah, USA) which is worth 4000 €, and 100 € more per probe. Another modern example of this kind of instruments is the Trase Systems (SoilMoisture Equipment Corp., Santa Barbara, California, USA). With these TDR measurement systems, the EC_a and θ calculations can be programmed to be made by the reflectometer, or either the TDR traces can be saved and calculations deferred. All these instruments are essentially research oriented.

There are even more compact instruments which constitute a second type of TDR systems. These instruments integrate reflectometer and probe and besides, a temperature sensor. Examples are the Trime-PICO 64/32 (Van Walt Ltd., Haslemere, Surrey, UK), which works at 1 GHz, and the CS615 and CS616 (Campbell Scientific Inc., Logan, Utah, USA), which operate at frequencies of 45 and 70 MHz, respectively [52], and thus are less expensive (less than 300 €) but also less accurate for θ estimation than the previous more complex TDR systems. These compact TDR probes make EC_a and θ calculations automatically and are adequate for practical applications in agriculture and soil studies.

A third type of TDR systems are profile probes. These devices are formed by several paired electrodes, which are usually equally separated on opposite sides of a non-conductive tube, which is vertically inserted into the soil. Examples are the Trime PICO-Profile (IMKO Micro-modultechnik GmbH, Ettlingen, Baden-Württemberg, Germany), and the Vector Probe (Aquaspy, San Diego, California, USA) with 12 sensors down to 1.2 m depth, which is worth less than 1000 €.

7.3.2. *Amplitude domain reflectometry sensors*

There are several commercial ADR sensors able to estimate θ such as the popular Theta Probe (Delta-T Devices Ltd., Cambridge, UK). However, to our knowledge, only the Hydra Probe (Stevens Water Monitoring Systems, Inc., Portland, Oregon, USA) is able to take both ε_a and EC_a measurements. This sensor forms its probe with three aligned stainless steel prongs, wired to an oscillator working at 50 MHz. It also measures soil temperature and is worth roughly 400 €.

7.3.3. *Frequency domain reflectometry sensors*

Commercial FDR sensors work at frequencies between 10 and 200 MHz, i.e. lower than TDR. There are two basic types of FDR sensors: single and profile probes. The degree in which the soil contributes to the dielectric medium in the soil sensing capacitors is markedly different in each kind of probe. For single-probe devices, the soil forms a good deal of the dielectric medium (**Figure 3**), while in profile probes, the soil medium is just an almost marginal part of it. In profile FDR probes, the soil sensing capacitors are attached to a non-conductive rod or plate which is introduced into an insulating access tube, and this in turn vertically into the soil. Therefore, the soil sensing capacitors are not in contact with the soil, and in fact only a small part of the electromagnetic field created by each one permeates the surrounding soil. This fringe field extinguishes rapidly from the capacitor. Accordingly, most of the sensitivity of FDR profile probes lies in the soil zone immediate adjacent to the access tube, which is the one more affected by soil drilling, thus raising concerns about representativeness [53].

There are several single-probe FDR instruments for both θ and EC_a estimations. One of the most used for agricultural applications is the WET sensor (Delta-T Devices Ltd., Cambridge, UK), which is worth roughly 1200 €. The WET sensor forms the soil sensing capacitor by means of three aligned 6-cm-long metal prongs, with the central rod acting as the plus plate, and the side prongs acting as the ground plates of the capacitor. Additionally, the WET sensor has a thermistor at the central rod tip that enables soil temperature measurements. Measurements of EC_a with the WET sensor are very similar to those carried out with ER techniques. However, the relatively low oscillation frequency in the WET sensor (20 MHz) makes ε_a measurements much too dependent on soil salinity and therefore, impairs the estimations of θ, and thus EC_p [41].

Although there are several commercial profile FDR instruments for θ estimations, to our knowledge, only the TriSCAN (Sentek Pty Ltd., Stepney, South Australia, Australia) estimates soil salinity in addition to θ. This probe can bear up to 16 pairs of electrodes no less than 10 cm apart, and it provides an estimation of soil salinity expressed as volumetric ion content (VIC). It works at two frequencies: over 100 Mhz for θ estimation and below 27 MHz for VIC assessment. The VIC is derived with a proprietary method and is related to EC_a, though it is not directly interchangeably with it [54]. Each TriSCAN probe is worth between 1000 and 1600 € depending on length (0.6–1.2 m).

8. Applications of EC measurements in agriculture

The capability of EC sensors to easily and quickly take high amounts of measurements at broad spatial scales that range from profile horizons to watersheds, and at no less wide time scales that range from seconds to years, permits the development of many applications for both mapping and monitoring of soil salinity, and also soil salt dynamics. Interestingly, EC measurement systems have been used even for crop yield estimation, due to the significant correlations between EC_a and yield found for different crops such as tomato [55], corn and soybean [56–58], sorghum [57, 58], and cotton [59, 60]. The inherent integration within EC_a of various soil properties on which plant development depends, such as θ, ϱ_b, clay content, and EC_p, explains such ability. The use of EC measurement systems is nowadays of paramount importance for irrigation, crop, and fertilizer management in a framework of precision agriculture (PA) in which management is adapted to the specific soil and crop characteristics as they change through space within a field, and throughout time within growing seasons.

PA is a farming management concept based on observe, measure, and thus, respond to variability in crop fields, both spatial and temporal [61]. The ultimate goal of PA is agricultural sustainability and efficiency. By matching agricultural inputs to needs, PA aims at simultaneously maximizing crop production and product quality, while minimizing environmental damage. In a PA framework, the obtaining of big data about soil properties within fields feds the decision-making process. Provided the capabilities of the EC_a sensors described in previous sections, they are crucial to acquire this information in many applications with the eventual aim of controlling soil salinity and additionally, improving nutrition of plants avoiding harmful side effects on the environment.

8.1. Soil salinity mapping

EC_a measurements have been widely used to characterize soil salinity at field scale. Modern mobile ER measurement systems based on a four-point array along with the use of GPS allow to elaborate 3D maps of EC_a in agricultural fields, giving rise to ER Imaging (ERI) in soil studies and agriculture [47, 62]. Good contact between the electrodes and the soil is maybe the most important constraint to have reliable EC_a measurements using ER. Depending on soil texture, water and coarse fragment contents, good contact could not be assured, and therefore, reliable measurements could neither be guaranteed with ER, overall with mobile instruments.

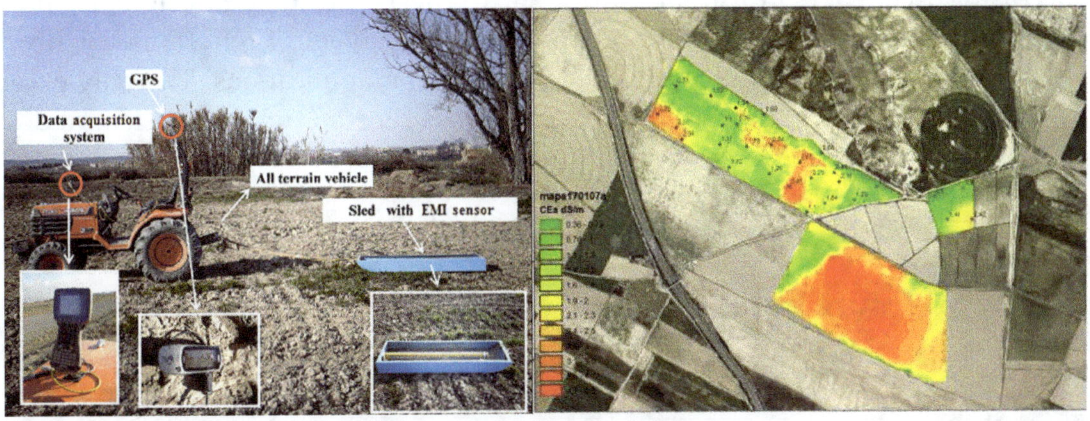

Figure 4. A mobile georeferenced electromagentic sensor (MGES) developed for the rapid carrying out of EC_a* surveys in agricultural plots [72].

Surveys for EC_a* taken with EMI sensors present several advantages over surveys for EC_a with ER techniques because EMI sensors do not require any contact with the soil. Therefore, EC_a* data can be more readily and reliably obtained on soils with stones and/or low water contents. Besides, with the specific aim of EC_a mapping, EMI techniques are usually overwhelmingly used instead of ERI, because EMI presents several advantages in addition to the previous such as i) the ability to make surveys on fields supporting growing crops, ii) the ability to make surveys on fields with beds and furrows, iii) the avoidance of soil alteration issues due to the low weight of EMI instruments, iv) the ability to survey faster because of the higher operating speeds of EMI instruments, and v) lower prices [63–65]. EMI methods present also some disadvantages, the most important of which is the more complex interpretation of EC_a* readings regarding EC_a.

The high volumes of EC_a* data obtained with EMI sensors are generally processed with the aid of geo-statistics [66, 67], multivariate statistics [68, 69], and GIS tools [70, 71]. EMI has been widely used for soil salinity mapping of agricultural plots, e.g. [72–75], by means of craft Mobile Georeferenced Electromagnetic Sensors (MGESs; **Figure 4**). Besides, EMI can be used along with remote sensing [76–78] to extend the capabilities of both techniques for soil salinity mapping at watershed scales.

In any case, since all EC_a measurements are affected by several other soil properties in addition to salinity, mainly water content and texture, EC_a data, and overall EC_a* data cannot be used alone for mapping of soil salinity. All EC_a surveys must be carried out along with traditional soil sampling and/or other sensor measurements and besides, other field observations must also be carried out.

8.2. Soil salinity monitoring

Since soil salinity is a dynamic property, many instruments able to automatically take EC_a measurements and additionally, to temporally save information and, to withstand the variable and tough outdoor conditions for long time spans, have been developed during the last decades. These instruments have been used mainly for agricultural water management and with increasing frequency, in a framework of PA. The EC_a technique originally used for monitoring was ER through four-point probes. Since the advent of reflectometry, however, TDR, ADR, and FDR have captured almost all monitoring applications. The only exception where ER still holds on is, maybe, the CC combined sensors, which are featured by an interesting price-quality ratio.

9. Present and future trends

The nowadays trends in EC_a sensor development are focused on improving accuracy, robustness, field installation ease, and data communication, while decreasing acquisition and maintenance expenses. All these improvements increase the applicability of EC_a sensors in agriculture, overall for irrigation and nutrient management in a PA framework. Nevertheless, while a large number of agricultural exploitations are using sensors for θ estimation to allow the subsequent adjustment of irrigation rates, the use of EC_a sensors is far less widespread. This fact is due to issues still not satisfactorily fixed about the correct interpretation of EC_a data under the ever-changing and diverse soil conditions of agricultural fields. Although many investigations have been carried out in order to interpret EC_a and, furthermore, to assess EC_p, far more research should be performed in this regard.

More accurate EC_a interpretations and EC_p assessments, and even $EC_{e,\ 25}$ estimations, from sensor measurements can proceed through three ways. First, it can proceed through the development of techniques able to separate water and salinity effects on sensor responses. Second, improvement of EC_a interpretations will take place through the development of calibrations based on more reliable models for EC_a. And third, improvement of EC_a interpretations will proceed through sensor fusion, in which data from different sensors will be used together in order to ascertain soil properties and/or crop status [79]. For the first requirement, reflectometry seems to be leading. Among the three different techniques, TDR has established itself as predominant. Nevertheless, the other two reflectometry techniques (ADR and FDR) are without doubt promising. On the one hand, in ADR, water and salinity effects have been claimed to be precisely separated [80] and, on the other hand, in FDR, progress can be made through the improved interpretation of soil dielectric spectra, i.e., soil permittivity against electromagnetic frequency up to 500 MHz [37]. Along with the improvement in the separation of water and salinity effects, the development of sensors less affected by temperature should also be performed so as to increase the robustness of measurement systems for EC_a interpretation.

Promising applications of reflectometry sensors for PA lie in the development of smart irrigation management systems (SIMS) (**Figure 5**). A SIMS has essentially two parts: i) the core, which is a decision support system (DSS) hosted in an Internet server, and ii) a field-deployed wireless sensor network (WSN), which feeds the DSS with e.g., EC_a and θ data. The WSN is made up of several probes distributed within the cropped field and wireless connected to one or more dataloggers which act as gateways to the WSN.. The dataloggers communicate data to the Internet server by means of GSM/GPRS, 3G, or 4G, for cloud computing by the DSS. The DSS is essentially a simulation model that runs on the data from the WSN and additionally, on other data, meteorological, soil, crop, water quality, fertilization, etc., and as a consequence, produces management recommendations which are sent to the farmer in real-time. A fully automatic system would include a third part. That is an actuator network, e.g., the simplest will be for irrigation hydrant control. Such fully automated SIMS including EC_a measurements seems to have not well been developed yet. However, remote irrigation management is possible by means of systems that use field θ sensor estimations [81, 82]. Further advances in cellular networks and cheaper and faster Internet communications will allow the spread of SIMS with fully farmer informed control through mobile telephony. This technology will benefit from the advances in EC_a data interpretation. One important way this will occur is through sensor fusion, which consists in the joint use of information originating from diverse sensors.

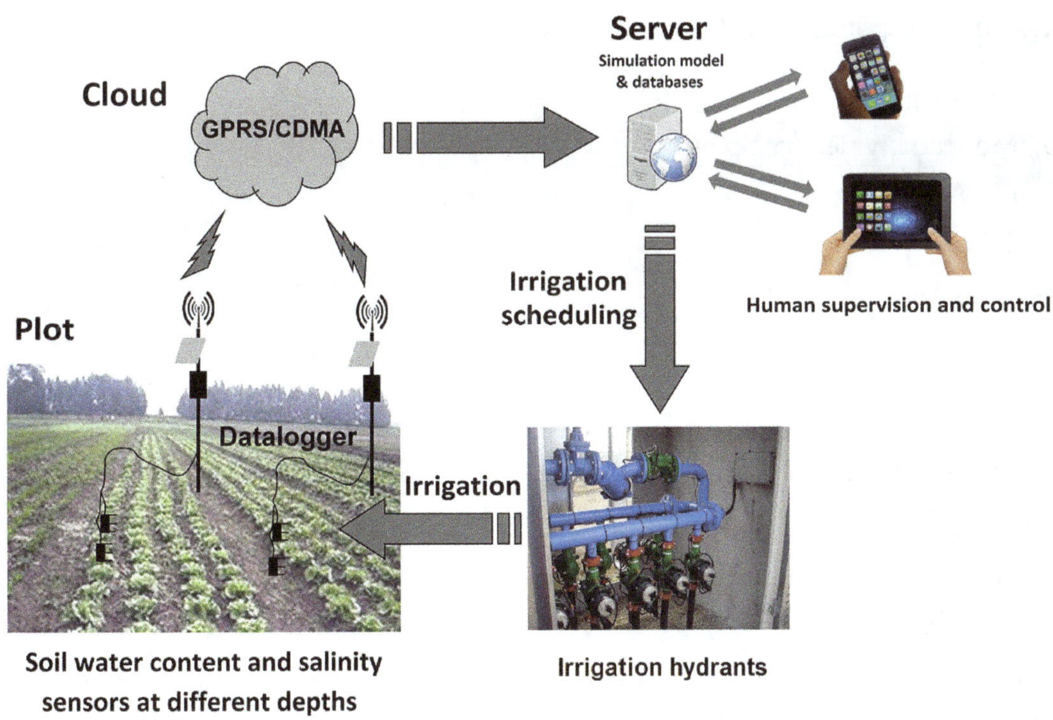

Figure 5. Scheme of a smart irrigation management system (SIMS).

Similarly to EC_a, which is an unselective factor on its own, i.e. it depends on various soil properties such as θ, ϱ_b, clay content, and EC_p, other proximal sensors as used in the field also provide unselective responses, e.g. hyperspectral, radiometric, mechanical, acoustic, pneumatic, and electrochemical [83]. Interestingly, under the concept of *sensor fusion*, all these unselective measurements including EC_a could be jointly used to generate selective information through the use of multivariate statistical methods such as MLR, PCR, and PLSR [69, 84].

Last, the use of EC_a measurements will remarkably benefit the fertilizer management for crop production in the next years. The conventional fertilizer management is generally based on monitoring the soil contents of nitrogen, phosphorous, and potassium by sampling and laboratory methods. However, since plant available N and K are in ionic form (NO_3^-, NH_4^+, and K^+), they contribute to EC_p and eventually EC_a. Therefore, the development of new sensors with higher capacity to separate water and salinity effects, in addition to sensor fusion, will contribute to the improvement of crop nutrition management, e.g. [85, 86]. Colburn [87] pioneering works on the secondary correlation of EC_a with soil nutrients pointed to the usage of EC_a measurements for fertilization management. However, more investigations are still needed on the application of EC_a sensors for fertilizer management [88].

Author details

Fernando Visconti* and José Miguel de Paz

*Address all correspondence to: fernando.visconti@uv.es

Valencian Institute for Agricultural Research (IVIA), Centre for Sustainable Agriculture (CDAS), Moncada, València, Spain

References

[1] Tanji, K. K. and Wallender, W. W. (2012). Nature and extent of agricultural salinity and sodicity. In W. W. Wallender and K. K. Tanji (Eds.), Agricultural salinity assessment and management (2nd Ed., pp. 1–25). Reston, VA: ASCE, EWRI.

[2] Ghassemi, F., Jakeman, A. J. and Nix, H. A. (1995). Salinisation of land and water resources: Human causes, extent, management and case studies. Wallingford: CAB International.

[3] Friedman, S. P. (2005). Soil properties influencing apparent electrical conductivity: A review. Computers and Electronics in Agriculture, 46(1–3 Spec. Iss.), 45–70.

[4] Wolt, J. D. (1994). Soil solution chemistry. New York: John Wiley & Sons, Inc.

[5] USSL Staff. (1954). Diagnosis and improvement of saline and alkali soils (Agriculture Handbook No. 60). Washington, DC: United States Department of Agriculture.

[6] Maas, E. V. and Hoffman, G. J. (1977). Crop salt tolerance – current assessment. ASCE J Irrig Drain Div, 103(2), 115–134.

[7] Shannon, M. C. and Grieve, C. M. (1998). Tolerance of vegetable crops to salinity. Scientia Horticulturae, 78(1–4), 5–38.

[8] Rhoades, J. D. (1996). Salinity: Electrical conductivity and total dissolved solids. In D. L. Sparks, et al. (Eds.), Methods of soil analysis part 3—Chemical methods (pp. 417–435). Madison, WI: SSSA, ASA.

[9] Ayers, R. S. and Westcot, D. W. (1985). Water quality for agriculture. Irrig drain paper 29, rev. 1. Rome: Food and Agriculture Organization of the United Nations.

[10] Corwin, D. L. and Lesch, S. M. (2005). Apparent soil electrical conductivity measurements in agriculture. Computers and Electronics in Agriculture, 46(1–3 Spec Iss.), 11–43.

[11] Ma, R., McBratney, A., Whelan, B., Minasny, B. and Short, M. (2011). Comparing temperature correction models for soil electrical conductivity measurement. Precision Agriculture, 12(1), 55–66.

[12] Johnson, P. (2007). The design of an integrated soil moisture sensor for agriculture. Brisbane, Queensland: Griffith University.

[13] Groom, D. (2008). Common misconceptions about capacitively-coupled resistivity (CCR) what it is and how it works. Environmental and Engineering Geophysical Society – 21st Symposium on the Application of Geophysics to Engineering and Environmental Problems, pp. 36–41.

[14] Wenner, F. (1915). A method of measuring earth resistivity. Bulletin of the Bureau of Standards, 12, 469–478.

[15] Loke, M. H., Chambers, J. E., Rucker, D. F., Kuras, O. and Wilkinson, P. B. (2013). Recent developments in the direct-current geoelectrical imaging method. Journal of Applied Geophysics, 95, 135–156.

[16] McNeill, J. D. (1980). Electrical conductivity of soils and rocks. Technical note TN-5. Mississauga, ON: Geonics Limited.

[17] Saey, T., Van Meirvenne, M., De Smedt, P., Neubauer, W., Trinks, I., Verhoeven, G., et al. (2013). Integrating multi-receiver electromagnetic induction measurements into the interpretation of the soil landscape around the school of gladiators at carnuntum. European Journal of Soil Science, 64(5), 716–727.

[18] Corwin, D. L., Lesch, S. M. and Lobell, D. B. (2012). Laboratory and field measurements. In W. W. Wallender and K. K. Tanji (Eds.), Agricultural salinity assessment and

management (2nd Ed., pp. 265–341). Reston, VA: American Society of Civil Engineers, Environmental and Water Resources Institute.

[19] Topp, G. C., Davis, J. L. and Annan, A. P. (1980). Electromagnetic determination of soil water content: Measurements in coaxial transmission lines. Water Resources Research, 16(3), 574–582.

[20] Savi, P., Maio, I. A. and Ferraris, S. (2011). Time-domain reflectometry (TDR) technique for the estimation of soil permittivity. In B. E. Özkaraova-Güngör (Ed.), Principles, application and assessment in soil science (pp. 351–370). Rijeka, Croatia: InTech.

[21] Muñoz-Carpena, R. (2012). Field devices for monitoring soil water content. BUL343. Gainsville, FL: Agricultural and Biological Engineering Department, Florida Cooperative Extension Service, Institute of Food and Agricultural Sciences, University of Florida.

[22] Noborio, K. (2001). Measurement of soil water content and electrical conductivity by time domain reflectometry: A review. Computers and Electronics in Agriculture, 31(3), 213–237.

[23] Tarantino, A., Ridley, A. M. and Toll, D. G. (2008). Field measurement of suction, water content, and water permeability. Geotechnical and Geological Engineering, 26(6), 751–782.

[24] Giese, K. and Tiemann, R. (1975). Determination of the complex permittivity from thin-sample time domain reflectometry improved analysis of the step response waveform. Advances in Molecular Relaxation Processes, 7(1), 45–59.

[25] Topp, G. C., Yanuka, M., Zebchuk, W. D. and Zegelin, S. (1988). Determination of electrical conductivity using time domain reflectometry: Soil and water experiments in coaxial lines. Water Resources Research, 24(7), 945–952.

[26] Muñoz-Carpena, R., Regalado, C. M., Ritter, A., Alvarez-Benedí, J. and Socorro, A. R. (2005). TDR estimation of electrical conductivity and saline solute concentration in a volcanic soil. Geoderma, 124(3–4), 399–413.

[27] Nadler, A. (2005). Methodologies and the practical aspects of the bulk soil EC (sa)-soil solution EC (sw) relations.

[28] Gaskin, G. J. and Miller, J. D. (1996). Measurement of soil water content using a simplified impedance measuring technique. Journal of Agricultural Engineering Research, 63(2), 153–160.

[29] Dean, T. J. (1994). The IH capacitance probe for measurement of soil water content. Wallingford, UK: Institute of Hydrology, Natural Environment Research Council.

[30] Robinson, D. A., Gardner, C. M. K., Evans, J., Cooper, J. D., Hodnett, M. G. and Bell, J. P. (1998). The dielectric calibration of capacitance probes for soil hydrology using an

oscillation frequency response model. Hydrology and Earth System Sciences, 2(1), 111–120.

[31] Seyfried, M. S. and Murdock, M. D. (2004). Measurement of soil water content with a 50-MHz soil dielectric sensor. Soil Science Society of America Journal, 68(2), 394–403.

[32] He, Y., DeSutter, T., Prunty, L., Hopkins, D., Jia, X. and Wysocki, D. A. (2012). Evaluation of 1:5 soil to water extract electrical conductivity methods. Geoderma, 185–186, 12–17.

[33] Rhoades, J. D., Chanduvi, F. and Lesch, S. (1999). Soil salinity assessment. Methods and interpretation of electrical conductivity measurements (FAO Irrigation and Drainage Paper 57). Rome, Italy: Food and Agricultural Organization of the United Nations.

[34] Kelleners, T. J. and Verma, A. K. (2010). Measured and modeled dielectric properties of soils at 50 megahertz. Soil Science Society of America Journal, 74(3), 744–752.

[35] Evett, S. R. (2007). Soil water and monitoring technology. In R. J. Lascano and R. E. Sojka (Eds.), Irrigation of agricultural crops (2nd Ed., pp. 25–84). Madison, WI: American Society of Agronomy, Crop Science Society of America, Soil Science Society of America.

[36] Malicki, M. A. and Walczak, R. T. (1999). Evaluating soil salinity status from bulk electrical conductivity and permittivity. European Journal of Soil Science, 50(3), 505–514.

[37] Wilczek, A., Szyplowska, A., Skierucha, W., Ciesla, J., Pichler, V. and Janik, G. (2012). Determination of soil pore water salinity using an FDR sensor working at various frequencies up to 500 MHz. Sensors (Switzerland), 12(8), 10890–10905.

[38] Hilhorst, M. A. (2000). A pore water conductivity sensor. Soil Science Society of America Journal, 64(6), 1922–1925.

[39] Kizito, F., Campbell, C. S., Campbell, G. S., Cobos, D. R., Teare, B. L., Carter, B., et al. (2008). Frequency, electrical conductivity and temperature analysis of a low-cost capacitance soil moisture sensor. Journal of Hydrology, 352(3–4), 367–378.

[40] Whalley, W. R. (1993). Considerations on the use of time-domain reflectometry (TDR) for measuring soil water content. Journal of Soil Science, 44(1), 1–9.

[41] Visconti, F., Martínez, D., Molina, M. J., Ingelmo, F. and De Paz, J. M. (2014). A combined equation to estimate the soil pore-water electrical conductivity: Calibration with the WET and 5TE sensors. Soil Research, 52(5), 419–430.

[42] Rhoades, J. D. (1981). Predicting bulk soil electrical conductivity versus saturation paste extract electrical conductivity calibrations from soil properties. Soil Science Society of America Journal, 45, 42–44.

[43] Rhoades, J. D., Shouse, P. J., Alves, W. J., Manteghi, N. A. and Lesch, S. M. (1990). Determining soil salinity from soil electrical conductivity using different models and

estimates. Soil Science Society of America Journal, 54, 46–54.

[44] Visconti, F. and De Paz, J. M. (2012). Prediction of the soil saturated paste extract salinity from extractable ions, cation exchange capacity, and anion exclusion. Soil Research, 50(7), 536–550.

[45] Rhoades, J. D. and van Schilfgaarde, J. (1976). An electrical conductivity probe for determining soil salinity. Soil Science Society of America Journal, 40(5), 647–651.

[46] Jakosky, J. J. (1938). Method and apparatus for electrical exploration of the subsurface. US-Patent No. 2.138.818.

[47] Pan, L., Adamchuk, V. I., Prasher, S., Gebbers, R., Taylor, R. S. and Dabas, M. (2014). Vertical soil profiling using a galvanic contact resistivity scanning approach. Sensors (Switzerland), 14(7), 13243–13255.

[48] Loke, M. H. (1999). Electrical imaging surveys for environmental and engineering studies. A practical guide to 2-D and 3-D surveys. Penang, Malaysia. http://moho.ess.ucla.edu/~pdavis/ESS135_2013/LITERATURE/%20LokeDCREsistivity.pdf. Accessed 13 April 2013.

[49] Urdanoz, V. and Aragüés, R. (2012). Comparison of geonics EM38 and dualem 1S electromagnetic induction sensors for the measurement of salinity and other soil properties. Soil Use and Management, 28(1), 108–112.

[50] Doolittle, J. A. and Brevik, E. C. (2014). The use of electromagnetic induction techniques in soils studies. Geoderma, 223–225(1), 33–45.

[51] Dalton, F. N. and Van Genuchten, M. T. (1986). The time-domain reflectometry method for measuring soil water content and salinity. Geoderma, 38(1–4), 237–250.

[52] Kelleners, T. J., Seyfried, M. S., Blonquist Jr., J. M., Bilskie, J. and Chandler, D. G. (2005). Improved interpretation of water content reflectometer measurements in soils. Soil Science Society of America Journal, 69(6), 1684–1690.

[53] Evett, S. R., Schwartz, R. C., Casanova, J. J. and Heng, L. K. (2012). Soil water sensing for water balance, ET and WUE. Agricultural Water Management, 104, 1–9.

[54] Buss, P., Dalton, M., Green, S., Guy, R., Roberts, C., Gatto, R., et al. (2004). Use of TriSCAN for measurement of water and salinity in the soil profile. Engineering Salinity Solutions: 1st National Salinity Engineering Conference 2004, Barton, ACT, pp. 206–211.

[55] Fortes, R., Prieto, M. H., Terrón, J. M., Blanco, J., Millán, S. and Campillo, C. (2014). Using apparent electric conductivity and NDVI measurements for yield estimation of processing tomato crop. Transactions of the ASABE, 57(3), 827–835.

[56] Zhu, Q., Lin, H. S. and Doolittle, J. A. (2013). Functional soil mapping for site-specific soil moisture and crop yield management. Geoderma, 200–201, 45–54.

[57] Kitchen, N. R., Sudduth, K. A. and Drummond, S. T. (1999). Soil electrical conductivity as a crop productivity measure for claypan soils. Journal of Production Agriculture, 12(4), 607–617.

[58] Kitchen, N. R., Drummond, S. T., Lund, E. D., Sudduth, K. A. and Buchleiter, G. W. (2003). Soil electrical conductivity and topography related to yield for three contrasting soil-crop systems. Agronomy Journal, 95(3), 483–495.

[59] Corwin, D. L., Lesch, S. M., Shouse, P. J., Soppe, R. and Ayars, J. E. (2003). Identifying soil properties that influence cotton yield using soil sampling directed by apparent soil electrical conductivity. Agronomy Journal, 95(2), 352–364.

[60] Guo, W., Maas, S. J. and Bronson, K. F. (2012). Relationship between cotton yield and soil electrical conductivity, topography, and landsat imagery. Precision Agriculture, 13(6), 678–692.

[61] Vanden-Heuvel, R. M. (1996). The promise of precision agriculture. Journal of Soil and Water Conservation, 51(1), 38–40.

[62] Lueck, E. and Ruehlmann, J. (2013). Resistivity mapping with GEOPHILUS ELECTRICUS—information about lateral and vertical soil heterogeneity. Geoderma, 199, 2–11.

[63] Corwin, D. L. and Lesch, S. M. (2005). Characterizing soil spatial variability with apparent soil electrical conductivity: Part II. Case study. Computers and Electronics in Agriculture, 46(1–3 Spec. Iss.), 135–152.

[64] Abdu, H., Robinson, D. A. and Jones, S. B. (2007). Comparing bulk soil electrical conductivity determination using the DUALEM-1S and EM38-DD electromagnetic induction instruments. Soil Science Society of America Journal, 71(1), 189–196.

[65] Serrano, J., Shahidian, S. and da Silva, J. M. (2014). Spatial and temporal patterns of apparent electrical conductivity: DUALEM vs. veris sensors for monitoring soil properties. Sensors (Switzerland), 14(6), 10024–10041.

[66] De Paz, J. M., Visconti, F. and Rubio, J. L. (2011). Spatial evaluation of soil salinity using the WET sensor in the irrigated area of the segura river lowland. Journal of Plant Nutrition and Soil Science, 174(1), 103–112.

[67] Ding, J. and Yu, D. (2014). Monitoring and evaluating spatial variability of soil salinity in dry and wet seasons in the werigan-kuqa oasis, china, using remote sensing and electromagnetic induction instruments. Geoderma, 235–236, 316–322.

[68] Rodrigues, F. A., Bramley, R. G. V. and Gobbett, D. L. (2015). Proximal soil sensing for precision agriculture: Simultaneous use of electromagnetic induction and gamma radiometrics in contrasting soils. Geoderma, 243–244, 183–195.

[69] Mouazen, A. M., Alhwaimel, S. A., Kuang, B. and Waine, T. (2014). Multiple on-line soil sensors and data fusion approach for delineation of water holding capacity zones for site specific irrigation. Soil and Tillage Research, 143, 95–105.

[70] Corwin, D. L. and Lesch, S. M. (2005). Characterizing soil spatial variability with apparent soil electrical conductivity: I. Survey protocols. Computers and Electronics in Agriculture, 46(1–3 Spec. Iss.), 103–133.

[71] Corwin, D. L., Lesch, S. M., Oster, J. D. and Kaffka, S. R. (2006). Monitoring management-induced spatio-temporal changes in soil quality through soil sampling directed by apparent electrical conductivity. Geoderma, 131(3–4), 369–387.

[72] Urdanoz, V., Amezketa Lizarraga, E., Clavería, I., Ochoa, V., Aragües Lafarga, R. (2008). Mobile and georeferenced electromagnetic sensors and applications for salinity assessment. Spanish Journal of Agricultural Research, 6(3), 469–478.

[73] Nogués, J., Robinson, D. A. and Herrero, J. (2006). Incorporating electromagnetic induction methods into regional soil salinity survey of irrigation districts. Soil Science Society of America Journal, 70(6), 2075–2085.

[74] Amezketa, E. (2006). An integrated methodology for assessing soil salinization, a pre-condition for land desertification. Journal of Arid Environments, 67(4), 594–606.

[75] Herrero, J., Ba, A. A. and Aragüés, R. (2003). Soil salinity and its distribution determined by soil sampling and electromagnetic techniques. Soil Use and Management, 19(2), 119–126.

[76] Metternicht, G. I. and Zinck, J. A. (2003). Remote sensing of soil salinity: Potentials and constraints. Remote Sensing of Environment, 85(1), 1–20.

[77] Yao, Y., Ding, J., Kelimu, A., Zhang, F. and Lei, L. (2013). Research on remote sensing monitoring of soil salinization based on measured hyperspectral and EM38 data. Spectroscopy and Spectral Analysis, 33(7), 1917–1921.

[78] Taghizadeh-Mehrjardi, R., Minasny, B., Sarmadian, F. and Malone, B. P. (2014). Digital mapping of soil salinity in Ardakan region, central Iran. Geoderma, 213, 15–28.

[79] Castrignanò, A., Landrum, C. and de Benedetto, D. (2015). Delineation of management zones in precision agriculture by integration of proximal sensing with multivariate geostatistics. examples of sensor data fusion. Agriculturae Conspectus Scientificus, 80(1), 39–45.

[80] Bellingham, K. (2013). Comprehensive Stevens Hydra Probe Users Manual. Rev. IV. Beaverton, OR: Stevens Water Monitoring Systems, Inc.

[81] Puerto, P., Domingo, R., Torres, R., Pérez-Pastor, A. and García-Riquelme, M. (2013). Remote management of deficit irrigation in almond trees based on maximum daily trunk shrinkage: Water relations and yield. Agricultural Water Management, 126, 33–45.

[82] Navarro-Hellín, H., Torres-Sánchez, R., Soto-Valles, F., Albaladejo-Pérez, C., López-Riquelme, J. A. and Domingo-Miguel, R. (2015). A wireless sensors architecture for efficient irrigation water management. Agricultural Water Management, 151, 64–74.

[83] Adamchuk, V. I., Hummel, J. W., Morgan, M. T. and Upadhyaya, S. K. (2004). On-the-go soil sensors for precision agriculture. Computers and Electronics in Agriculture, 44(1), 71–91.

[84] De Benedetto, D., Castrignanò, A., Rinaldi, M., Ruggieri, S., Santoro, F., Figorito, B., et al. (2013). An approach for delineating homogeneous zones by using multi-sensor data. Geoderma, 199, 117–127.

[85] Fuentes, S., Rogers, G., Jobling, J., Conroy, J., Camus, C., Dalton, M., et al. (2008). A soil-plant-atmosphere approach to evaluate the effect of irrigation/fertigation strategy on grapevine water and nutrient uptake, grape quality and yield. Acta Horticulturae, (792), 297–303

[86] Rogers, G., Shuttleworth, L., Fox, M., Fuentes, S., Dalton, M., and Conroy, J. (2008). Evaluation of a combined soil EC and moisture sensor and its use to co-manage soil moisture and vine nitrogen in grapevines (cv. shiraz) under deficit irrigation. Acta Horticulturae, 792, 543–549.

[87] Colburn, J. W. (1999). Soil doctor multi-parameter, real-time soil sensor and concurrent input control system. In P. C. Robert, R. H. Rust & W. E. Larson (Eds.), Precision agriculture (pp. 1011–1021). Madison, WI: American Society of Agronomy, Crop Science Society of America, Soil Science Society of America.

[88] Paul, W. (2002). Prospects for controlled application of water and fertiliser, based on sensing permittivity of soil. Computers and Electronics in Agriculture, 36(2–3), 151–163.

9

Force Measuring System for High-Precision Surface Characterization under Extreme Conditions

Roman Nevshupa and Marcello Conte

Abstract

Force measuring is used in various surface characterization techniques such as indentation, scratch tests, tribological analysis, determination of gas content, etc. The main problems related with force measurement under extreme conditions have been analysed. A strategy that should be followed to solve these problems has been discussed and several examples of successive solutions that recently were developed by the authors are presented. The need to carry out the characterization under extreme conditions poses serious problems for the designers of the measuring systems that may include the incompatibility of the sensors with the test conditions, undesirable interactions with other components, stability, precision and uncertainty issues, the measurement range, etc. Resolving these problems must be based on a global approach in which the characterization system is considered as a whole, while the designer must analyse and solve the possible conflicts between the subsystems. The way how an appropriate force measuring system can be selected is described. The proposed method is illustrated by an example in which an indirect force measurement using optical fibre displacement sensor was used. Another example describes measuring system developed for vacuum high-temperature nanoindentation. At high temperature, proper heat management based on non-contact heating and laminar flow cooling system is mandatory to avoid experimental data being affected by external noise and thermal drift.

Keywords: Force sensor, Uncertainty, Modelling, Extreme conditions, Vacuum, High-Temperature Nanoindentation

1. Introduction

Like aerospace industry itself, vacuum tribology rapidly developed in the middle of the twentieth century pursuing to cover the needs in this growing sector in reliable durable materials and lubricants suitable for operating under vacuum and rarefied gas atmosphere [1–3]. By the 1970s, when some of the problems had been resolved and the limitations of other had been defined, further development of vacuum tribological techniques slowed down [4]. The situation changed in 1990s driven by the advancements in various sectors including semiconductor, energy, transport, etc. that issued new challenges for development of materials with controlled micro- and nanostructure for extreme operation conditions. For example, advanced tribological materials for space applications must comply with very strict requirements on strength, toughness, and wear resistance and also have low outgassing, emission of volatiles, high chemical stability, low secondary electron emission, etc. [3,5,6]. Moreover, in semiconductor industry gaseous and particular contamination associated with tribochemical degradation and wearing of materials in tribo-contacts of mechanical components is one of the main problems that interferes development of fully automatic robotized technological lines with enhanced production yield.

Surface mechanical characterization in vacuum such as precise tribometry, indentation, scratching, etc. is now enjoining a renaissance with various new techniques developed in the last three decades [4,7–9]. These new techniques have been evolved towards both enhancement of their features and integration with other techniques. Combination of mechanical , physical and chemical characterization [4] has offered the opportunity to study complex processes of gas-phase lubrication [10], mechanically stimulated gas emission [4,11–13], triboplasma [14–16], triboelectrification [17], triboluminescence [15,16,18], emission of charged particles [19,20], tribochemical reactions [21–25], and so on. On the other hand, nanoindentation has become widespread as a method to study the mechanical response of materials [26], particularly the measurement of hardness [27,28], elastic modulus [27,28], hardening exponents [29], creep parameters [30], and residual stresses [31]. For all of these purposes, however, nanoindentation testing has most commonly been conducted at room temperature. This is in spite of the fact that micro-materials and devices are often employed at elevated temperatures, and deformation physics are usually thermally activated. The use of nanoindentation to study materials and thin films at high temperature is of high interest and undergo through several limitation due to material oxidation, thermal drift, and machine stability [32].

Therefore, development of precise and reliable measuring systems for force and displacement that must be compatible with extreme operating conditions and with other instruments for physico-chemical characterization is crucial for overall performance of the techniques. In this work we summarize some recent advances in development of force measuring systems for high-temperature nanoindentation and ultrahigh vacuum tribometry.

2. Problems and challenges of measuring systems under extreme conditions: resolving the conflicts

2.1. Ultrahigh vacuum tribology

Force transducers suitable for ultrahigh vacuum applications must comply not only with the requirements on rated capacity, non-linearity, combined error, repeatability, reproducibility, creep, long-term stability, frequency response, fatigue life, etc. but also with a series of specific provisions [33]. Among others these provisions include possibility of outgassing baking, reduced emission rate of volatiles and gas desorption, appropriate thermal management and heat sinks in absence of convection and heat conduction through the air, low emission of electromagnetic interferences that could affect other neighbour analytic tools, low emission of gases and volatiles, and so on. For example, strain gauges, capacitive, and inductive transducers can cause crosstalk in charge detectors, electron multipliers, antennas and other sensitive electromagnetic devices. Force transducers that utilize visible light may alter the measurements of triboluminescence by introducing undesirable background radiation.

In general, selection and development of the transducers for measuring system being used under vacuum must be done using holistic approach in which a satisfactory compromise between the requirements set by various subsystems of the entire test rig upon the environmental and operational conditions, relative position of the components and their compatibility must be found. A design of the tests rig in which such satisfactory compromise is achieved is considered as optimal. Recently it was shown that an experimental system for combined tribological and , physical and chemical characterization normally comprises seven subsystems one of which is the measurement system [4]. Therefore, optimal design must involve matching the measuring systems with other six subsystems including vacuum, mechanical, sample handling, environmental control, loading, and kinematics and physico-chemical characterization (**Figure 1**).

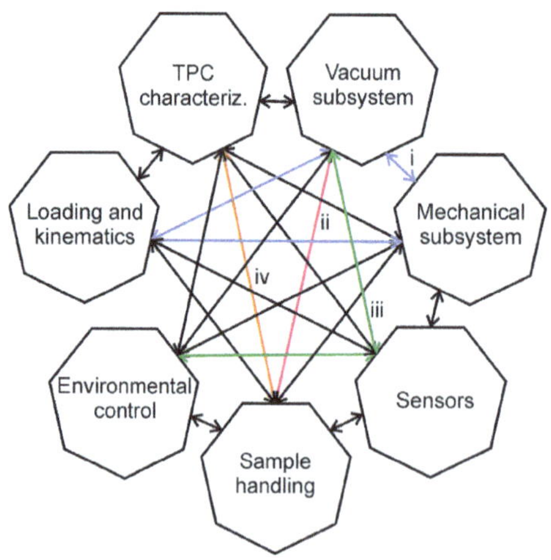

Figure 1. Subsystems of a vacuum test rig for complex tribological and tribo-physico-chemical (TPC) characterization of materials. Some of the critical relationships and constrictions between the subsystems are shown by arrows: i (blue) – limitations between Vacuum, Mechanical, and Loading and Kinematics subsystems; ii (red) – limitations between Vacuum and Sample handling subsystems; iii (green) – limitations between Sensors, Environmental control, and Vacuum subsystems; iv (orange) – limitations between Sample handling and TPC subsystems.

2.2. Nanoindentation at high and low temperatures

In normal ambient conditions most of metallic materials, semiconductors, metal hydrides, etc., when heated up, experience oxidation. In many cases an oxide layer at elevated temperature can be as thick as the penetration depth of a tip. This poses serious problem for material characterization through nanoindentation since the measured hardness value may correspond to the oxide layer rather than to the substrate. To avoid oxidation the measurement should be done in inert environment or in high vacuum. The latter is preferable because of the higher surface cleanliness. Also, in absence of convection in vacuum more precise and efficient heating control can be achieved. Vacuum is also desirable when indentation must be carried out at cryogenic temperatures. In this case, vacuum is needed to avoid condensation and to cut down thermal flow towards cold surfaces through conductivity and convection. Nevertheless, an increase the complexity of the experimental system by adding subsystems, e.g. pumping means, vacuum measurement, etc. can introduce important external noise or limit the system functionality. For example, the materials being tested must be vacuum compatible, i.e. to have sufficiently low pressure of saturated vapours at given temperature in order to reach the required vacuum degree. Another important problem is related with deformations of the system components due to differences in thermal expansion of construction materials. The components of the measurement system may be subjected to important temperature gradients when a pin and material at the indentation zone are heated up or cooled down, while some sensitive elements including electronic circuits situated in proximity to the indentation zone have to be maintained at room temperature. When temperature distribution in the components of the measuring system is non-stationary it can lead to thermal drift. Therefore, the collected data, in general, is a complex function of the mechanical behaviour of material under study, fluctuations and noise induced due to operation of subsystems, and thermal drift. Measurement uncertainties and errors associated with fluctuations, noise and thermal drift can be significantly reduced by correct designing of the measurement system. Analysis of some problems that must be solved to achieve the optimal design and possible solutions are discussed in Section 3.

3. Nanoindentation at high temperature

Development of turbomolecular pumping system with a magnetically suspended rotor (maglev) and oil-less primary pump has offered a cost-effective and clean solution for those applications where vibration is a serious problem. By combining opportune materials with maglev pumping system it is possible to significantly dump vibrations in nanoindentation systems. Nevertheless, it is not enough when vibrations due to thermal expansion happen. An interesting solution was presented in [34,35]. It is based on an active top referencing configuration which eliminates almost entirely the problem of noise and, furthermore, reduces to negligible values the instrument frame compliance. Such substantial improvement has been achieved by the use of a unique principle of load and displacement measurements. So far all existing nanoindentation systems have been based on only one actuator and one sensor. The ultra nanoindentation method uses two separated actuators and three separated sensors, which provide real measurements of depth and load, as well as a feedback loop that allows continuous and accurate control of the applied load (**Figure 2**). In particular, each axis has its own actuator, displacement and load sensors. For both axes, the displacement is applied via piezo actuators A1 and A2. The load on the indenter and the reference is obtained from the displacement of the springs K1 and K2, measured with capacitive sensors C1 and C2. The displacement of the indenter is measured relative to the reference through the differential capacitive sensor C3.

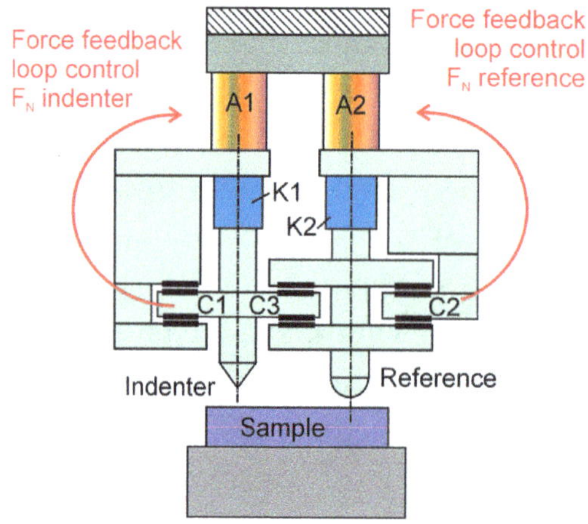

Figure 2. Schematic design of Ultra Nanoindentation head. F_N – normal load; A1 – Indenter's piezo actuator; A2 – Reference's piezo actuator; K1 – Indenter's load cell spring; K2 – Reference's load cell spring; C1 – Indenter's load cell capacitive sensor; C2 – Reference's load cell capacitive sensor; C3 – Penetration depth differential capacitive sensor.

Continuous control of normal force on both the indenter and the reference is ensured by precise feedback loops. The crucial components used in the measurement head are made of Zerodur®, a material with extremely low coefficient of thermal expansion (0.01×10^{-6} K^{-1} over the range 0 °C – 100 °C).

Considering it as a starting point, addressing the thermal drift problem is the next step before performing high precision nanoindentation tests. The need to reduce the thermal flow at the contact point can be achieved in two ways called passive heating and active heating, respectively. The former is based on holding the indenter in contact with the sample surface for a time long enough to equilibrate the temperature of the involved bodies. The latter consist of heating up the involved bodies independently. This was demonstrated to be the most appropriate way, especially for temperature higher than 400 °C, that is, when incandescence becomes important [8]. In the same work, the effect of thermal drift on the elastic modulus and hardness measurements on standard reference materials has been analysed.

Heating the sample to be characterized is not a difficult task as any contact/non-contact heating system can be used. However, heating the indentation tip is not easy as any contact acts as a spring and its compliance can be read out by the force and displacement sensors. A non-contact heating system is then preferred.

In [9] a novel patent pending non-contact heating system based on IR radiation was presented (**Figure 3**). The main advantage of IR heaters is their almost zero thermal inertia. Such heating system located in high vacuum environment uses the active top referencing system previously illustrated for high temperature nanoindentation. The reference and indenter shafts are irradiated independently by infrared emitters and their temperatures are read out by means of thermocouples embedded in the tips. When the sample approaches the measurement device, an IR bath is established allowing a fast and precise regulation of the sample surface temperature. Opportune reflective coatings and water cooling management in the measurement area allow reaching accuracy of temperature about 0.1 °C, that is the accuracy of the thermocouples at temperatures as high as 800 °C.

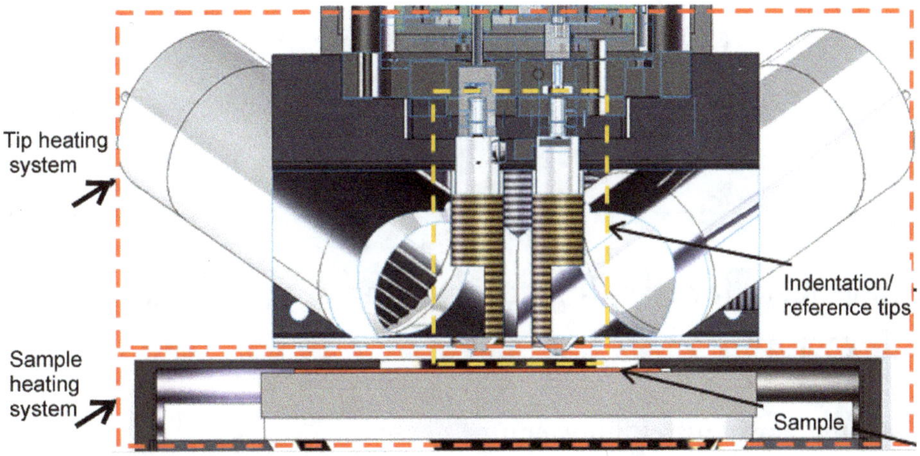

Figure 3. Schematic drawing of the sample and tip heating systems.

Thermal drift rate was measured on oxygen free high conductivity copper (OFHC Cu) by loading up to 100 mN in at least 30 s, holding the maximum load for few seconds and unloading to 5 mN, that is less than 10% of the maximum load to avoid material creep (**Figure 4**). The indenter was hold in contact for 1 minute and the deviation of the penetration depth was measured along this period. A thermal drift less than 1 nm/min was recorded for all the measurements up to 600 °C (**Figure 5**).

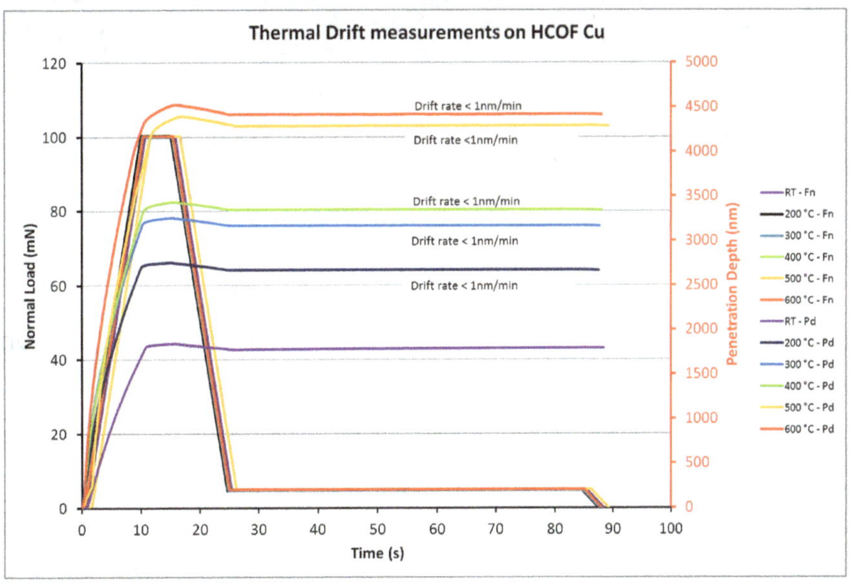

Figure 4. Thermal drift of the measuring system registered under various loads and temperatures.

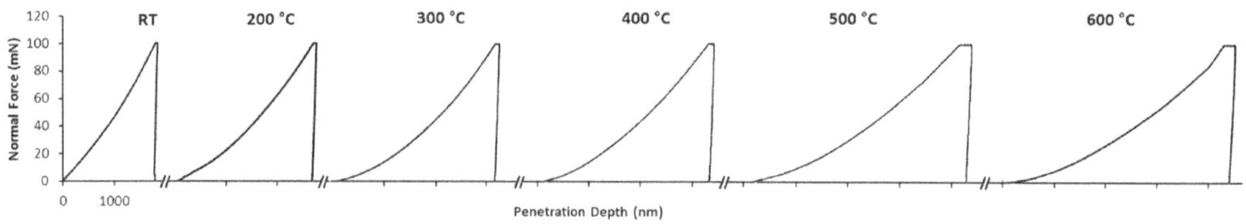

Figure 5. Indentation tests on oxygen free high conductivity copper at various temperatures.

The extrapolation of Elastic Modulus and Hardness from the uncorrected data is then an easy task.

4. Force transducers for tribological and physicochemical characterization in ultrahigh vacuum

4.1. Force measurement in ultrahigh vacuum

The force transducer is a device that is subjected to the force to be measured and changes it into another measurable physical quantity through a known relationship.There are several groups of force transducers that use different physical relationships. These transducers are based on piezoelectric phenomenon, light pressure [37], magneto-elasticity [38], vibration, surface waves, gyroscopic effect, electromagnetic watt balance [39], etc. Many force transducers employ some form of elastic load-bearing element with known load-deformation behaviour.Elastic devices represent the most common type of a force transducer, while the most frequent method is to make measurements of the longitudinal and lateral strain. When a force is applied to the elastic element it deflects and the deformation is measured by a displacement transducer. Displacement means moving from one position of the object to another for a specific distance or angle referencing to its own prior position rather than to an external reference [36].

Elastic devices are made of materials which exhibit a linear relationship between the stress and strain with low hysteresis and low creep in the working range such as metal alloys (beryllium copper, phosphor bronze, Monel, Inconel, high-carbon and alloyed steel, amorphous metals, etc.), silicon, borosilicate glass, etc. and are perfectly compatible with ultrahigh vacuum conditions. The shape of elastic elements depends on a number of factors including the range of force to be measured, required performance, sensitivity to misalignments and buckling of the displacement transducer coupled to the elastic elements, etc.

Among different geometries of elastic elements of force transducers the most commonly used are the bending beam, cantilever or leaf spring arranged perpendicularly to the direction of acting force to be measured (**Figure 6**) [34,40]. A cantilever employed in Atomic Force Microscopy is an example of a single beam that is used for measuring of forces in two directions. **Figure 7** shows a set-up developed for microtribological measurements and consisting of two parallel beams [41–43].

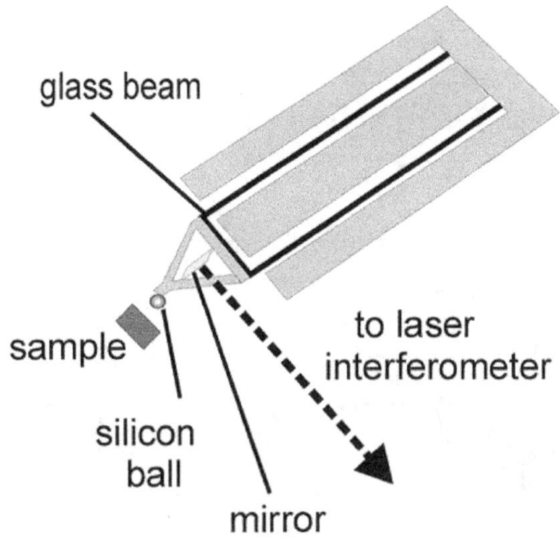

Figure 6. Schematic drawing of a double-leaf spring made of glass and used for simultaneous measurement of normal and tangential forces in a microtribometer.

The important disadvantage of a single beam configuration is that under applied load the free end of the beam undergoes both displacement and tilting (**Figure** 7). In some cases tilting is undesirable since it can lead to systematic errors in some displacement transducers.

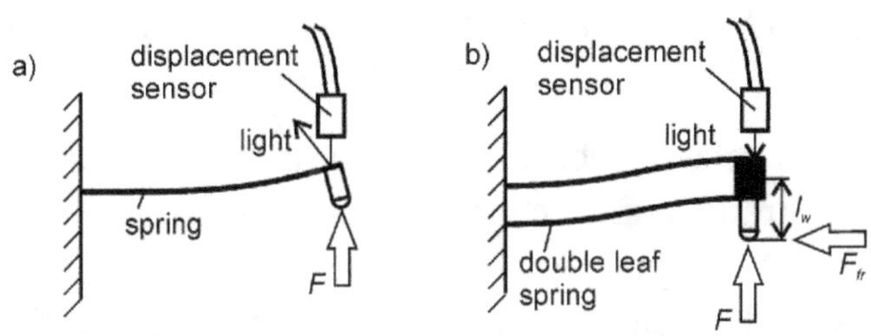

Figure 7. Schematic drawing of a force sensor with a single leaf spring (a) and double-leaf spring (b).

Double-leaf spring with two or more parallel beams with their free ends rigidly connected to a block are nearly free of this disadvantage and follows approximately straight displacements [44]. Further improvements of the spring linearity and reducing tilting were achieved by using symmetrical double-leaf configuration with two couples of springs facing each other on both sides of a point, where the force is applied. **Figure 8** shows a complex elastic element designed for measuring of two perpendicular forces [45] using four compliant units (A-D) and their respective mirrored compliant units (A′-D′). The operation of the unit has been described as the following: *When the normal force is applied the Flexure B, B′ and D, D′ would bend to give the desired displacement and Flexure A, A′ and C, C′ are in tensile/compressive load. Similarly, when the lateral force is applied flexure A, A′ and C, C′ deflects to give the desired motion and Flexure B, B′ and D, D′ are in tensile/compressive load. Any parasitic errors due to bending of compound flexures are compensated by the secondary motion stage. Furthermore, this force measuring mechanism is relatively insensitive to thermal disturbances and manufacturing errors due to its symmetry.*

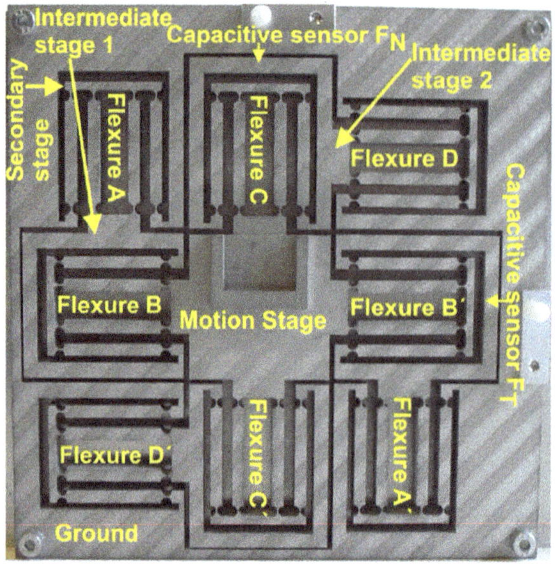

Figure 8. Elastic element for measuring two perpendicular forces using four compliant symmetric double-leaf units. (Reprinted from [45]).

A number of force transducers employing elastic elements in the form of parallel double-leaf spring have been developed and patented so far for atmosphere and vacuum tribometry [41,43,46–50]. A different approach was proposed to measure 3D forces by a single elastic element having a shape of a hollow elastic dome [51]. Deformation of the dome is sensed by a photodiode matrix situated at the circular baseplate of the dome.

Sensing displacement of the elastic element in vacuum can be done by almost any kind of displacement transducer. Strain gauge technology has been the most prominent on the market since its inception in 1938. Other types of displacement transducers include capacitive [34,52, 53], laser interferometry [54], Doppler sensor [55], linear variable differential transformer, optical fibre strain gauge, interference-optical load cell, and so on. An optical displacement transducers is preferable when some sensitive devices are used for physic-chemical characterization. A fibre-optic displacement sensor based on the principle of reflective light intensity modulation is a simple, contactless and compact transducer suitable for operating under vacuum and extreme environments [41,42,56–60]. The basic principles and uncertainty analysis of this type of the sensors are discussed in the following section.

4.2. Intensity modulated bundle fibre-optic displacement sensor

Intensity modulation fibre-optic displacement sensor (IMFODS) consists of one or several transmitting and one or several receiving step-index optic fibres packed in a bundle. At the proximal end the transmitting fibres are illuminated by a light source. Light passed through the transmitting fibres exits their distal end and is reflected from the surface, a distance to which is being measured (**Figure 9**). It is assumed that the reflection is pure specular; thus, the irradiance distribution function at the image plane is the same as for the radiant emittance at the exit from the transmitting fibre. Then the reflected light is coupled to the receiving fibre and directed to a photodiode detector to measure its power. Displacement of the reflecting surface towards or backwards the distal end of the bundle will modulate the radiant power of light entering the receiving fibres, thus producing variable electrical signal on the photodiode [61,62]. A thorough explanation of the operating principles of the intensity modulation optic fibre displacement sensors is given in the literature [56,62–64].

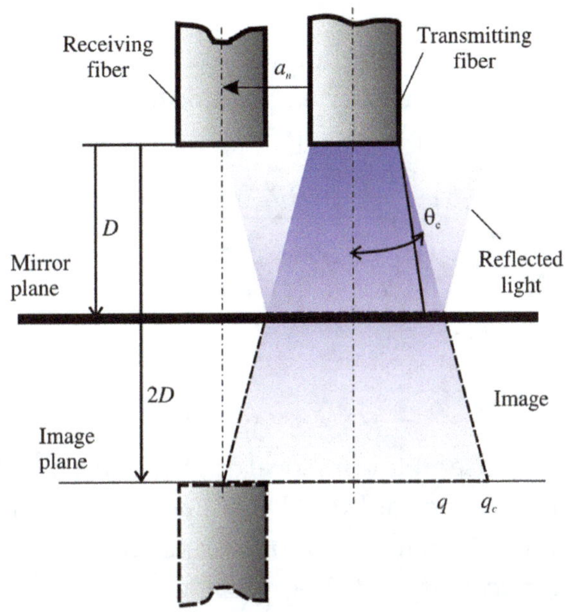

Figure 9. Geometrical considerations of a two-fibre sensor.

The irradiance distribution function at the image plane and the configuration of the receiving and transmitting fibres within a bundle provide the necessary information to determine the sensor modulation characteristic, i.e., the electrical signal at the photodiode detector vs. displacement.

Normally, IMFODSs have two measurement regions: one at near side (front slope) and another at far side (back slope) of the modulation characteristic (**Figure 10**).

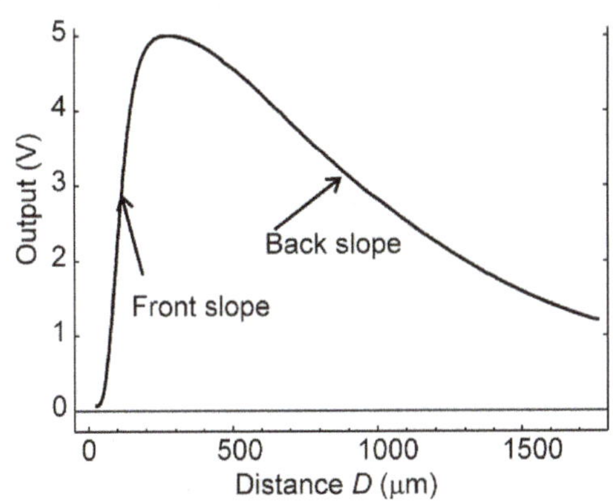

Figure 10. Typical characteristic of IMFODS.

It is gener0ally assumed that both regions are linear. So, when the slope *K*, of a linear region is known, displacement of the target surface, Δl, can be easily determined from the difference in the output voltage at the final, U_f, and initial, U_i, positions of the target surface [65]:

$$\Delta l = K\left(U_f - U_i\right) \tag{1}$$

Assuming that the measurements of output voltage at the initial and final positions are independent and have the same uncertainty, u_u, the uncertainty in the displacement measurement is found from the following expression [65]:

$$u_{\Delta l} = \Delta l \sqrt{\frac{u_K^2}{K^2} + 2\frac{u_U^2}{\left(U_f - U_i\right)^2}} \tag{2}$$

where u_K is the uncertainty of *K*.

Although linear assumption is useful for practical applications, the measurement uncertainty is quite large since the characteristics of IMFODS are not perfectly linear in both ranges [65]. In addition, linear regions are quite narrow as compared with the total available measurement range. All these limit the applicability of this approach.

Another approach is based on the displacement determination directly from the modulation characteristic without any assumption on its form [65]:

$$\Delta l = l\left(U_f\right) - l\left(U_i\right) \tag{3}$$

where $l(U_z)$ is the distance between the sensor and the target surface corresponding to the output voltage U_z.

In this approach an inverse modulation characteristic $l=g(U)$ is used. $g(U)$ is a monotonous function with two existence domains corresponding to front and back slopes of normal modulation characteristic.

The corresponding uncertainty is [65]:

$$u_{\Delta l} = \sqrt{u_{l(U_f)}^2 + u_{l(U_i)}^2} \tag{4}$$

where $u_{l(U)}$ is the uncertainty in the determination of distance $l(U)$.

Measurement uncertainty can be found experimentally in each point of the modulation characteristics or from the uncertainty of the normal modulation characteristic, $u_{f(l)}$, using the following transformation [65]:

$$u_{l(U)} = \frac{\sqrt{u_{f(l)}^2 - u_U^2}}{df(l)/dl} \tag{5}$$

In most cases the uncertainty in the voltage reading is much below the uncertainty of the modulation characteristic, so (5) can be simplified:

$$u_{l(U)} \approx \frac{u_{f(l)}}{\left|df(l)/dl\right|} \tag{6}$$

These formulae indicate that knowledge of the modulation characteristic is required for displacement measurement and determination of its uncertainty.

So far, several models were developed for modulation characteristic of a single pair of fibres and different irradiance distribution function: uniform [56,62,66,67], Gaussian [61,64,68], modified Gaussian [69,70], uniangular [63], theoretical [56]. Though these models showed good correlation with the experimental results, most of them have not been studied for multiple bundled fibres. Cao and co-workers [64] developed a model for the optical fibre bundle displacement sensor for several bundle configurations assuming quasi-Gaussian irradiance distribution function. The quasi-Gaussian, i.e., truncated Gaussian, distribution was adopted by these authors assuming that it coincides best with the actual facts, though, no argument for this assumption were presented. He and Cuomo [56] developed a more realistic theoretical irradiance distribution function for the lossless step-index multimode fibre illuminated with a Lambertian light source. This model is based on the assumption that each light ray propagating through the optical fibre core at different angles with respect to the fibre axis carries the same power. The model was validated for a bundle of seven fibres (six receiving fibres surrounding a transmitting one) with core radius 50 μm and the displacement range between 0 and 900 μm.

Before getting deeper into modelling the coordinates and geometrical considerations must be defined. It is assumed that light exits the transmitting fibre with a conical shape with a critical angle θ_c (**Figures 9** and **11**). The radius of a light spot at the image plane is a function of distance D between the distal end of the bundle and the reflecting surface:

$$q_c = x_0 + 2D\tan\theta_c \tag{7}$$

and corresponding dimensionless critical radius:

$$k_c = \frac{q_c}{x_0} = 1 + \frac{2D}{x_0}\tan\theta_c \tag{8}$$

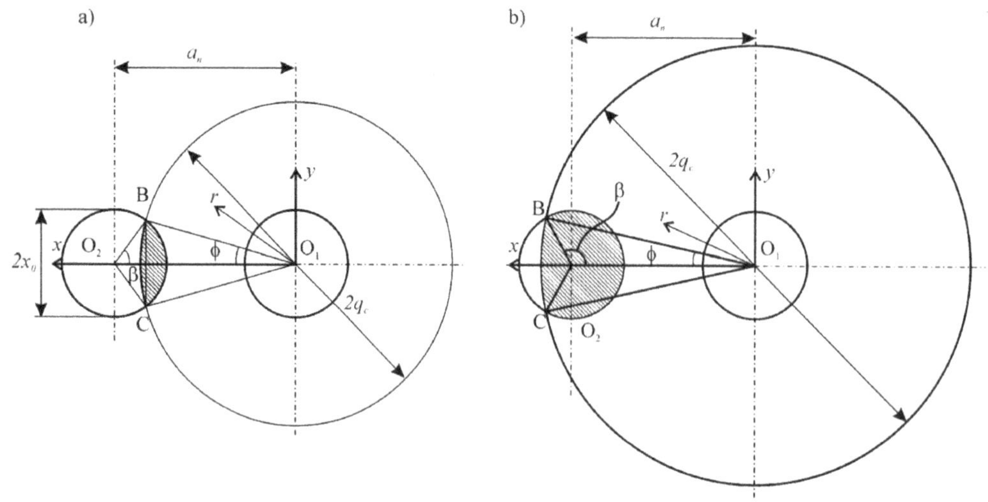

Figure 11. Geometry for determination of subtended area for a pair of fibres placed at a_n distance an between their centres.

The distance between the centres of adjacent fibres depends on the core diameter and the cladding thickness: $a_a = 2(c_l + x_0)$, whereas for any other pair of fibres within a bundle the distance between the fibre centres a_n is a function of the bundle configuration. The corresponding dimensionless distance is $m = a_n/x_0$. For simulation purpose, m was varied proportionally to the dimensionless distance between the centres of the adjacent fibres that in our case was 20/9.

The reflected light collected by a receiving fibre depends on the subtended area, S_i, and the irradiance distribution function, I_i, there on. The subtended area shown by the shadowed region in **Figure 11** is determined as function of only geometrical parameters from the following expression:

$$S_i(D,\theta_c) = x_0^2\left(k_c^2\varphi + \sin^{-1}(k_c \sin\varphi) - mk_c \sin\varphi\right),\ k_c \le \sqrt{m^2+1} \tag{9a}$$

$$S_i(D,\theta_c) = x_0^2\left(k_c^2\varphi + \pi - \sin^{-1}(k_c \sin\varphi) - mk_c \sin\varphi\right), \sqrt{m^2+1} < k_c < m+1 \tag{9b}$$

where

$$\varphi = \cos^{-1}\left(\left(k_c^2 + m^2 - 1\right)/\,2k_c m\right). \tag{10}$$

The above model is a further development of more simple models that account only for certain fixed distance between the fibres [63,67,69,70]. In addition, this model corrects some errors present in [63].

The received radiant power is calculated by integration the irradiance distribution function over the subtended area (**Figure 11**):

$$P_{ig} = \int_{S_i} I_i dS = x_0^2 \int_{m-1}^{k_{\max}} I_i k \cos^{-1} \frac{k^2 + m^2 - 1}{2km} dk \quad (11)$$

where $k_{max} = m + 1$ or $k_{max} = k_c$ whichever is smaller.

In the simplest case the irradiance distribution function is assumed uniform over the light spot. Then, assuming the reflection losses negligible, the irradiance at the image plane is determined as follows:

$$I_{iu} = I_0 \frac{x_0^2}{q_c^2} \quad (12)$$

where I_0 is mean radiant emittance at the exit from the transmitting fibre.

The radiant flux directed to the photodiode detector is given by the following expressions in dimensionless form:

$$\frac{P_i}{I_0 x_0^2} = k_c^{-2}\left(k_c^2\varphi + \sin^{-1}\left(k_c \sin\varphi\right) - mk_c \sin\varphi\right),\ k_c \le \sqrt{m^2+1} \quad (13a)$$

$$\frac{P_i}{I_0 x_0^2} = k_c^{-2}\left(k_c^2\varphi + \pi - \sin^{-1}\left(k_c \sin\varphi\right) - mk_c \sin\varphi\right),\ \sqrt{m^2+1} < k_c < m+1 \quad (13b)$$

$$\frac{P_i}{I_0 x_0^2} = \pi k_c^{-2} k_c > m+1$$

In more complex model the irradiance distribution is simulated by Gaussian function [61]:

$$I_{ig}\left(k\right) = \frac{I_0}{k_c^2} \exp\left(-\frac{k^2}{k_c^2}\right) \quad (14)$$

After substitution (14) in (11) the received radiant power can be obtained by integration in polar coordinates:

$$\frac{P_{ig}}{I_0 x_0^2} = 2k_c^{-2} \int_{m-1}^{k_{max}} \exp\left(-\frac{k^2}{k_c^2}\right) k \cos^{-1} \frac{k^2 + m^2 - 1}{2km} dk \tag{15}$$

Finally, the most realistic theoretical irradiance function was developed by He and Cuomo [56] as a piecewise-defined smooth function composed of eight formulas. These formulas combine according to the values of the parameters k and k_c. Five of these formulas correspond to the interval $k \geq 1$:

$$I_{it}(k) = \frac{I_0 d}{2\theta_c (d^2 - 1)} \left(\tan^{-1}(k_c - 1) - \tan^{-1}(k - 1) + d^{-1} \left(\tan^{-1} \frac{k-1}{d} - \tan^{-1} \frac{k_c - 1}{d} \right) \right), 1 \leq k_c \leq 2 \text{ and } 1 \leq k \leq k_c \tag{16a}$$

$$I_{it}(k) = \frac{I_0 d}{2\theta_c (d^2 - 1)} \left(\frac{\pi}{4} - d^{-1} \tan^{-1} d^{-1} - \tan^{-1}(k - 1) + d^{-1} \tan^{-1} \frac{k-1}{d} \right) + \frac{I_0}{8\theta_c d} \ln \frac{(k_c - 1)^2 (1 + \mathrm{d}^{-2})}{1 + \mathrm{d}^{-2} (k_c - 1)^2}, k_c > 2, 1 \leq k \leq 2 \text{ and } k \geq k_c - 2; \tag{16b}$$

$$I_{it}(k) = \frac{I_0 d}{2\theta_c (d^2 - 1)} \left(\frac{\pi}{4} - d^{-1} \tan^{-1} d^{-1} \tan^{-1}(k - 1) + d^{-1} \tan^{-1} \frac{k-1}{d} \right) + \frac{I_0}{8\theta_c d} In \frac{(k - 1)^2 (1 + d^{-2})}{1 + d^{-2} (k - 1)^2}, \mathrm{k}_c > 2, 1 \leq k \leq 2 \text{ and } k < k_c - 2; \tag{16c}$$

$$I_{it}(k) = \frac{I_0}{8\theta_c d} \ln \frac{(k_c - 1)^2 \left(1 + d^{-2} (k - 1)^2\right)}{(k - 1)^2 (1 + d^{-2} (k_c - 1)^2}, k_c > 2, k > 2 \text{ and } k \geq k_c - 2; \tag{16d}$$

$$I_{it}(k) = \frac{I_0}{8\theta_c d} \ln \frac{(k + 1)^2 \left(1 + d^{-2} (k - 1)^2\right)}{(k - 1)^2 \left(1 + d^{-2} (k + 1)^2\right)}, k_c > 2, k > 2 \text{ and } k < k_c - 2; \tag{16e}$$

where $d = 2D/x_0$.

The radiant power received by the receiving fibre was calculated numerically in dimensionless form for three irradiation distribution function. The results are given in **Figure 12**.

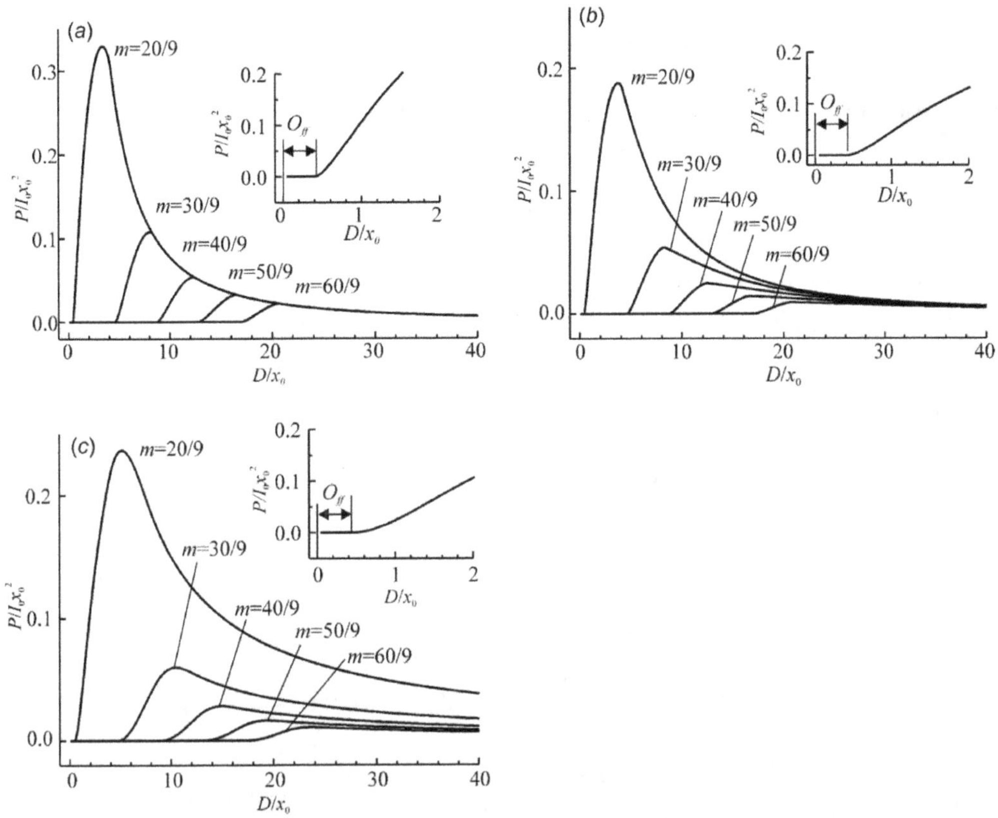

Figure 12. Elemental two-fibre modulation characteristics for various light intensity distribution functions: a) uniform, b) Gaussian, c) He-Cuomo theoretical distribution. Insets show the offset (blind) regions.

The graphs of dimensionless received radiant power vs. D/x_0 for uniform irradiance function show a horizontal offset at the origin, O_{ff}, that increases with increasing m (**Figure 12a**). The offset in dimensionless form can be determined from the following expression:

$$O_{ff} = \frac{m-2}{2\tan\theta_c} \tag{17}$$

For adjacent fibres $O_{ffa} = \frac{c_l}{2\tan\theta_c}$, i.e., is proportional to the cladding thickness. When m increases the value at the maximum decreases and the peak gets less sharp. The falling edges of all peaks converge to the function $(D/x_0)^{-2}$.

For Gaussian irradiance distribution function, the falling edge of all graphs is much more gradual as compared with the uniform irradiance model and all the graphs converge at larger displacements than for the uniform irradiance model (**Figure 12b**).

The graphs for He-Cuomo irradiance model are much less sharp as compared to other models considered above. After reaching the maximum the graphs decrease but do not converge in the studied range of D. For all irradiance distribution functions, the offset of the graphs is the same since it depends only on the cladding thickness; however, for He-Cuomo model the graph raises very smoothly just after the offset, thus creating illusion of a bigger offset.

To compare the resulting graphs the plots for m=20/9 were normalized dividing by the corresponding maximum values (**Figure 13**). The position of the maximum shifts to larger D/x_0 in the following order: Uniform ,*U*, Gaussian, *G*, He-Cuomo, *H-C*, models. Also, the slopes of the front and falling edges of the peaks decrease following the same order. It should be mentioned that for He-Cuomo model the falling edge of the peak is much higher at larger distances than for other two models.

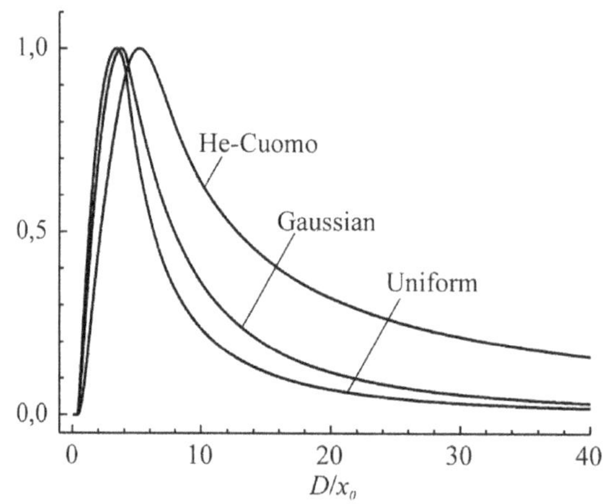

Figure 13. Normalized received radiant power for Uniform, Gaussian and He-Cuomo irradiance functions.

In many cases the transducers have more than one of each transmitting and receiving fibres assembled in a bundle. Independently of the order of the fibres in the bundle, the resulting signal of the photodiode detector can be found as a superposition of the signals from the individual receiving fibres. In turn, the signal on a receiving fibre is proportional to a sum of the received radiant power from each transmitting fibre. Therefore, the total voltage at the photodiode detector can be determined by summation [65]:

$$U(D) = \sum_{l=1}^{l=N_r} U_r(D) \propto \sum_{l=1}^{l=N_r} \sum_{j=1}^{j=N_t} P_i(l,j,D) \tag{18}$$

where $P_i(l,j,D)$ is the radiant power from the j-th transmitting fibre subtended by the l-th receiving fibre at a distance D from the reflecting surface, N_t is the number of transmitting fibres and N_r is the number of receiving fibres.

The output voltage of the sensor depends not only on the distance to the reflecting surface and the irradiance function but also on $a_n(l,j)$, i.e., on the configuration of the fibres within a bundle.

A summation model for equally spaced fibres with 1×2 periodicity was developed in [64]. It is worth mentioning that the adjacent fibres contribute most significantly to the rising front of the resulting modulation characteristics of the sensor.

The effect of bundle configuration was modelled for a bundle containing 85 fibres with a fibre core radius x_0 = 22.5 μm and cladding thickness c_l = 2.5. Geometry and enumeration of individual fibres in the bundle are shown in **Figure 14**.

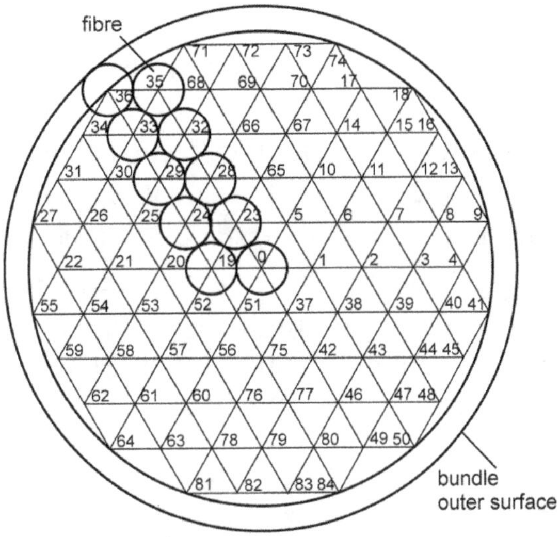

Figure 14. Three bundle configurations were modelled The numbers denote the centres of individual fibres.

The following three configurations of a bundle were modelled: alternating linear configuration of fibres (a line of transmitting fibres followed by a line of receiving fibres and so on), semicircle, and random configuration. The results for alternating linear configuration and for the semicircular one are shown in **Figure 15**.

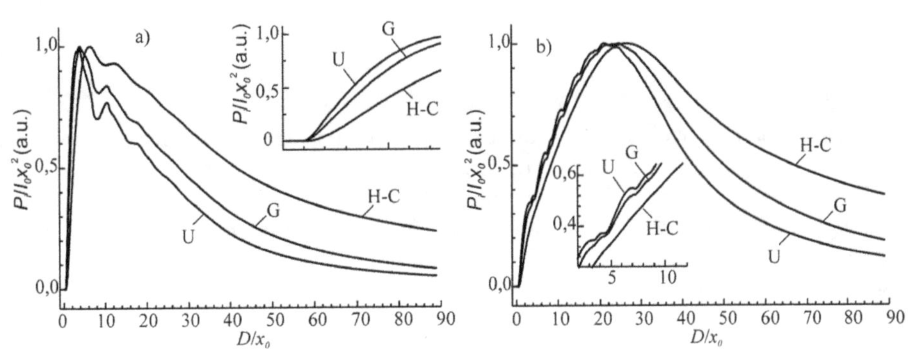

Figure 15. Modulation characteristics of a IMFODS with a) alternating linear configuration of transmitting and receiving fibres; b) semicircle configuration of emitting and receiving fibres. U – uniform irradiance function; G – Gaussian irradiance model; H-C – He-Cuomo theoretical irradiance function.

For the alternating configuration of fibres, there is a large number of transmitting fibres situated at a short distance from the receiving ones. Therefore, for all three irradiance distribution functions the plots have sharp peaks at the falling edge resulting from the contribution of the nearby transmitting fibres. However, for He-Cuomo function these peaks are less important than for other functions. A similar effect can also be observed for the semicircle configuration (**Figure 15b**). Whereas the sensor is being retracted from the reflecting surface new rows of receiving and transmitting fibres enter synchronously into the light spot at the image plane producing sharp steps on the modulation characteristic. Again, for He-Cuomo model the graphs are smoother than for other irradiance distribution functions.

Earlier Cao et al. [64] simulated modulation characteristics of an IMFODS using various bundle configurations including the "random" one. In fact, the configuration they referred to as "random" was an ordered one with a definite pattern (see **Figure 2** in [64]). In this work, real random distribution of the fibres was obtained using a random number generator. Five different random configurations were created. **Figure 16** a shows mean modulation characteristic determined as a mean of five particular modulation characteristics corresponding to five random bundle configurations. In addition, standard deviation and error of mean were determined for all models. Despite the use of random configuration followed by averaging of particular modulation characteristics, the graphs in **Figure 16** for uniform and Gaussian models have important corrugation in the region near to the maximum. However, He-Cuomo model gives very smooth modulation curve without any significant peaks and corrugation.

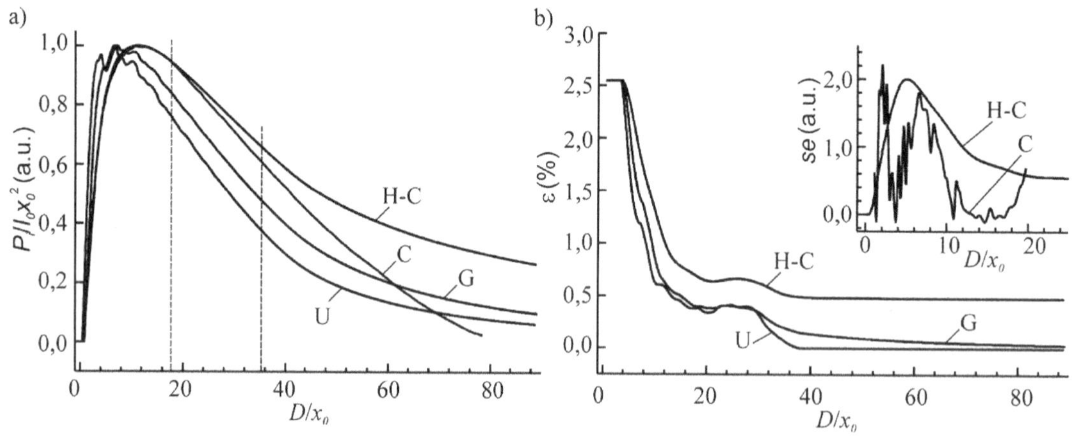

Figure 16. a) Mean modulation characteristic of a bundle optical fibre displacement sensor with random configuration of transmitting and receiving fibres. Portion of graphs fitted by linear function is shown by vertical dashed lines. C – experimental data, U – uniform model, G – Gaussian model, H-C – He-Cuomo theoretical model. b) Relative standard error for random configuration of a bundle and different irradiation distribution functions. Inset: Absolute dimensionless value of standard error for random configuration and He-Cuomo theoretical irradiation distribution. Graph C is the difference between the mean modulation characteristic and the experimental data.

Graph C in **Figure 16** a represents experimental calibration graph of a bundle optical fibre sensor. This graph is well fitted by the He-Cuomo modulation characteristic within the interval of dimensionless displacements from 0 to 20. This finding is consistent with the results published earlier by He and Cuomo [3] in the same range of dimensionless distances. The absolute difference between the two graphs (C) and the standard error for He-Cuomo model (H-C) are plotted in the inset in **Figure 16b**. The standard error reflects dispersion of the mean modulation characteristic related to variation of the bundle configuration. Within the interval of displacements from 3 to 20 the absolute difference between the theoretical and experimental graphs does not exceed standard error, thus indicating that this difference is due to mismatch between the theoretical and real bundle configurations. A sharp peak at the beginning can be attributed to the experimental error related to limited measurement capacity of the experimental test rig. At the displacements larger than 18 graph C sharply increases indicating significant divergence between two graphs. The slope of the H-C graph decreases, whereas the slope of the experimental graph remains almost constant (**Figure 16a**). Surprisingly, the experimental calibration curve cannot be fitted on the entire range of the displacement using only He-Cuomo theoretical model.

In the range of D/x_0 between 18 and 35 all the plots, both calculated and the experimental, are almost linear. All graphs were fitted in this range by a linear function using a least-squares method and mean slope and the standard error (se) were determined for these linear functions (see **Table 1**). Adjusted coefficient of determination for all fittings was larger than 0.99. The mean slope of the experimental calibration graph is very close to that of the Gaussian model, although the difference between the mean slopes is statistically significant.

Symbol	Uniform	Gaussian	He-Cuomo	Calibration
Mean slope	-2.22×10^{-2}	-2.08×10^{-2}	-1.63×10^{-2}	-1.91×10^{-2}
se	3.64×10^{-5}	2.08×10^{-5}	1.76×10^{-5}	7.36×10^{-5}
Sample volume	426	426	426	32

Table 1. Mean slope of fitted linear functions, standard error of the slope (*se*) and the sample volume for calculated modulation characteristics and experimental calibration graph.

Seemingly, the reason for the discrepancy between the calculated and the experimental modulation characteristics should be sought in some discrepancy between the real irradiance distribution function and the function used in He-Cuomo model. To fit the experimental data, the real irradiance distribution function should conceivably combine the characteristics of both Gaussian and He-Cuomo functions: the real irradiance distribution function should have a sharp peak at the origin as He-Cuomo function, but this central part should decrease with increasing D/x_0 as rapidly as Gaussian function.

It was reported that the experimental graph of the measurement uncertainty vs. distance is almost identical to the simulated one obtained by varying the bundle configuration [65]. Bearing in mind that the bundle configuration was constant during the test it was suggested that small imperfections on the mirror surface could be responsible. In fact, presence of spots that reduce reflectivity can alter particular contributions from certain pairs of receiving and transmitting fibres that is equivalent to changing of the bundle configuration. The imperfections should have the size beyond a certain critical value, which is comparable with the fibre core diameter.

Another important conclusion that has been drawn from this study is that, in contrast to implicit expectations [71], the derivatives of the modulation characteristics have no flat features at the near side that suggests absence of nominally linear region. Thus, using the linear approach at the near side will lead to systematic errors related with the deviation from linearity of the modulation characteristic. The values of these errors depend on the displacement range as well as on the initial and final positions. However, at the far side the derivatives of both theoretical and experimental modulation characteristics have definite flat region, especially at D/x_0 between 30 and 50. Therefore, the using of the linear approach is justified at this side.

5. Concluding remarks

Technological advances in development of precision surface characterization techniques have been notable in the last 20 years offering a wide range of solutions to the research community. Several improvements were achieved thanks to the better understanding of limitations of devices and to the development of reliable measurement techniques pushing the envelope to extreme working conditions. This led to resolving conflicts between different subsystems by means of a holistic approach involving different physic-chemical phenomena. As for complex tribological characterization and indentation, an indirect measurement of the involved forces through displacement measurements is needed, fibre optical and capacitive sensors were found to be the most suitable. The limitations for the former are mainly related to reflectivity loss, critical dependences of the measurement uncertainty on the minute defects on reflecting surfaces, limited temperature range and the non-linearity of its working characteristic; for the latter, a differential capacitive sensor is preferable leading to the need of an active reference. In addition, use of capacitance sensor can be limited when it is used together with other sensitive techniques due to electromagnetic interferences. However, when these kinds of sensors are implemented in the device, their accuracy is so high that the noise floor is recorded and overlaps the measurements. A proper damping is than needed as well as the right models taking into account the estimated instrument cumulated uncertainty. It should then be highlighted the need of subsystem influencing as less as possible these sensors and materials with stable mechanical characteristics over a wide range of temperatures. It is then easy to understand how only non-contact heating device, such as laser-based, inductive or radiation-based heat sources, can be used to such accurate machines, and how important is to keep the measuring device free from external noises and change of temperature.

The results presented in this chapter show that such recommendation are not utopic although a deep experience is needed in interpreting the machine response and proper calibrations should be carried out. As an example, **Figure 17** shows the force measuring head designed by the authors for an ultrahigh vacuum system for tribological characterization of materials and lubricants [4]. The patented head allows measuring two forces in perpendicular directions with very low crosstalk using two pairs of leaf springs. The displacement of the leaf springs is measured by IMFODS mounted on corresponding micropositioning motorized stages. These stages permit not only fine adjustment of the initial position of the IMFODS that is necessary to reduce the measurement uncertainty but also switching from back to front slopes during the experiment. Since the sensitivity of the IMFODS at the front slope is almost tenfold greater than on the back slope, flexibility in selection the range in vacuum offers the opportunity to significantly increase the measurement range.

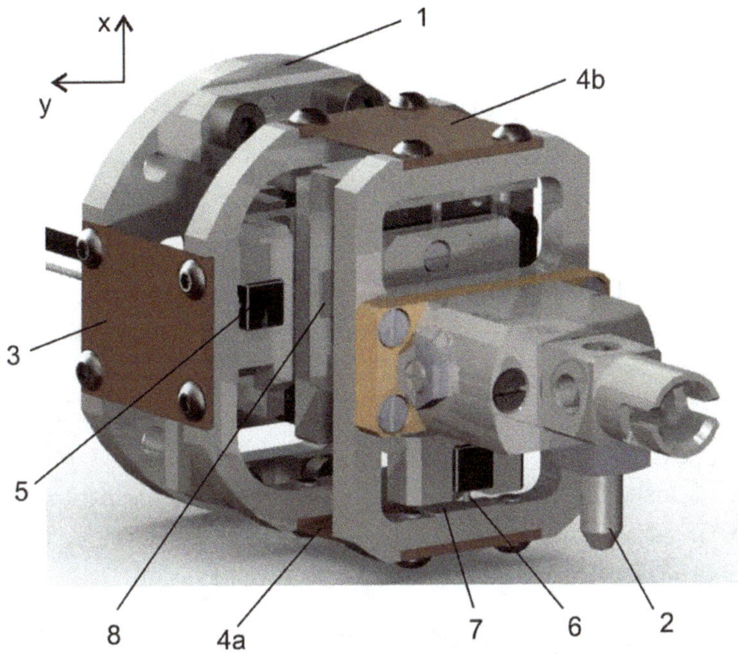

Figure 17. UHV compatible 2-axes force sensor. 1 – the base plate; 2 – the pin; 3 – one of the two leaf-springs used for measuring y component of the force ;4a and 4b – the leaf-springs for measuring x component of the force; 5,6 – IM-FODS; 7 – mirror of the x-stage; and 8 – microactuator for positioning of the optical sensor of the y-stage.

Author details

Roman Nevshupa[1*] and Marcello Conte[2*]

*Address all correspondence to: r.nevshupa@csic.es and marcello.conte@anton-paar.com

1 Spanish National Research Council, Institute of Constructional Sciences "Eduardo Torroja" (IETCC-CSIC), Madrid, Spain

2 Anton Paar Tritec SA, Peseux, Switzerland

References

[1] Jones WR, Jansen MJ. Space tribology. Hanover, MD: NASA Center for Aerospace Information; 2000. p. 33.

[2] Miyoshi K. Solid lubricants and coatings for extreme environments: State-of-the-art survey. Technical memorandum. Cleveland, Ohio: National Aeronautics and Space Administration, Glenn Research Center; 2007. p. 16.

[3] Buckley DH. Friction, wear, and lubrication in vacuum. Washington, D.C.: Scientific and Technical Information Office, National Aeronautics and Space Administration; 1971.

[4] Nevshupa RA, Conte M, Igartua A, Roman E, de Segovia JL. Ultrahigh vacuum system for advanced tribology studies: Design principles and applications. Tribology International. 2015;86:28-35. DOI: 10.1016/j.triboint.2015.01.020

[5] ESAE. Space product assurance. Thermal vacuum outgassing test for the screening of space materials. Noordwijk: ECSS Secretariat, ESA-ESTEC; 2008. p. 45.

[6] Fischer D. Untersuchung von reibungsvorschungen mit hlfe der exoelektronenemission. Berlin: Berlin Technical University; 1970.

[7] Vanhulsel A, Velasco F, Jacobs R, Eersels L, Havermans D, Roberts EW, et al. Dlc solid lubricant coatings on ball bearings for space applications. Tribology International. 2007;40:1186-1194. DOI: 10.1016/j.triboint.2006.12.005

[8] Roberts EW. Space tribology: Its role in spacecraft mechanisms. Journal of Physics D: Applied Physics. 2012;45:503001.

[9] Roberts EW, Todd MJ. Space and vacuum tribology. Wear. 1990;136:157-167. DOI: 10.1016/0043-1648(90)90078-O

[10] Martin J-M, Bouchet M-IDB, Matta C, Zhang Q, Goddard WA, Okuda S, et al. Gas-phase lubrication of ta-c by glycerol and hydrogen peroxide. Experimental and computer modeling. The Journal of Physical Chemistry C. 2010;114:5003-5011. DOI: 10.1021/jp909940j

[11] Řepa P. Mechanically induced desorption. Vacuum. 1992;43:367-371. DOI: 10.1016/0042-207X(92)90039-Y

[12] Nevshupa RA, Roman E, de Segovia JL. Origin of hydrogen desorption during friction of stainless steel by alumina in ultrahigh vacuum. Journal of Vacuum Science & Technology A. 2008;26:1218-1223. DOI: 10.1116/1.2968682

[13] Nevshupa RA, Roman E, de Segovia JL. Contamination of vacuum environment due to gas emission stimulated by friction. Tribology International. 2013;59:23-29. DOI: 10.1016/j.triboint.2012.07.009

[14] Nakayama K, Nevshupa RA. Plasma generation in a gap around a sliding contact. Journal of Physics D: Applied Physics. 2002;35:L53-L56.

[15] Nevshupa RA. Effect of gas pressure on the triboluminescence and contact electrification under mutual sliding of insulating materials. Journal of Physics D: Applied Physics. 2013;46:185501.

[16] Nevshupa R, Hiratsuka K. Triboluminescence. In: Gnecco E, Meyer E, editors. Fundamentals of friction and wear on the nanoscale: Springer International Publishing; 2015. p. 57-77.

[17] Williams MW. Triboelectric charging of insulating polymers-some new perspectives. Aip Advances. 2012;2. DOI: 010701 10.1063/1.3687233

[18] Walton AJ. Triboluminescence. Advances in Physics. 1977;26:887-948. DOI: 10.1080/00018737700101483

[19] Nakayama K, Ikeda H. Triboemission characteristics of electrons during wear of amorphous carbon and hydrogenated amorphous carbon films in a dry air atmosphere. Wear. 1996;198:71-76. DOI: 10.1016/0043-1648(96)06934-7

[20] Evdokimov VD. Specific features of exoelectron emission during friction of metals. Soviet Physics Journal. 1968;11:11-13. DOI: 10.1007/BF01890910

[21] Mahrova M, Conte M, Roman E, Nevshupa R. Critical insight into mechanochemical and thermal degradation of imidazolium-based ionic liquids with alkyl and mPEG side chains. Journal of physical chemistry C. 2014;118:22544–22552. DOI: 10.1021/jp504946h

[22] Nevshupa R, Ares JR, Fernández JF, del Campo A, Roman E. Tribochemical decomposition of light ionic hydrides at room temperature. The Journal of Physical Chemistry Letters. 2015;6:2780-2785. DOI: 10.1021/acs.jpclett.5b00998

[23] Rusanov A, Nevshupa R, Fontaine J, Martin J-M, Le Mogne T, Elinson V, et al. Probing the tribochemical degradation of hydrogenated amorphous carbon using mechanically stimulated gas emission spectroscopy. Carbon. 2015;81:788-799. DOI: 10.1016/j.carbon. 2014.10.026

[24] Le Mogne T, Martin J-M, Grossiord C. Imaging the chemistry of transfer film in aes/xps analytical UHV tribotester. In: Dowson D, editor. Lubrication at the frontier: The role of the interface and surface layers in the thin film and boundary regime. Amsterdam: Elsevier; 1999. p. 413-422.

[25] Kajdas C. General approach to mechanochemistry and its relation to tribochemistry Tribology in engineering Zagreb: InTech; 2013. p. 209-240.

[26] Schuh CA. Nanoindentation studies of materials. Materials Today. 2006;9:32-40. DOI: 10.1016/S1369-7021(06)71495-X

[27] Bhushan B. Handbook of micro/nano tribology. 2nd ed: Taylor & Francis, 1998. 880 p.

[28] Oliver WC, Pharr GM. Measurement of hardness and elastic modulus by instrumented indentation: Advances in understanding and refinements to methodology. Journal of Materials Research. 2004;19:3-20. DOI: 10.1557/jmr.2004.19.1.3

[29] Dao M, Chollacoop N, Van Vliet KJ, Venkatesh TA, Suresh S. Computational modeling of the forward and reverse problems in instrumented sharp indentation. Acta Materialia. 2001;49:3899-3918. DOI: 10.1016/S1359-6454(01)00295-6

[30] Storåkers B, Larsson P-L. On brinell and boussinesq indentation of creeping solids. Journal of the Mechanics and Physics of Solids. 1994;42:307-332. DOI: 10.1016/0022-5096(94)90012-4

[31] Suresh S, Giannakopoulos AE. A new method for estimating residual stresses by instrumented sharp indentation. Acta Materialia. 1998;46:5755-5767. DOI: 10.1016/S1359-6454(98)00226-2

[32] Trenkle JC, Packard CE, Schuh CA. Hot nanoindentation in inert environments. Review of Scientific Instruments. 2010;81:073901. DOI: 10.1063/1.3436633

[33] Guide to the measurement of force. London: Weight and Force Measuring Panel, The Institute of Measurement and Control, 2013. p.

[34] Woirgard J, Bellaton B, Consiglio R. Measuring head for nanoindentation instrument and measuring method using same. International patent number US7685868 B2. Priority date: 23.12.2004

[35] Nohava J, Randall NX, Conte N. Novel ultra nanoindentation method with extremely low thermal drift: Principles and experimental results. Journal of Materials Research. 2009;24:873-882. DOI: 10.1557/jmr.2009.0114

[36] Fraden J. Position, displacement, and level. Handbook of modern sensors: Springer New York; 2010. p. 279-326.

[37] Smith SB, Cui Y, Bustamante C. Optical-trap force transducer that operates by direct measurement of light momentum. Methods in enzymology: Academic Press; 2003. p. 134-162.

[38] Ştefănescu D. Magnetoelastic force transducers. Handbook of force transducers. Berlin Heidelberg: Springer 2011. p. 165-184.

[39] Richard LS, Edwin RW, David BN, Ruimin L. Towards an electronic kilogram: An improved measurement of the planck constant and electron mass. Metrologia. 2005;42:431.

[40] Stachowiak G, Batchelor AW. Experimental methods in tribology: Elsevier Science, 2004. p.

[41] Liu H, Bhushan B. Adhesion and friction studies of microelectromechanical systems/nanoelectromechanical systems materials using a novel microtriboapparatus. Journal of Vacuum Science & Technology A. 2003;21:1528-1538. DOI: 10.1116/1.1560711

[42] Nevshupa RA, Scherge M, Ahmed SIU. Transitional microfriction behavior of silicon induced by spontaneous water adsorption. Surface Science. 2003;526:209-209. DOI: 10.1016/S0039-6028(02)02067-8

[43] Scherge M, Schaefer JA. Microtribological investigations of stick/slip phenomena using a novel oscillatory friction and adhesion tester. Tribology Letters. 1998;4:37-42. DOI: 10.1023/a:1019182500536

[44] Brouwer DM, Meijaard JP, Jonker JB. Large deflection stiffness analysis of parallel prismatic leaf-spring flexures taking into account shearing, constrained warping and anticlastic curving effects. Precision engineering. 2013;37:505 - 521.

[45] Yaqoob MA, M.B. R, Schipper DJ. Design of a vacuum based test rig for measuring micro adhesion and friction force. In: De Wilde WP, Brebbia CA, Hernández S, editors. High performance structures and materials vi. Southampton: WIT press; 2012. p. 261-274.

[46] Block B. Solid state force transducer and method of making same. Patent number CA1114644 A1. Priority date: 9 Feb 1976

[47] Uchikawa H, Munekata M, Motohashi H. Kinetofrictional force testing apparatus. Patent number US 5212657. Priority date: 16.01.1990

[48] Hegde SG, Praino AP, Root SJ, Sri-Jayantha M. Tunable feedback transducer for transient friction measurement. Patent number US 5115664. Priority date: 25.06.1990

[49] Börner H, Frindt F, Mollenhauer O, Spaltmann D. Ultrahochvakuum-tribometer. Patent number DE10390125B4. International patent number WO2003060487A1. Priority date: 18.01.2002

[50] Dellah A, Wild PM, Moore TN, Shalaby M, Jeswiet J. An embedded friction sensor based on a strain-gauged diaphragm. Journal of Manufacturing Science and Engineering. 2002;124:523-527. DOI: 10.1115/1.1461839

[51] TarÁS, Cserey GG, Veres J. Sensor device. International patent number WO2013072712 A1. Priority date: 17.11.2011

[52] Bonin WA. Multi-dimensional capacitive transducer. Patent number US 5661235. Priority date: 20.12.1995

[53] Bellaton B, Consiglio R, Woirgard J. Measuring head for nanoindentation instrument and measuring method. Patent application WO2014202551 A1. Priority date: 17.06.2013

[54] Kosinskiy M, Ahmed SI-U, Liu Y, Schaefer JA. A compact reciprocating vacuum microtribometer. Tribology International. 2012;56:81-88. DOI: 10.1016/j.triboint.2012.06.019

[55] Holton C, Ahmadian M. Doppler sensor for the derivation of torsional slip, friction and related parameters. Patent number US7705972 B2. Patent number US7705972 B2. Priority date: 20.06.2006

[56] He G, Cuomo FW. A light intensity function suitable for multimode fiber-optic sensors. Journal of Lightwave Technology. 1991;9:545-551. DOI: 10.1109/50.76670

[57] Mollenhauer O, Scherge M, Karguth A. Device for examining friction conditions. Patent number US 6666066. Priority date: 03.04.1998

[58] Spaltmann D, Boerner H, Frindt F, Mollenhauer O. Tribometer. Patent number WO 03060487. International patent number WO 03060487. Priority date: 20.01.2003

[59] Roman E, Nevshupa R, de Segovia JL, Konovalov PI, Menshikov IP. Method and apparatus for analysis of gas content in solids and surface coatings. Patent number WO2007ES70216 20071220. Priority date: 23.02.2007

[60] Nevshupa R, Conte M, Delgado A, Igartua A, Egaña F, Aranzabe A. Dispositivo de medida de fuerzas. Patent number WO 2011080355. Priority date: 29.12.2009

[61] Weik MH. Fiber optics standard dictionary. New York: Springer US, 1989. 366 p. DOI: 10.1007/978-1-4757-1177-6

[62] Cuomo FW. The analysis of a three-fiber lever transducer. Proceedings of SPIE Symposium. 1984;478:28-32. DOI: 10.1117/12.942653

[63] Cook RO, Hamm CW. Fiber optic lever displacement transducer. Appl Opt. 1979;18:3230-3241. DOI: 10.1364/AO.18.003230

[64] Cao H, Chen Y, Zhou Z, Zhang G. Theoretical and experimental study on the optical fiber bundle displacement sensors. Sensors and Actuators A: Physical. 2007;136:580-587. DOI: 10.1016/j.sna.2006.12.010

[65] Nevshupa R, Conte M, Van Rijn C. Measurement uncertainty of a fibre-optic displacement sensor. Measurement Science and Technology. 2013;24:035104. DOI: 10.1088/0957-0233/24/3/035104

[66] Lawson CM, Tekippe VJ. Fiber-optic diaphragm-curvature pressure transducer. Opt Lett. 1983;8:286-288. DOI: 10.1364/OL.8.000286

[67] Faria JB. A theoretical analysis of the bifurcated fiber bundle displacement sensor. IEEE Transactions on Instrumentation and Measurement. 1998;47:742-747. DOI: 10.1109/19.744340

[68] Samian FRP, Pramono YH, Rohedi AY, Rusydi F, Zaidan AH. Theoretical and experimental study of fiber-optic displacement sensor using multimode fiber coupler. Journal of Optoelectronics and Biomedical Materials. 2009;1:303-308.

[69] Libo Y, Jian P, Tao Y, Guochen H. Analysis of the compensation mechanism of a fiber-optic displacement sensor. Sensors and Actuators A: Physical. 1993;36:177-182. DOI: 10.1016/0924-4247(93)80190-r

[70] Lin C-S, Chang R-S. Fiber optic displacement sensors for the measurement of a vibrating object. Precision Engineering. 1994;16:302-306. DOI: 10.1016/0141-6359(94)90008-6

[71] Buchade PB, Shaligram AD. Influence of fiber geometry on the performance of two-fiber displacement sensor. Sensors and Actuators A: Physical. 2007;136:199-204. DOI: 10.1016/j.sna.2006.11.020

Permissions

 Every chapter published in this book has been scrutinized by our experts. Their significance has been extensively debated. The topics covered herein carry significant findings which will fuel the growth of the discipline. They may even be implemented as practical applications or may be referred to as a beginning point for another development.

The contributors of this book come from diverse backgrounds, making this book a truly international effort. This book will bring forth new frontiers with its revolutionizing research information and detailed analysis of the nascent developments around the world.

We would like to thank all the contributing authors for lending their expertise to make the book truly unique. They have played a crucial role in the development of this book. Without their invaluable contributions this book wouldn't have been possible. They have made vital efforts to compile up to date information on the varied aspects of this subject to make this book a valuable addition to the collection of many professionals and students.

This book was conceptualized with the vision of imparting up-to-date information and advanced data in this field. To ensure the same, a matchless editorial board was set up. Every individual on the board went through rigorous rounds of assessment to prove their worth. After which they invested a large part of their time researching and compiling the most relevant data for our readers.

The editorial board has been involved in producing this book since its inception. They have spent rigorous hours researching and exploring the diverse topics which have resulted in the successful publishing of this book. They have passed on their knowledge of decades through this book. To expedite this challenging task, the publisher supported the team at every step. A small team of assistant editors was also appointed to further simplify the editing procedure and attain best results for the readers.

Apart from the editorial board, the designing team has also invested a significant amount of their time in understanding the subject and creating the most relevant covers. They scrutinized every image to scout for the most suitable representation of the subject and create an appropriate cover for the book.

The publishing team has been an ardent support to the editorial, designing and production team. Their endless efforts to recruit the best for this project, has resulted in the accomplishment of this book. They are a veteran in the field of academics and their pool of knowledge is as vast as their experience in printing. Their expertise and guidance has proved useful at every step. Their uncompromising quality standards have made this book an exceptional effort. Their encouragement from time to time has been an inspiration for everyone.

The publisher and the editorial board hope that this book will prove to be a valuable piece of knowledge for researchers, students, practitioners and scholars across the globe.

List of Contributors

Andrzej B. Dobrucki, Piotr Pruchnicki, Przemysław Plaskota, Piotr Staroniewicz, Stefan Brachmański and Maciej Walczyński
Department of Acoustics and Multimedia, Faculty of Electronics, Wrocław University of Technology, Wrocław, Poland

Chuxiong Hu and Yu Zhu
State Key Lab of Tribology, Department of Mechanical Engineering, Tsinghua University, Beijing, China
Beijing Key Lab of Precision/Ultra-precision Manufacturing Equipment and Control, Tsinghua University, Beijing, China

Leopoldo Angrisani
DIETI - Department of Information Technology and Electrical Engineering, CeSMA - Center of Advanced Measurement Services, University of Napoli Federico II, Napoli, Italy

Giovanni Cavallo and Annalisa Liccardo
DIETI - Department of Information Technology and Electrical Engineering, University of Napoli Federico II, Napoli, Italy

Gian Paolo Papari and Antonello Andreone
DF - Department of Physics, University of Napoli Federico II, Napoli, Italy

Xiaohai Cui and Yong Li
National Institute of Metrology, Beijing, China

Yu Song Meng and Yueyan Shan
National Metrology Centre, Agency for Science, Technology and Research (A*STAR), Singapore

Daniel Discini Silveira
Federal University, Ufjf, Juiz de Fora, Minas Gerais, Brazil

Marcos Paulo de Souza Silva
São Paulo's Electricity Company, CESP, São Paulo, Brazil

Marcel Veloso Campos
State University, Unimontes, Montes Claros, Minas Gerais, Brazil

Maurício Silveira
Federal University, UFSCar, São Carlos, São Paulo, Brazil

Laura Rossi
Altran S.p.A., Bologna, Italy

Paulo Pereira
Quality Management Department, Portuguese Institute of BloodTransplantation, Lisbon, Portugal

Fernando Visconti and José Miguel de Paz
Valencian Institute for Agricultural Research (IVIA), Centre for Sustainable Agriculture (CDAS), Moncada, València, Spain

Roman Nevshupa
Spanish National Research Council, Institute of Constructional Sciences "Eduardo Torroja" (IETCC-CSIC), Madrid, Spain

Marcello Conte
Anton Paar Tritec SA, Peseux, Switzerland

Index

www.ingramcontent.com/pod-product-compliance
Lightning Source LLC
LaVergne TN
LVHW080951210926
848485LV00004B/818